Embedded Microcontroller Interfacing for M·CORE Systems

Academic Press Series in Engineering

Series Editor
J. David Irwin
Auburn University

Designed to bring together interdependent topics in electrical engineering, mechanical engineering, computer engineering, and manufacturing, the Academic Press Series in Engineering provides state-of-the-art handbooks, textbooks, and professional reference books for researchers, students, and engineers. This series provides readers with a comprehensive group of books essential for success in modern industry. A particular emphasis is given to the applications of cutting-edge research. Engineers, researchers, and students alike will find the Academic Press Series in Engineering to be an indispensable part of their design toolkit.

Published books in the series:
Industrial Controls and Manufacturing, 1999, E. Kamen
DSP Integrated Circuits, 1999, L. Wanhammar
Time Domain Electromagnetics, 1999, S. M. Rao
Single- and Multi-Chip Microcontroller Interfacing for the Motorola 68HC12, 1999, G. J. Lipovski
Control in Robotics and Automation, 1999, B. K. Ghosh, N. Xi, T. J. Tarn
Soft Computing and Intelligent Systems, 1999, N. K. Sinha, M. M. Gupta
Introduction to Microcontrollers, 1999, G. J. Lipovski
Control of Induction Motors, 2000, A. M. Trzynadlowski
Embedded Microcontroller Interfacing for M·CORE Systems, 2000, G. J. Lipovski

Embedded Microcontroller Interfacing for M·CORE Systems

G. Jack Lipovski
Department of Electrical and Computer Engineering
University of Texas
Austin, Texas

ACADEMIC PRESS

A Harcourt Science and Technology Company

San Diego   San Francisco   New York   Boston
London   Sydney   Tokyo

ACADEMIC PRESS
A Harcourt Science and Technology Company
525 B Street, Suite 1900, San Diego, CA 92101-4495, USA
http://www.academicpress.com

Academic Press
Harcourt Place, 32 Jamestown Road, London NW1 7BY, UK
http://www.academicpress.com

Library of Congress Catalog Card Number: 00-102018
ISBN: 0-12-451832-X

Printed in the United States of America

00 01 02 03 04 05 IP 9 8 7 6 5 4 3 2 1

*Dedicated to my wife
Isabelle Lipovski*

Contents

Preface

The embedded microcontroller industry is moving towards inexpensive micro-controllers with significant amounts of ROM and RAM, and some user-designed hardware that is put on a single microcontroller chip. In these microcontrollers, the majority of the design cost is incurred in the writing of software that will be used in them. The memory available in such microcontrollers permits the use of real-time operating systems. Further, C++ compilers permit the use of classes to encapsulate the function members, their data members, and their hardware, in an object. Both of these techniques reduce software design cost. This book aims to give the principles of and concrete examples of design, especially software design, of the Motorola MMC2001, a particular M·CORE embedded microcontroller.

The first four chapters of the book provide background. The first chapter is aimed at the high-level programmer who will need to acquire a reading knowledge of assembler language to be able to debug his or her high-level language programs. The second chapter is aimed at the hardware designer, who will need to know enough C and C++ programming to be able to write the programs in an embedded micro-controller. The third chapter introduces the real-time operating system, including the use of device drivers. The fourth chapter provides information for programmers who need to understand the issues involved in hardware design, including the design of ASIC modules that are implemented in an M·CORE chip. While many readers will be familiar with one or more of these topics, the designer of embedded micro-controllers needs to be familiar with all of them. These chapters bring the reader to an adequate level of background needed for embedded microcontroller design.

The next three chapters are the core of this book. The fifth chapter discusses the alternatives to the parallel port, and ways to program interfaces to control them. The sixth chapter describes alternatives to interrupts, and ways to program interrupt and other synchronization interfaces. The seventh chapter highlights the techniques for and problems with time slice operation of embedded microcontrollers. A simple multi-threaded time sharing system is introduced, followed by an object-oriented time sharing system. The use of real-time operating systems multitasking is then discussed.

Chapter 8 shows how to design additional hardware to be added into the MMC2001 chip. It gives an ASIC design example, and describes a processor architecture that is suitable for special-purpose designs. The last two chapters provide some examples of system design. Chapter 9 discusses communication techniques and shows several programming approaches to the MMC2001 UART device. The tenth chapter shows the programming of display and storage systems.

This book provides a concrete understanding of hardware-software tradeoffs, high-level languages, and embedded microcontroller operating systems. Because these very practical areas should be understood by many if not all computer engineering graduate students, this book is written as a textbook for a graduate level course. However, it will also be very useful to practitioners, especially those who will work with the Motorola M·CORE embedded microcontroller. It is therefore also written for engineers who need to understand and use these microcontrollers.

List of Figures

List of Tables

Acknowledgments

The author would like to express his deepest gratitude to everyone who contributed to the development of this book. Special thanks are due to Jim Thomas, who initiated the development of this book, and Greg Watkins, who coordinated its development with M·CORE personnel. I also acknowledge extensive and helpful proofreading from several of these personnel, especially Steve Sobel, Kirby Kyle, and Howard Owens at Motorola, and Phil Walsh and Alan Anderson at Microware.

About the Author

G. Jack Lipovski has taught electrical engineering and computer science at The University of Texas since 1976. He is a computer architect internationally recognized for his design of the pioneering data-base computer, CASSM, and the parallel computer, TRAC. His expertise in microcomputers has brought international recognition—he has served as a director of Euromicro and an editor of IEEE Micro. Dr. Lipovski has published more than 70 papers, largely in the proceedings of the International Symposium on Computer Architecture (ISCA), the IEEE Transactions on Computers and the National Computer Conference. At the 25th ISCA, Dr. Lipovski was noted as having written more papers at this prestigious symposium than any other author. He holds 12 patents, generally in the design of logic-in-memory integrated circuits for database and graphics geometry processing. He has authored nine books and edited three. He has served as chairman of the IEEE Computer Society Technical Committee on Computer Architecture, member of the Computer Society Governing Board, and chairman of the Special Interest Group on Computer Architecture of the Association for Computer Machinery. He has been elected Fellow of the IEEE and a Golden Core Member of the IEEE Computer Society. He received his Ph.D. degree from the University of Illinois, 1969, and has taught at the University of Florida, and at the Naval Postgraduate School, where he held the Grace Hopper chair in Computer Science. He has consulted for Harris Semiconductor, designing a microcomputer, and for the Microelectronics and Computer Corporation, studying parallel computers. He founded Linden Technology Ltd., and is the chairman of its board. His current interests include parallel computing, data-base computer architectures, artificial intelligence computer architectures, and microcomputers.

1

Microcomputer Architecture

Microcomputers, microprocessors, and microprocessing are at once quite familiar and a bit fuzzy to most engineers and computer scientists. When we ask the question: "What is a microcomputer?" we get a wide range of answers. This chapter aims to clear up these terms. Also, the designer needs to be sufficiently familiar with the microcomputer instruction set to be able to read the object code generated by a C compiler. Clearly, we have to understand these concepts to be able to discuss and design I/O interfaces. This chapter contains essential material on microcomputers and microprocessors needed as a basis for understanding the discussion of interfacing in the rest of the book.

We recognize that the designer must have a comprehensive knowledge about basic computer architecture and organization. But the goal of this book is to impart enough knowledge so the reader, on completing it, should be ready to design good hardware and software for microcomputer interfaces. We have to trade material devoted to basics for material needed to design interface systems. There is so much to cover and so little space, that we will simply offer a summary of the main ideas. If you have had this material in other courses or absorbed it from your work or from reading those fine trade journals and hobby magazines devoted to microcomputers, this chapter should bring it all together. Some of you can pick up the material just by reading this condensed version. Others should get an idea of the amount of background needed to read the rest of the book.

For this chapter, we assume the reader is fairly familiar with some kind of Assembly Language on a large or small computer or is able to pick it up quickly. In this chapter, he or she should learn about the software view of microcomputers and embedded systems in general, and the M·CORE embedded processor in particular.

1.1   An Introduction to the Microcomputer

Just what is a microcomputer and a microprocessor, and what is the meaning of microprogramming — which is often confused with microcomputers? This section will survey these concepts and other commonly misunderstood terms in digital systems design. It describes the architecture of digital computers and gives a definition of architecture. Note that all *italicized* words are in the index and are listed at the end of each chapter; these serve as a glossary to help you find terms that you may need later.

Because the microcomputer is much like other computers except that it is smaller and less expensive, these concepts apply to large computers as well as micro-computers. The concept of the computer is presented first, and the idea of an instruction is scrutinized next. The special characteristics of microcomputers will be delineated last.

1.1.1   Computer Architecture

Actually, the first and perhaps the best paper on computer architecture, "Preliminary discussion of the logical design of an electronic computing instrument," by A. W. Burks, H. H. Goldstein, and J. von Neumann, was written 15 years before the term was coined. We find it fascinating to compare the design therein with all computers produced to date. It is a tribute to von Neumann's genius that this design, originally intended to solve nonlinear differential equations, has been successfully used in business data processing, information handling, and industrial control, as well as in numeric problems. His design is so well defined that most computers — from large computers to microcomputers — are based on it, and they are called *von Neumann computers*.

In the early 1960s a group of computer designers at IBM — including Fred Brooks — coined the term "architecture" to describe the "blueprint" of the IBM 360 family of computers, from which several computers with different costs and speeds (for example, the IBM 360/50) would be designed. The *architecture* of a computer is, strictly speaking, its instruction set and the input/output (I/O) connection capabilities. More generally, the architecture is the view of the hardware as seen by the programmer. Computers with the same architecture can execute the same programs and have the same I/O devices connected to them. Designing a collection of computers with the same "blueprint" or architecture has been done by several manufacturers. This definition of the term "computer architecture" applies to this fundamental level of design, as used in this book. However, outside of this book the term "computer architecture" has become very popular and is also rather loosely used to describe the computer system in general, including the implementation techniques and organization discussed next.

The *organization* of a digital system like a computer is usually shown by a block diagram which shows the registers, busses, and data operators in the computer. Two computers have the same organization if they have the same block diagram. For instance, Motorola manufactures several computers having the same architecture but different organizations to suit different applications. Incidentally, the organization of a computer is also called its *implementation*. Finally, the *realization* of the computer is its actual hardware interconnection and construction. It is entirely reasonable for a company to change the realization of one of its computers by replacing the hardware in a block of its block diagram with a newer type of hardware, which might be faster or cheaper. In this case the implementation or organization remains the same while the realization is different. In this book we will name the component by its full part number, like PMC2001HDCPU34 when we want to discuss an actual realization. However, we are usually interested only in the organization or the architecture only. In these cases, we will refer to an organization as a partial name without the suffix, such as MMC2001 without HDCPU34, and refer to the architecture as an M·CORE architecture or a number 6812. This should clear up any ambiguity, while also being a natural, easy-to-read shorthand.

The architecture of von Neumann computers is disarmingly simple, and the following analogy shows just how simple. (For an illustration of the following terms, see Figure 1.1) Imagine a person in front of a mailbox, with an adding machine and window to the outside world. The mailbox, with numbered boxes or slots, is analogous to the *primary memory*; the adding machine, to the *data operator* (arithmetic-logic unit); the person, to the *controller*; and the window, to *input/output* (I/O). The person's hands *access* the memory. Each slot in the mailbox has a paper that has a string of, say, 8 1s and 0s (*bits*) on it. A string of 8 bits is a *byte*, and four bits is a *nibble*. A string of 16 bits is called a *halfword*, and 32 bits is called a *word*.

The primary memory may be in part a *random access memory* (RAM) (so-called because the person is free to access its data in any order at random, without having to wait any longer for data because it is in a different location). RAM may be *static ram* — *SRAM* — if bits are stored in flip-flops, or *dynamic ram* — *DRAM* — if bits are stored as charges in capacitors. Memory that is normally written at the factory, never to be rewritten by the user, is called *read-only memory* — *ROM*. A *programmable read-only memory* — *PROM* — can be written once by a user, by blowing fuses to store bits in it. An *erasable programmable read-only memory* — *EPROM* — can be erased by ultraviolet light, and then written electrically by a user. An *electrically erasable programmable read-only memory* — *EEPROM* — can be erased and then written by a user, but erasing and writing words in EEPROM takes several milliseconds. A variation of this memory, called *flash*, is less expensive but can not be erased one word at a time.

With the left hand the person takes out a word from slot or box *n*, reads it as an instruction, and replaces it. Bringing a word from the mailbox (primary memory) to the person (controller) is called *fetching*. The hand that fetches a word from box *n* is

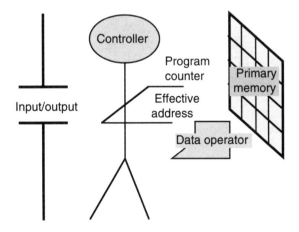

Figure 1.1. Analogy to the von Neumann Computer

analogous to the *program counter*. It is ready to take the word from the next box, box $n + 1$, when the next instruction is to be fetched.

An instruction in the M·CORE processor is a *binary code* such as 01001100. Consistent with the notation used by Motorola, binary codes are denoted in this book by a 0b (zero bee), followed by 1s or 0s. (Decimal numbers, by comparison, will not use any special symbols.) Since all those 1s and 0s are hard to remember, a convenient format is often used, called *hexadecimal notation*. In this notation, a 0x (zero ex) is written (to designate that the number is in hexadecimal notation), and the bits, in groups of 4, are represented as if they were "binary coded" digits 0 to 9 or letters A, B, C, D, E, and F to represent values 10, 11, 12, 13, 14, and 15, respectively. For example, %0100 is the binary code for 4, and %1100 is the binary code for 12, which, in hexadecimal notation, is represented as 0xC. The binary code 01001100, mentioned previously, is represented as 0x4C in hexadecimal notation. Whether the binary code or the simplified hexadecimal code is used, instructions written this way are called *machine-coded* instructions because that is the actual code fetched from the primary memory of the machine, or computer.

However, this is too cumbersome. So a *mnemonic* (which means a memory aid) is used to represent the instruction. All instructions in the M·CORE are entirely described by one 16-bit halfword. The M·CORE instruction 0x6001 actually puts a one into register r1, so it is written as

```
movi r1, #1
```

(The M·CORE registers such as r1 are described in §1.2[1]. The mnemonic movi is described in §1.2.1. Strictly speaking, M·CORE mnemonics should be written in lower case to conform with Motorola's Applications Binary Interface Standards Manual MCOREABISM/AD.)

As better technology becomes available, and as experience with an architecture reveals its weaknesses, a new architecture may be crafted that includes most of the old instruction set and some new instructions. Programs written for the old computer should also run, with little or no change, on the new one, and more efficient programs can perhaps be written using new features of the new architectures. Such a new architecture is *upward compatible* from the old one if this property is preserved. If an architecture executes the same machine code the same way, it is fully upward compatible, but more generally, if it executes the same mnemonic instructions, even though they may be coded as different machine codes, then the architecture is *source code upward compatible*. The 6812 architecture is source code upward compatible from the 6811.

An *assembler* is a program that converts mnemonics into machine code so the programmer can write in convenient mnemonics and the output machine code is ready to be put in primary memory to be fetched as an instruction. The mnemonics are therefore called *assembly-language* instructions. A *compiler* is a program that converts statements in a *high-level language* either to assembly language, to be input

---

[1] § means "Section."

to an assembler, or to machine code, to be stored in memory and fetched by the controller.

While a lot of interface software is written in assembly language and many examples in this book are discussed using this language, most will be written in the high-level language C. However, quick fixes to programs are occasionally even written in machine code. Moreover, an engineer should want to know exactly how an instruction is stored and how the controller understands it. Therefore, in this chapter we will show the assembly language and machine code for some assembly-language instructions.

Now that we have some ideas about instructions, we resume the analogy to illustrate some things an instruction might do. For example, an instruction may direct the controller to clear register r1 and write this word from r1 to a box, where the address is the sum of a register r2 plus 20. In the M·CORE architecture an instruction to store a word from r1 into the word at the location indicated by r2 plus twenty, is fetched as:

$$0x9152$$

where each byte essentially represents one of the instruction's parameters, and is represented by mnemonics as

```
st.w r1, (r2, 20)
```

in assembly language. The main operation — writing a word into the mailbox (primary memory) from the adding machine (data operator) — is called *memorizing* data. The right hand is used to get the word; it is analogous to the *effective address.*

As with instructions, assembly language uses a shorthand to represent locations in memory. A *symbolic address*, which is actually some address in memory, is a name that means something to the programmer. For example, ALPHA might be the twenty. Then the assembly-language instruction above can be written as follows:

```
st.w r1, (r2, ALPHA)
```

Other symbolic addresses and other locations can be substituted, of course. A symbolic address is just a representation of a number, which usually happens to be the numerical address in primary memory, or an offset of the word in primary memory relative to a register pointer. As a number, it can be added to other numbers, doubled, and so on. In particular, the instruction

```
st.w r1, (r2, ALPHA+4)
```

will store the word from register r1 into the 24th location below that pointed to by r2.

Before going on, we point out a feature of the von Neumann computer that is easy to overlook, but is at once von Neumann's greatest contribution to computer architecture and yet a major problem in computing. Because instructions *and* data are stored in the primary memory, there is no way to distinguish one from the other except by which hand (program counter or effective address) is used to get the data. We can conveniently use memory not needed to store instructions — if few are to be stored — to store more data, and vice versa. It is possible to modify an instruction as if it were data, just before it is fetched, although a good computer scientist would shudder at the thought. However, through an error (*bug*) in the program, it is possible to start fetching data words as if they were instructions, which produces strange results fast.

Generally, after such an instruction has been executed, the left hand (program counter) is in position to fetch the next instruction in box $n + 1$. For instance, if the pair of words shown below are in consecutive locations, they are executed sequentially:

0x6001

0x9152

These instructions are indicated in assembly-language source code in successive lines:

```
movi    r1, 0
st.w    r1, (r2, 20)
```

A *program sequence* is a sequence of instructions fetched from consecutive locations one after another. The program sequence given here cleared the word that is five words below the word whose address is in r2. Unless something is done to change the left hand (program counter), a sequence of words in contiguously numbered boxes will be fetched and executed as a program sequence. For example, a sequence of load and store instructions can be fetched and executed to copy a collection of words from one place in the mailbox into another place. However, when the controller reads the instruction, it may direct the left hand to move to a new location (load a new number in the program counter). Such an instruction is called a *jump*, which is an example of a *control instruction*. Such instructions will be discussed further in §1.2.3, where concrete examples using the M·CORE instruction set are described. To facilitate the memory access functions, the effective address can be computed in a number of ways, called *addressing modes*. M·CORE addressing modes will be explained in §1.2.1.

1.1.2   The Instruction

In this section the concept of an instruction is described from different points of view. The instruction is discussed first with respect to fetching, decoding, and executing them. Then the instruction is discussed in relation to hardware-software trade-offs. Some concepts used in choosing the best instruction set are also discussed.

The controller fetches a word or a couple of words from primary memory and sends commands to all the modules to execute the instruction. An instruction, then, is essentially a complex command carried out under the direction of a single word or a couple of words fetched as an inseparable group from memory.

The bits in the instruction are broken into several fields. These fields may be the bit code for the instruction or for options in the instruction, or for an address in primary memory or an address for some registers in the data operator. For example, the instruction ST.W r1,(r2,5) may look like the hexadecimal pattern 0x9152 when it is completely fetched into the controller. The leftmost nibble - 9 - tells the computer that this is a store word instruction. Each instruction must have a different *opcode* bit sequence, like the first nibble 9, so the controller knows exactly which instruction to execute just by looking at the instruction word. The second nibble from the left - 1 - may identify the register that is to be stored. The third nibble from the left - 5 - may indicate the scaled number to be added to get the address to access the word to be stored. Finally, the last nibble - 2 - may indicate the register to be added to get the address to access the word to be stored. Generally, options, registers, addressing modes, and primary memory addresses differ for different instructions. It is necessary to decode the opcode - 9 - in this example before it can be known that the next nibble - 1 - is a register, the next - 5 - is a number, and the last - 2 - is a register, and so on.

The instruction can be executed by the controller as a sequence of small steps, called *microinstructions*. As opposed to instructions, which are stored in primary memory, microinstructions are usually stored in a small fast memory called *control memory*. A microinstruction is a collection of data transfer *orders* that are simultaneously executed; the *data transfers* that result from these orders are movements of, and operations on, bytes of data as these bytes are moved about the machine. While the control memory that stores the microinstructions is normally ROM, in some computers it can be rewritten by the user. Writing programs for the control memory is called *microprogramming*. It is the translation of an instruction's required behavior into the control of data transfers that carry out the instruction.

The entire execution of an instruction is called the *fetch-execute cycle* and is composed of a sequence of microinstructions. Access to primary memory being rather slow, the microinstructions are grouped into *memory cycles*, which are fixed times when the memory fetches an instruction, memorizes or recalls data, or is idle. A *memory clock* beats out time signals, one clock pulse per memory cycle. The fetch-execute cycle is thus a sequence of memory cycles. The first cycle is the *fetch cycle* when the instruction code is fetched. If the instruction is n bytes long, the first n memory cycles are usually fetch cycles. In some computers, the next memory cycle is a *decode cycle* when the instruction code is analyzed to determine what to do next. The M·CORE processor does not need a separate cycle for this. The next cycle may be for *address calculations*. Then the instruction's main function is done in the *execute cycle*. Finally, the data may be memorized in the last cycle, the *memorize cycle*, or data may be read from memory to the data operator in a *recall cycle*. This fetch-execute sequence is repeated indefinitely as each instruction is fetched and executed.

An instruction may be designed to execute a very complicated operation. In other computers, a sequence of instructions can perform the same thing. It is also generally possible to fetch and execute a sequence of simple instructions to carry out the same net operation. In the M·CORE architecture, a memory word is cleared by the two-instruction sequence movi r1,0 st.w r1,(r2,5). If a useful operation is not performed in a single instruction, but in a sequence of simpler instructions such as the program sequence already described, such a sequence is either a *macro(instruction)* or may be done in a subroutine.

It is a macro if, every time in a program that the operation is required, the complete sequence of instructions is written. It is a subroutine if the instruction sequence is written just once, and a jump to the beginning of this sequence is written each time the operation is required. In many ways macroinstructions and subroutines are similar techniques to get an operation done by executing a sequence of instructions. Perhaps one of the central issues in computer architecture design is this: What should be created as instructions or included as addressing modes, and what should be left out, to be carried out by macros or subroutines? At one extreme, it has been proven that a computer with just one instruction can do anything any existing computer can. It may take a long time to carry out an operation, and the program may be ridiculously long and complicated, but it can be done. On the other extreme, programmers might find complex machine instructions that enable one to execute a high level (for example, C) language statement desirable. Such complex instructions create undesirable side effects, however, such as long latency time for handling interrupts (see the end of §1.2.2). However, the issue is overall efficiency. A computer's instructions are selected on the basis of which can be executed most quickly (speed) and which enable the programs to be stored in the smallest room possible (program density) without sacrificing low I/O latency (time to service an I/O request — see §1.2.2). (The related issue of storing data as efficiently as possible is discussed in §2.2.)

The choice of instructions is complicated by the range of requirements in two ways. Some applications need a computer to optimize speed while others need their computer to optimize program density. For instance, if a computer is used like a desk calculator and the time to do an operation is only 0.1 s, there may be no advantage to doubling the speed because the user will not be able to take advantage of it, while there may be considerable advantage to doubling the program density because memory cost may be halved and machine cost may drop substantially. In another instance, if a computer is used in a computing center with plenty of memory, doubling the speed may permit twice as many jobs to be done, so that the computer center income is doubled, while doubling the program density is not significant because there is plenty of memory available. Moreover, the different applications demanded of computers require different proportions of speed and density.

No known computer is best suited to every application. Therefore, there is a wide variety of computers with different features, and there is a problem picking the computer that best suits the operations for which it will be used. Generally, to choose the right computer from among many, a collection of simple well-defined programs pertaining to the computer's expected use, called *benchmarks*, are available. Some benchmarks are: multiply two unsigned 16-bit numbers, move some words from one location in memory to another, and search for a word in a sequence of words. Programs are written for each computer to effect these benchmarks, and the speed and program density are recorded for each computer. A weighted sum of these values is used to derive a figure of merit for each machine. If storage density is studied, the weights are proportional to the number of times the benchmark (or programs similar to the benchmark) is expected to be stored in memory, and the figure of merit is called *static efficiency*. If speed is studied, the weights are proportional to the number of times the benchmark (or similar routines) is expected to be executed, and the figure of merit is called *dynamic efficiency*. These figures of merit, together with computer rental or purchase cost, available software, reputation for serviceability, and other factors, are used to select the machine.

In this chapter and throughout the subject of software interface design, the issues of efficiency and I/O latency (see the end of §1.2.2) continually appear in the selection instructions for "good" programs. The currently popular RISC (*Reduced Instruction Set Computer*) architectural philosophy exploits the concept of using many very simple instructions to execute a program most efficiently. The M·CORE architecture has a RISC instruction set, with additional instructions that are very useful in handling I/O. The CISC (*Complex Instruction Set Computer*) architectural philosophy uses more complex instructions to execute a program most efficiently. Readers are strongly encouraged to develop the skill of using the most efficient techniques. They should try to select instructions that execute the program the fastest, if dynamic efficiency is prized, or that can be stored in the least number of bytes, if static efficiency is desired.

1.1.3 Microcomputers

One can regard microcomputers as similar to the computers already discussed, but which are created with inexpensive technology. If the controller and data operator are on a single LSI integrated circuit, such a combination of data operator and controller is called a *microprocessor*. If memory and I/O module are added, the result is called a *microcomputer*. If the entire microcomputer (except the power supply and some of the hardware used for I/O) is in a single chip, we have a *single-chip microcomputer*. A *personal computer*, whether small or large, is any computer used by one person at a time, but a microcomputer intended for industrial control rather than personal computing is generally called a *microcontroller*. A microcontroller can be a single-chip or multiple-chip microcomputer. An *embedded* microcomputer or microcontroller is one that is so embedded or integrated into a system as to be indistinguishable from the system; for instance, an embedded microcomputer for an automobile has input/output devices such as a gas peddle, a speedometer, and a spark plug, rather than a printer and a modem.

However, the prefix "micro" is now superfluous, since essentially all von Neumann computers are implemented with VLSI, and it is almost impossible to find a processor that is not a microprocessor, and so on. Herein, we refer to the M·CORE processor as the data operator and controller we study, the M·CORE architecture as the programmer's view of it, and the M·CORE embedded processor as the integrated circuit containing it.

Ironically, this superstar of the 1970s through the 1990s, the microcomputer, was born of a broken marriage. At the dawn of that period, we were already putting fairly complicated calculators on LSI chips. So why not a computer? Fairchild and Intel made the PPS-25 and 4004, which were almost computers, but were not von

Neumann architectures. Datapoint Corporation, a leading and innovative terminal manufacturer and one of the larger users of semiconductor memories, talked both Intel and Texas Instruments into building a microcomputer they had designed. Neither Intel nor Texas Instruments was excited about such an ambitious task, but Datapoint threatened to stop buying memories from them, so they proceeded. The resulting devices were disappointing — both too expensive and much too slow. As a recession developed, Texas Instruments dropped the project, but did get the patent on the microcomputer. Datapoint decided they would not buy it after all, because it did not meet specs. For some time, Datapoint was unwilling to use microcomputers. Once burned, twice cautious. It is ironic that two of the three parents of the microcomputer disowned the infant. Intel was a new company and could not afford to drop the project altogether. So they marketed it as the 8008, and it sold. It is also ironic that Texas Instruments has the patent on the Intel 8008. The 8008 was incredibly clumsy to program and took so many additional support-integrated circuits that it was about as large as a computer of the same power that didn't use microprocessors. Some claim it set back computer architecture at least 10 years. However, it was successfully manufactured and sold. It was in its way a triumph of integrated circuit technology because it proved a microcomputer was a viable product by creating a market where none had existed. The Intel Pentium, designed to be upward compatible to this 8008, is one of the most popular microcomputers in the world.

We will study the M·CORE processor and its architecture in this book because the MMC2001 has 256K bytes of ROM and 32K bytes of SRAM. A single-chip implementation can support a real-time operating system where we can explore the writing of device drivers. Nevertheless, other microcomputers have demonstrably better static and dynamic efficiency or economy for certain applications. Even if they have comparable (or even inferior) performance, they may be chosen because they cost less, have a better reputation for service and documentation, or are available, while the "best" chip does not meet these goals. The reader is also encouraged to be prepared to use other microcomputers if warranted by the application.

The microcomputer has unleashed a revolution in computer engineering. As the cost of microcomputers approaches ten dollars, computers become mere components. They are appearing as components in automobiles, kitchen appliances, toys, instruments, process controllers, communication systems, and computer systems. They replace larger computers in process controllers much as fractional horsepower motors replaced the large motor and belt shaft. They are "fractional horsepower" computers. This aspect of microcomputers will be our main concern through the rest of the book, since we will focus on how they can be interfaced to appliances and controllers. However, there is another aspect we will hardly have time to study, but which will become equally important: their use in conventional computer systems. We are only beginning to appreciate their significance in computer systems. Microcomputers continue to spark startling innovations; however, the features of microcomputers, minicomputers, and large computers are generally very similar. In the following subsections the main features of the M·CORE architecture, a von Neumann RISC architecture, are examined in greater detail. Having learned basic principles on an M·CORE processor, you will be prepared to work with other similar microcontrollers.

1.2 The M·CORE Instruction Set

This section describes the M·CORE instruction set. The M·CORE *Reference Manual*, available from Motorola (document MCORERM/AD), can be used as a more thorough reference to this instruction set. A typical machine has six types of instructions and several addressing modes. Most M·CORE instruction set addressing modes apply to specific instructions. We will describe the M·CORE instructions grouped according to the instruction type, and as we meet new addressing modes, we will discuss them in conjunction with the instructions they apply to. Before we discuss these instructions, we introduce the registers and memory organization of the M·CORE architecture.

The M·CORE processor has a user and a supervisor mode. Typically, the operating system executes in supervisor mode, and user programs run in either mode. The user mode has 16 general purpose registers r0 to r15 while the supervisor mode has these, and an additional set of alternative registers r0' to r15', which can be used for interrupts only, to reduce latency (see the end of §1.2.2). The supervisor mode has 13 additional control registers cr0 to cr12. See Figure 1.2a. Generally, the instructions that use these registers will work with any of them, but a few instructions only work with specific registers. The user or supervisor uses the *program counter* PC to fetch instructions, and a *condition code bit* C to control branching. The condition code bit is in fact the least significant bit of the *program status register* PSR, which is control register cr0, and the most significant bit of this register is the *supervisor bit* S, which is 1 if the processor is running in the supervisor mode, and 0 if in the user mode. Figure 1.3 shows the M·CORE memory. It can be addressed in 8-bit bytes, 16-bit halfwords, and 32-bit words. The least significant bit of a byte, or a halfword, or a word, is bit 0.

1.2.1 M·CORE Data Operator Instructions

The simplest class of instructions is the *move* class, such as load and store. These instructions move data to or from a controller or data operator register, from or to memory. Typically, a third of the program instructions are moves. If an architecture has good move instructions, it will have good efficiency for many benchmarks. (Table 1.1 lists the M·CORE processor's move instructions.)

The simplest instruction of this class, mov can move any GPR to any GPR; it uses an addressing mode called *register addressing* to indicate the register used. The destination register is the register specified first, on the left of the source register. The instruction

```
mov r3,r7
```

will move the 32-bit contents of general purpose register r7 to general purpose register r3. Here and in the following examples, r3 and r7 represent any of the GPRs. Similarly, data can be moved to or from control registers from or to general purpose registers, but only in the supervisor mode. The instruction

```
mfcr r3,cr7
```

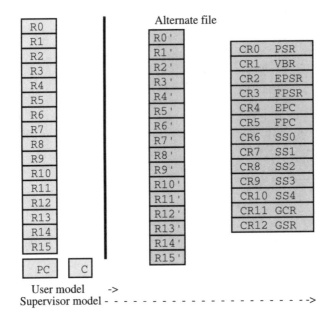

a. The Machine State

31	30	29	28	27	26	25	24	23	22	21	20	19	18	17	16
S	0	Sp		U				0	VEC						

15	14	13	12	11	10	9	8	7	6	5	4	3	2	1	0
TM		TP	TC	0	SC	MM	EE	IC	IE	0	FE	0	0	AF	C

b. Breakdown of the PSR Register (cr0)

Figure 1.2. M·CORE Registers

will move the 32-bit contents of control register cr7 to general purpose register r3, and

```
mtcr r7,cr3
```

will move the full contents of general purpose register r7 to control register cr3. Here and in following the examples, cr3 represents any of the control registers. Six moves use the C bit. The instruction

```
mvc r3
```

puts the C bit into the least significant bit of general purpose register r3, filling the remaining bits of r3 with zeros. The instruction

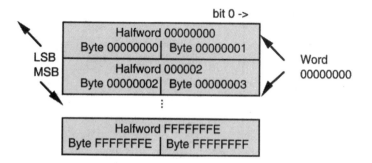

Figure 1.3. M·CORE Memory

```
mvcv r3
```

puts the complement of the C bit into the least significant bit of general purpose register r3, filling the other 31 bits of r3 with zeros. The instruction

```
movt r3,r7
```

will move the 32-bit contents of general purpose register r7 to general purpose register r3 if (and only if) the C bit is 1 (true), and the instruction

```
movf r3,r7
```

will move the full contents of general purpose register r7 to general purpose register r3 if the C bit is 0. Similarly,

```
clrf r3
```

clears r3 if the condition bit C is 0 (false), and the instruction

```
clrt r3
```

clears r3 if the condition bit C is 1 (true). In all the preceeding examples, any general purpose register may be used in place of r3 and r7.

Four moves transfer multiple registers to or from memory. The instruction

```
ldm r13-r15,(r0)
```

will load the 32-bit contents of general purpose registers r13 to r15 from consecutive locations in memory, in decreasing significance from ascending memory

Table 1.1. M·CORE Processor's Move Instructions

mov	mfcr	mtcr
mvc	mvcv	
movt	movf	
ldm	stm	
ldq	stq	
movi	clrf	clrt
ld. [b, h, w]	st. [b, h, w]	
lrw		

locations, beginning at the address given in general purpose register r0. GPR r13 can be replaced by any other GPR, to store all registers from it to GPR r15, inclusive. The instruction

```
stm r13-r15,(r0)
```

will store the 32-bit GPRs r13 to r15 into consecutive locations in memory, in exactly the reverse operation to LDM. Any register, except r0 and r15, may be chosen in place of r13, but the data from the indicated register up to r15, inclusive, are moved. Note that high-numbered registers are more easily saved and restored, using ldm and stm. They are intended to hold a subroutine's local variables. The instruction

```
ldq r4-r7,(r11)
```

will load the 32-bit contents of general purpose registers r4 to r7 from consecutive locations in memory, in increasing significance from ascending memory locations, beginning at the address given in general purpose register r11. Any register, except r4, r5, r6, or r7, may be chosen in place of r11, but the data from register r4 to r7, inclusive, will be moved. The instruction

```
stq r4-r7,(r11)
```

stores the words in general purpose registers r4 to r7 into consecutive locations in memory, performing the inverse of the ldq instruction. Any register, except r4, r5, r6, or r7, may be chosen in place of r11, but the data from register r4 to r7, inclusive, will be moved.

These instructions use *implied addressing*, in which the instruction always deals with the same memory word or register so that no instruction bits specify it. In ldm r13 and stm r13, the contents of general purpose register r0 are implied as the address in memory where the contents of the range of registers are loaded from or stored to. In ldq r11 and stq r11, general purpose registers r4 to r7 are implied as the range of registers loaded from or stored into memory.

Another "nonaddressing" addressing mode is called *immediate addressing*. Herein, part of the instruction is the actual data, not the address of data. For example

```
movi r3,127
```

can write a 7-bit unsigned immediate number such as 127 into a GPR such as r3 shown here. This form of addressing has also been called *literal addressing*.

The instruction, ld.b can load any general purpose register (GPR) with a byte from memory, at an effective address, which is the sum of a GPR and a 4-bit unsigned constant multiplied by the data size. For example

```
ld.b r3,(r7,13)
```

can add a 4-bit unsigned number such as 13 into a GPR such as r7 shown here to get an address, and load the byte at that address into r3 shown here. The number, 13 shown here, is called the *offset*. ld.h similarly loads a 16-bit halfword, but the offset is always even. For example

```
ld.h r3,(r7,26)
```

can add the offset, such as 26, to a GPR, such as r7, to get an address. It loads the halfword at that address (and the next higher address) into r3 shown here. ld.w similarly loads a 32-bit word, but the offset is always a multiple of four. For example

ld.w r3,(r7,52)

can add the offset such as 52 into a GPR such as r7 shown here to get an address, and load the word at that address (and the three next higher addresses) into r3 shown here. Other general purpose registers can be substituted for r3 and r7 shown here. These load instructions load the right bits of the GPR, filling the other bits with zeros. Similarly, st.b, st.h, and st.w store the rightmost 8, 16, or 32 bits of a GPR into memory using ld.b's address mode. These instructions use the mode *index addressing.*

Relative addressing uses a page offset to put a 32-bit constant into a GPR; it is the only way to load an arbitrarily chosen constant into a GPR. This addressing mechanisms is best introduced with an example (see in what follows). Suppose the constant 0x00001004 is at location 0x0000F0B2 and the instruction, lrw r3,[*+20], begins at 0x0000F09C. It loads r3 with 0x00001004. Four times the offset, which is the instruction's least significant byte, 0x05, is added to the PC, which is the address of the next instruction, 0x0000F09E, and then the two least significant bits are cleared, to get the effective address, which is 0x0000F0B0. The instruction then reads the 32-bit word of data there, which is 0x0000, followed by 0x1004, which is 0x00001004, into r3.

location	opcode operand	comment
0000F09C 7305	lrw r3, [*+20]	$0000F0B0 address of *d*
0000F0B0 0000	· · ·	high bytes of *d*
0000F0B2 1004		low bytes of *d*

Different assemblers write a page relative address in different ways. In current HI-WARE C++'s embedded assembly language, which we also use in disassembled code throughout this book, the lrw instruction has the relative address written between square brackets, as shown here. In other assemblers, the lrw instruction has the data at that address in it. For instance, the preceding instruction is written:

lrw   r3,0x00001004

In these other assemblers, the assembler directive .literal causes the 32-bit constant to be written out. In this book, we concentrate on the syntax used in the disassembler.

The M·CORE instruction set has *arithmetic instructions* to be used with 32-bit registers. These instructions add, subtract, multiply, or divide the value of a GPR with the value of another GPR or a constant. See Table 1.2.

The basic 32-bit addu instruction can add any GPR to any GPR. The instruction

addu r3,r7

adds r7 to r3. An unsigned 5-bit immediate operand can be added to any GPR.

addi r3,31

Table 1.2. M·CORE Arithmetic Instructions

addc	addi	addu	
rsub	rsubi	subc	subi
subu			
mult	divs	divu	
cmphs	cmplt	cmplti	tstnbz
cmpne	cmpnei		
decf	dect	incf	inct
decgt	declt	decne	
abs	ffl	ixh	ixw

adds 31 to GPR r3. Neither addu nor addi change the condition code C bit. However, the instruction

addc r3,r7

adds the (former) C bit to r3 and r7, putting the sum in r3, and the carry out into the (updated) C bit. The basic 32-bit subtract instructions subu and rsub can subtract any GPR from any GPR.

subu r3,r7
rsub r3,r7

subu subtracts r7 from r3 putting the result in r3. rsub subtracts r3 from r7 putting the result in r3. A 5-bit immediate operand can be used in place of the source register.

subi r3,31
rsubi r3,31

subi subtracts 31 from r3 putting the result in r3. rsubi subtracts r3 from 31 putting the result in r3. These instructions do not change the condition code C bit. The instruction

subc r3,r7

subtracts the complement of the (former) C bit and r7 from r3, putting the difference in r3, and the borrow out into the (updated) C bit. If the borrow is 0, the C bit is 1.

Instructions can multiply or divide any GPR by any GPR.

mult r3,r7
divs r3,r1
divu r3,r1

The first multiplies r3 by r7 putting the low-order 32 bits of the product into r3. The numbers can be signed or unsigned. The two divide instructions divide any GPR by GPR r1. divs executes signed division, while divu executes unsigned division. A remainder is not produced by either instruction.

Compare instructions can compare any GPR to any GPR to change the condition code C bit. For instance, in the instructions

```
cmphs r3,r7
cmpne r3,r7
cmplt r3,r7
```

cmphs sets C if the unsigned value of r3 is greater than or equal to the unsigned value of r7, cmpne sets C if the value of r3 is not equal to the value of r7, cmplt sets C if the signed value of r3 is less than the signed value of r7. A 5-bit immediate operand can be used in place of the second GPR in the last two instructions shown above:

```
cmpnei  r3,31
cmplti r3,-16
```

cmpne sets C if the value of r3 is not equal to 31 and cmplt sets C if the signed value of r3 is less than −16.

A compare-like instruction is provided that permits testing of register data for the presence of zero bytes. The tstnbz instruction will check each byte of a register. If any byte is all zeros, the condition bit C is cleared, otherwise it is set.

Increment and decrement instructions change a GPR depending on the C bit. In

```
incf r3
inct r3
decf r3
dect r3
```

incf increments register r3 if the C bit is 0 (false), inct increments register r3 if the C bit is 1 (true), decf decrements register r3 if the C bit is 0 (false), and dect decrements register r3 if the C bit is 1 (true). Other decrement instructions change C. In

```
decgt r3
declt r3
decne r3
```

decgt decrements r3 and loads C with 1 if the result left in r3 is greater than zero, otherwise it clears C. Similarly declt decrements r3, loading C with the test: final r3 less than zero, and decne decrements r3 and loads C bit with the test: final r3 not zero.

Four rather unusual instructions are provided. In

```
abs r3
ff1 r3
```

abs puts the absolute value of r3 into r3 and ff1 puts the bit location of the leftmost 1 bit of r3 into r3, where bit 0 is the left (sign) bit. In

```
ixh r3,r7
ixw r3,r7
```

ixh adds twice the value of r7 into r3 and ixw adds four times the value of r7 to r3. These instructions are very useful in indexing into vectors and arrays. They add a scaled value of r7 into a base address in r3.

Addition and subtraction are unsigned, there being no condition code bit available for a signed overflow check. But since data moved into a GPR can be sign-extended using sextb or sexth as will be shown later, and addition and subtraction are 32-bit operations, a 32-bit signed overflow is unlikely. Before a store such as st.b or st.h, the high bits, which are not stored, can be checked to see if they are all zeros or all ones.

The reader should observe that the M·CORE architecture has unusually extensive logic and edit instructions. These instructions are valuable for I/O operations. However, there are comparatively fewer arithmetic and move instructions in this RISC processor.

The *logic instructions* (see Table 1.3) are similar to arithmetic instructions except that they operate logically on corresponding bits of two GPRs, or a GPR and an immediate operand. The instruction:

```
and r3,r7
```

will logically "and," bit by bit, the contents of r7 into r3. For example, if the low-order bits of r3 were 01101010 and those of r7 were 11110000, then after such an instruction is executed, the low-order bits of the result in r3 would be 01100000. In

```
andi    r3,31
andn    r3,r7
tst     r3,r7
```

andi will AND the 5-bit unsigned value 31 into r3, andn will AND the negated value of r7 into r3, and tst sets C if the AND of r3 and r7 is nonzero. Only tst changes the C bit. In

```
or   r3,r7
xor  r3,r7
not  r3
```

or will OR r7 into r3, xor will exclusive-OR r7 into r3, and the complement instruction not will complement each bit in r3. None of these instructions change the C bit.

Bit-oriented instructions permit the setting and testing of individual bits. In the instructions:

```
bclri   r3,31
bseti   r3,31
btsti   r3,31
```

Table 1.3. M·CORE Logic Instructions

and	andi	andn
or	xor	not
bclri	bseti	btsti
bmaski	bgeni	bgenr
tst		

```
bmaski r3,31
bgeni  r3,31
bgenr  r3,r7
```

bclri will clear bit 31 in r3, bseti will set bit 31 in r3, and btsti will copy bit 31 in r3 into the C bit. bmaski will set all the bits to the right of bit 31 in r3, bgeni will set bit 31 (like bseti) but also clear all the other bits of r3. Other immediate operands less than 31 can be used in these instructions. bgenr sets the r7th bit of r3, clearing all the other bits of r3. Note that movi can be used to generate any value less than 127, so bgeni and bmaski may not be used to generate such values.

The next class of instructions — the *edit* instructions (see Table 1.4) — rearrange the data bits without changing their meaning. The M·CORE edit instruction

```
asr r3,r7
```

shifts r3 right arithmetically (filling with sign bits) a number of bits specified by r7. The C bit is not affected. The instruction

```
asrc r3
```

shifts r3 right arithmetically one bit, putting the bit shifted out into C

```
asri r3,31
```

shifts r3 right arithmetically 31 bits. Similar instructions lsr, lsrc, and lsri shift right logically (filling with zeros) and lsl, lslc, and lsli shift left, in similar manner. The instruction

```
rotli r3,31
```

is a circular left shift of register r3 by 31 bit positions. The instruction

```
xsr r3
```

rotates a GPR 1-bit right circular in the 33-bit register consisting of r3 and the C bit. The instruction

```
brev r3
```

reverses the bits in r3, so that bits 0 and 31 are exchanged, bits 1 and 30 are exchanged, and so on.

Table 1.4. M·CORE Edit Instructions

```
asr asrc asri
lsl lsr lslc lsrc lsli
lsri
brev  rotli xsr
sextb sexth zextb zexth
xtrb0 xtrb1 xtrb2 xtrb3
```

sextb r3

sign extends r3 from the low-order 8 to the full 32 bits and

sexth r3

sign extends r3 from 16 to 32 bits. Similarly,

zextb r3

zero extends r3 from 8 to 32 bits and

zexth r3

zero extends a GPR from 16 to 32 bits. The instruction

xtrb0 r3

extracts byte zero (least significant byte) of r3 to the least significant byte of GPR register r1, filling remaining bytes with zero and setting C if that byte is zero. xtrb1 similarly extracts byte one of any GPR, xtrb2 extracts byte two, and xtrb3 extracts byte three. In all these cases, the register r3 may be any GPR but the resulting byte is always put into the least significant byte of r1, and the remaining bits in r1 are cleared.

The next class of instructions is the I/O group for which a wide variety of approaches is used. In most computers, there are 8-bit and 16-bit registers in the I/O devices and control logic in the registers. In other computers there are instructions to transfer a byte or 16-bit word from the accumulator to the register in the I/O device; to transfer a byte or 16-bit word from the register to the accumulator; and to start, stop, and test the device's control logic. In the M·CORE architecture, there are no special I/O instructions; rather, I/O registers appear as words in primary memory (*memory mapped* I/O). The ld.b or ld.h or ld.w instructions serve to input a byte, halfword or word from an input port, and st.b or st.h or st.w serves to output a byte, halfword or word to an output port.

1.2.2   M·CORE Control Instructions

A final instruction group is the *control* group of instructions that affects the program counter. (See Table 1.5.) Next to move instructions, control instructions are most common, so their performance has a strong impact on a computer's performance. In addition, microcomputers with an instruction set missing such operations as floating point arithmetic, multiple word shifts, and high-level language (e.g., C) operations, implement these "instructions" as subroutines rather than macros, to save memory space. These control instructions are now scrutinized.

Table 1.5. M·CORE Control Instructions

br	bf	bt	jmp	jmpi	loopt
bsr	jsr	jsri	trap	rte	rfi
bkpt	wait	doze	stop	sync	

The simplest M·CORE control instruction:

br ALPHA

has encoded in it an 11-bit signed relative address, which is doubled and then added to the program counter PC. Branching can be conditional. bt will branch if C is true and bf will branch if C is false. The instruction

jmp r3

copies r3 into the PC. The jmpi instruction uses essentially the same mechanism as the lrw instruction described in §1.2.1. An example of the jmpi instruction is

location	opcode	operand	comment
3000104A 7004 jmpi	[*+16]		indirect address
		...	
3000105C 3000			high bytes of d
3000105E 1000			low bytes of d

This instruction's execution adds four times the displacement, which is the low byte of the instruction, 0x04, to the current program counter, the address of the next instruction, 0x3000104C, and clears the low-order two bits of this sum. It puts the 32-bit data there, 0x30001000, into the PC. This is generally called *relative indirect addressing*.

If the (previous value of) C is 1,

loopt r3,ALPHA

decrements the GPR r3, and sets the C bit if r3 is positive; then it branches backwards up to 32 byte locations to implement a loop. Otherwise it decrements the GPR and continues to execute the instruction below it. The instruction's offset is doubled, and then added to the program counter PC minus 32, which is put into the PC.

If we move the program intact from one address in memory to another, their relative address remains unchanged. You may use relative addressing of a br in place of register or indirect addressing used in a jmp or jmpi instruction. If a program does not use direct addressing in jump instructions but rather uses branch instructions, we say it has *position independence*. This means a program can be located anywhere in memory, and it will run without change, thus simplifying program loading. This also means that a ROM can be loaded with the program and the same ROM will work wherever it is addressed. Position independence permits ROMs to be usable in a larger range of multiple chip microcontrollers where the ROMs are addressed at different places to avoid conflicts with other ROMs, so they can be sold in larger quantities and will therefore cost less. Relative branch instructions simplify position independence.

Subroutines can be called by three instructions. bsr L0 is like br L0 except that the return address is saved in r15. The second subroutine call is jsr r3. It saves the PC in GPR register r15, and copies r3 into the PC. The last instruction, jsri saves the PC in GPR register 15, and uses relative indirect addressing like jmpi to go to the subroutine. The following example shows how jsri appears in disassembled programs.

location	opcode operand	comment
3000104A 7F03	JSRI [*+12]	relative address
	...	
30001058 3000		high bytes of subroutine addr.
3000105A 1000		low bytes of subrotuine addr.

Incidentally, note that a jsr instruction can copy the PC to r15, so that its value can be used in an expression that computes a relative address to effect position independence. The calculated address is put in the register used by a jmp instruction.

Subroutines that do not call other subroutines are *leaf subroutines*; other subroutines are *nonleaf subroutines* (Figure 1.4). Leaf subroutines (Sub2, Sub3, Sub4, and Sub5) can merely leave the return address in r15, so that jmpr 15 returns to the caller.

For nonleaf subroutines (Main and Sub1) to call other subroutines, to implement *nesting of subroutines*, the programmer has to explicitly save and restore the calling program's return address, which is left in GPR r15 by a jsr instruction, to make room for the subroutine return address, when it calls another subroutine. The programmer has to push the nonleaf subroutine's return address onto a *stack*. In the M·CORE processor, GPR r0 is reserved as a stack pointer and points to the stack's top byte. At the beginning of subroutine A, main's return address in r15 is pushed onto the stack, on top of (in lower memory words than) the other return address. The first instructions in subroutine A can be

```
subi    r0,4
st.w r15,(r0,0)
```

When the nonleaf subroutine completes, it pulls a word from the stack and copies it to the PC to return to the main program, using the following instruction sequence:

```
ld.w    r15,(r0,0)
addi    r0,4
jmp     r15
```

```
subi        r0,12
stm r13-r15,(r0)
```

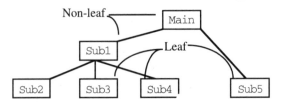

Figure 1.4. Leaf and Nonle   broutines

The subroutine may have local variables. According to Motorola's *Application Binary Interface Standard*, the first seven local variables should be stored in GPR r8 to r14. These can be saved when the return address is saved, and restored when the return address is restored. For instance, if the subroutine has two 32-bit local variables in r13 and in r14, then they are saved by:

```
subi       r0,12
stm r13-r15,(r0)
```

and they are restored using:

```
ldm r13-r15,   (r0)
addi           r0,12
jmp            r15
```

The *stack* fills out, starting at high addresses and building toward lower addresses, in the stack buffer. If it builds into addresses lower than the stack buffer, a *stack overflow* error occurs, and if it is pulled too many times, a *stack underflow* occurs. If no such errors occur, then the last word pushed onto the stack is the first word pulled from it, a property that sometimes labels a stack a LIFO (last in, first out). Overflow or underflow often causes data stored outside of the stack buffer to be modified. This bug is hard to find. You should push some number of bytes on the stack and pull the same number from the stack, never pulling more bytes than you push, to *balance* it.

The stack pointer r0 must be treated with respect. It should be initialized to point to the high address end of the stack buffer in RAM as soon as possible, right after power is turned on, and should not be changed except by incrementing or decrementing it to effectively push or pull words from it. Words above (at lower addresses relative to) the stack pointer must be considered garbage and may not be read after they are pulled.

The instruction

```
trap #3
```

having a 2-bit immediate operand such as 3 (called the trap number) saves the PC and PSR, and then loads the PC with the address stored at 0x40 plus the trap number times four. Hardware interrupts operate essentially the same as trap, but there are *normal interrupts* and *fast interrupts*, as we discuss in Chapter 6. Such interrupts and instructions as trap are called *exceptions*. Normal exceptions save the PSR in cr2 and the PC in cr4, and fast exceptions save the PSR in cr3 and the PC in cr5. In an exception handler, execution is in the supervisor mode. The instruction rte returns from an exception, and rfi returns from a fast interrupt exception, restoring the saved PC and PSR. These instructions generally return execution to the supervisor/user mode in effect before the exception occurred.

The instruction bkpt causes a breakpoint exception; it loads the PC with the address stored at 0x1C. It can be used to stop a program so that the debugger can examine memory or registers, and resume. An illegal instruction can also be useful as a convenient subroutine call to execute I/O operations. Its handler's address is put at location 0x10. Other hardware accelerator "instructions," like floating point add, are serviced by a handler whose address is at location 0x30. Some illegal instructions can be used for special subroutine calls to a monitor, to input or output data from or to a terminal or a personal computer that the designer uses to debug his or her design.

Supervisor mode instructions can cause the processor to stop execution. wait causes the processor to enter low-power wait mode, in which all peripherals continue to run, and doze causes the processor to enter low-power doze mode, in which some peripherals continue to run; stop causes the processor to enter low-power stop mode. The MMC2001, a first implementation of the M·CORE processor family, normally uses 40 ma. at 3.3 V. Both wait and doze modes have current drain of 3 ma. and the stop mode has current drain of only 60 μa. sync causes the processor to suspend fetching new instructions until all previously fetched instructions complete execution.

The (*hardware or I/O*) *interrupt* is an architectural feature that is very important to I/O interfacing. Basically, it is evoked when an I/O device needs service, either to move some more data into or out of the device, or to detect an error condition. *Handling* an interrupt stops the program that is running, causes another program to be executed to service the interrupt, and then resumes the main program exactly where it left off. The program that services the interrupt (called an *interrupt handler or device handler*) is very much like a subroutine, and an interrupt can be thought of as an I/O device tricking the computer into executing a subroutine. An ordinary subroutine called from an interrupt handler is called an *interrupt service routine*. However, a handler or an interrupt service routine should not disturb the current program in any way. The interrupted program should get the same result no matter when the interrupt occurs.

The PSR, PC, and other registers in the controller and data operator are collectively called the *machine state* and are saved and restored whenever an interrupt occurs. Hardware saves and restores PC and PSR. ldm, stm, ldq, and stq can be used for saving and restoring the remaining machine state.

If a subroutine is currently being executed, and the same subroutine is called from within an interrupt handler, or an interrupt service routine, data from the program that was interrupted could get mixed up with data used in the subroutine called from the handler or interrupt service routine, producing errors. If this is avoided, then the subroutine is said to be *reentrant* because it can be entered again, even when it is entered and not finished. Reentrancy is important in designing software for interfaces. Related to it is *recursion* — a property whereby a subroutine can call itself as many times as it wants. While recursion is a nice abstract property and useful in working with some data structures to be discussed in §2.2, it is not generally useful in interfacing; however, recursive subroutines are usually reentrant, and that is important. If the subroutine is reentered, the local data for the subroutine's first execution are saved on the stack as new local data are pushed on top of them and the new data are used by the subroutine's second execution. When the first execution is resumed, it uses the old data. Keeping all local data on the stack this way simplifies implementation of reentrancy.

I/O devices may request an interrupt in any memory cycle. The time from when an I/O device requests an interrupt until data that it wants moved is moved, or the error condition is reported or fixed is called the *latency time*. Fast I/O devices require low latency interrupt service; the fast exception mechanism is provided so that devices requiring low latency can be serviced without having to save and restore registers used by normal exceptions, such as cr2 and cr4. Fast exceptions save the PSR and PC in cr3 and cr5. Additionally, the interrupt can use the alternate register file (Figure 1.2) so that the interrupted program may leave its data in the normal GPRs.

   An omission from the instruction set is a test and set instruction. Such an instruction is used to implement semaphores (§3.2.4). It first tests a memory variable, whether it is zero or not, and then sets it to 1. A zero indicates a resource is available, and setting the memory variable reserves the resource for use. A key aspect of this operation is that it is indivisible. In an improper implementation of this operation, one program might test the memory variable, then another program might test and set it, and finally the original program might finally set it. This operation is improper because both programs testing the variable would see it clear, and conclude that they both had access to the resource. A test-and-set operation can be made indivisible by disabling interrupts before the test, and enabling them after the set to make the test-and-set operation indivisible. However, when multiple processors share a common memory, the test-and-set operation is not made indivisible when interrupts are turned off. But if the test-and-set operation were made in a single memory cycle, the hardware can be designed to make the operation indivisible for each program that runs in any of the processors. So a test-and-set that executes in a single memory cycle would simplify semaphore operations in multicomputer systems.

1.2.3   M·CORE Alias Instructions

Several often-used operations are just special cases of other more general instructions. Because they are often used they are called *alias* instructions, and are listed in Table 1.6. We call these alias instructions because they are merely different names for previously introduced instructions. Motorola's documentation calls these instructions pseudo-instructions.

   The move class instruction

```
                          clc
```

clears the condition code bit (C). It is a special case of cmpne r0, r0. Similarly

```
                          setc
```

which is equivalent to cmphs r0, r0. sets the condition code bit (C). The arithmetic class instruction

```
                        neg  r3
```

is a special case of rsubi r3, 0; it negates the value in r3. The alias instruction

Table 1.6. Alias Instructions for the M·CORE Architecture

clc	setc	cmplei	cmpgt
jbsr	jbr	jbf	jbt
neg	rotic	rotri	
rts	tstle	tstlt	tstne

cmplei   r3,30

is a signed comparison, setting C if r3 is less or equal to 5-bit unsigned constant 30; it is an alias of cmplti r3, 31. Similarly

cmpls   r3,r7

is an alias of cmphs r7, r3. It is an unsigned comparison, setting C if r3 lower or the same as r7, and

cmpgt   r3,r7

which is equivalent to cmplt r7, r3, is a signed comparison, setting C if r3 is greater than r7.

Further,

tstle   r3

which is an alias of cmplti r3, 1, test for a negative or zero value in r3,

tstlt   r3

which is an alias of btsti r3,1, tests for a negative value in r3, and

tstne   r3

which is an alias of cmpnei r3,0, test for a nonzero value in r3. The edit class instruction

rotic   r3,1

which is an alias of addc r3,r3, rotates r3 with carry left by one bit, and

rotri   r3,7

which is equivalent to rotli r3,25, rotates r3 right 7 (i.e., 32 - 25) bits.

A control class instruction

rts

is the return from subroutine; it is an alias of jmp r15. Four alias instructions provide two alternatives, dependent on whether the label is within the range of the shorter branch instruction.

jbsr   label

is either jsr label or bsr label. Similarly

jbr   label

is either jmpi label or br label, and

jbfl   abel

is either bf label or bt .+4 jmpi label. The meaning of .+4 will be explained at the beginning of the next section. Similarly,

jbt   label

is either bt label or bf .+4 jmpi label.

1.3   Assembly-Language Programming

The purpose of this book is to show how C and C++ programs can interface an
M·CORE processor. What assembly language we use will be embedded in these C
and C++ programs. We do not develop a separate section for writing source code
for assemblers in this book. You can try the assembly-language examples and problems in this and later chapters by writing a C-language `main` procedure which
immediately inserts a statement asm { and that ends in a symbol }, which enclosed
the assembly language program you wish to run. The syntax for the embedded
assembly language is described in §2.3.5.

1.4   Organization of M·CORE Microcontrollers

In this section, we describe the block diagram of the MMC2001's hardware — particular implementations of the M·CORE architecture — considering its general implementation and its particular input/output hardware. Because we will present
hardware descriptions in a style similar to the block diagrams programmers commonly
see, we will also call our descriptions *block diagrams*. After then discussing the memory
and I/O organization, we introduce the memory map, which explains the location of
memory and I/O devices so programmers can write instructions to access them.

1.4.1   Notation for Block Diagrams

A block diagram is used to describe hardware organization from the programmer's
point of view (see §1.1.1). It is especially useful for showing how IC's work so a
programmer can focus on the main ideas without being distracted by details unrelated to the software. In this memory-mapped I/O architecture, a register is a
location in memory that can be read or written as if it were a word in memory. A
block diagram shows modules and registers as rectangles, with the most important
inputs and outputs shown around the perimeter. The effects of software instructions
can be shown nicely on a block diagram; for instance, if r2 is 0x4000, the ld.h r1,
(r2, 0) instruction reads a 16-bit word from a certain input port or module; this is
shown as in Figure 1.5. The instruction and the arrow away from the module show
the port can be read (a readable port). If an instruction like st.h r0, (r2, 0)
appears there and an arrow is shown into the module, the port can be written (a
writable port). And if both are shown (ld.h/st.h and a double arrow), the port is
a read-write port. Similarly, an instruction like ld.b indicates an 8-bit input port,
st.b indicates an 8-bit output port, ld.w indicates a 32-bit input port, and st.w
indicates a 32-bit output port.

Figure 1.5. Block Diagram Showing the Effect of an Instruction

1.4.2   M·CORE Microcontroller I/O and Memory Organization

The MMC2001, whose photomicrograph is shown in Figure 1.6, can operate as a single-chip microcontroller, or I/O devices can be added to the address/data/control bus as illustrated in Figure 1.7. The MMC2001 can be the only chip in a system, for it is self-sufficient. The processor, memory, controller, and I/O are all in the chip. The controller and data operator execute the M·CORE instruction set discussed earlier. The memory consists of 32K bytes of static random access memory(SRAM), and 256K bytes of read-only memory (ROM). The I/O devices include an *interrupt controller*, a *keypad*, a *timer*, a *pulse width modulator* (PWM), an *interval mode serial peripheral interface* (ISPI), and a *universal asynchronous receiver/transmitter* (UART). There is also an *on-chip emulation* circuit (OnCE).

Figure 1.6. Photomicrograph of the MMC2001 Chip

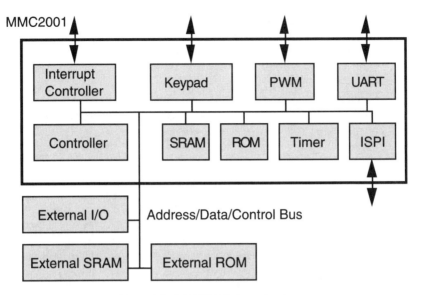

Figure 1.7. MMC2001 Organization

Internally, the MMC2001 can use the 32 address bits, the 32-bit data bus, and control signals, to send the address and send and receive data to or from modules within the chip. Externally, the MMC2001 can use the low-order 20 of the 32 address bits, the 16-bit data bus, and control signals. Memory such as RAM, ROM, or PROM, or I/O devices, can be added to this expanded bus.

1.4.3   The MMC2001 Memory Map

A *memory map* is a description of the memory showing what range of addresses is used to access each part of memory or each I/O device. Figure 1.8 presents a memory map for the MMC2001. At the lowest address, ROM appears, I/O is above 0x10000000, and RAM is above 0x30000000. The memory range from 0x20000000 to 0x2FFFFFFF is intended for external memory. In all these cases, memory is assigned the low-addressed part of each block; using higher addresses than those that are implemented may cause a *transfer error*, which terminates the program.

1.5   Conclusions

In this chapter, we have surveyed the background in architecture needed for microcomputer interfacing. The first section covered bare essentials about von Neumann computers, instructions and what they do, and microcomputers. You will find this background helpful as you begin to learn precisely what happens in an interface.

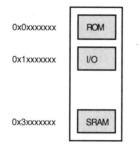

0x0xxxxxxx

0x1xxxxxxx

0x3xxxxxxx

Figure 1.8. Memory Map of the MMC2001

The middle section covered addressing modes and instructions that you may expect in any microcomputer, discussing those in the M·CORE processor in more detail. The general comments there should help if you want to learn about another machine. And the M·CORE architecture comments should help you read the examples and do some of the experiments suggested in the book. A short section provides additional information needed for reading and modifying assembly-language programs that are generated by a compiler.

The final section described a programmer's perspective of some of the hardware used in interfacing. You need to know this perspective to write interfacing programs.

Do You Know These Terms?

All italicized words in this chapter are listed below. Check them to be sure that you recognize their meaning. You can look up those that you do not understand in the index.

von Neumann	static ram (SRAM)	fetching	bug
computer	dynamic ram	program counter	program sequence
architecture	(DRAM)	binary code	jump
organization	read-only memory	hexadecimal	control instruction
implementation.	(ROM)	notation	addressing modes
realization	programmable read-	machine-coded	opcode
primary memory	only memory	mnemonic	microinstruction
data operator	(PROM)	upward compatible	control memory
controller	erasable program-	source code upward	orders
input/output	mable read-only	compatible	data transfers
access	memory	assembler	microprogram
bits	(EPROM)	assembly-language	fetch-execute cycle
byte	electrically erasable	instruction	memory cycle
nibble	programmable	compiler	memory clock
halfword	read-only	high-level language	fetch
word	memory	memorize	decode
random access	(EEPROM)	effective address	address calculations
memory (RAM)	flash	symbolic address	execute cycle

memorize cycle
recall cycle
macro(instruction)
benchmark
static efficiency
dynamic efficiency
Reduced Instruction
 Set Computer
 (RISC)
Complex Instruction
 Set Computer
 (CISC)
microprocessor
microcomputer
single-chip
 microcomputer
personal computer
microcontroller
embedded micro-
 computer
embedded
 microcontroller

program counter
condition code bit
program status
 register
supervisor bit
register addressing
implied addressing
immediate
 addressing
literal addressing
offset
index addressing
relative addressing
arithmetic
 instructions
logic instructions
edit instructions
memory mapped I/O
control
relative indirect
 addressing

position
 independence
leaf subroutines
nonleaf subroutine
nesting of subroutine
stack
stack overflow
stack underflow
balance
exception
interrupt
hardware interrupt
I/O interrupt
Handling an inter-
 rupt
interrupt handler
device handler
interrupt service
 routine
machine state
reentrant
recursion

latency time
alias instructions
block diagram
interrupt controller
keypad
timer
pulse width mod-
 ulator (PWM)
interval mode serial
 peripheral
 interface (ISPI)
universal
 asynchronous
 receiver/
 transmitter
 (UART)
on-chip emulation
memory map
transfer error
move
main procedure

Problems

Problems 1 to 3 in this chapter and many problems in later chapters are paragraph correction problems. We use the following guidelines for all these problems.

The paragraph correction problems have been found useful in helping students understand concepts and definitions. The paragraph has some correct and some erroneous sentences. Your task is to rewrite the paragraph so the whole paragraph is correct without any sentences that do not fit into the paragraph's theme. However, if a sentence is correct, you should not change it, and you cannot use the word "not" or its equivalent to correct the sentence. Consider the first sentence in problem 1. "The architecture is the block diagram of a computer." This is incorrect. It can be made correct by changing "architecture" to "organization," or by changing "block diagram" to either "programmer's view" or "instruction set and I/O connection capabilities." Any of these corrections would be acceptable. The second sentence is correct, however, and should not be rewritten. Try to complete the problems without referring to the chapter, then check your answers by looking up the definitions. If you get a few sentences wrong, you are doing fine. But if you have more trouble, you should reread the sections the problem covers. Paragraph correction problems are marked with an asterisk (*) so you will not confuse them with other problems that are true statements.

1. * The architecture is the block diagram of a computer. Von Neumann invented the architecture used on microcomputers. In it, the controller is analogous to the adding machine. We "recall" words from primary memory into the controller, using the program counter (left hand). Symbolic addresses are used in assembly languages to represent locations in this memory. A macro is a program in another part of memory that is called by a program, so that when the macro is done, the calling program resumes execution at an instruction below the jump to macro. An (I/O) interrupt is like a subroutine that is requested by an I/O device. The latency time is the time needed to completely execute an interrupt. To optimize the speed of execution, choose a computer with good static efficiency. A microcomputer is a controller and data operator on a single LSI chip, or a few LSI chips.

2. * Addressing modes are used by the most common class of instructions, the arithmetic class. Direct addressing has the operand data in a part of the instruction called the immediate value, which is 8-bits long and is sign-extended, for the instruction movi. The M·CORE processor initializes a register with an arbitrarily chosen 32-bit value, that is read using a relative address. Index addressing is especially useful for jumping to nearby locations. The M·CORE processor always accesses arbitrarily chosen memory bytes, words, or halfwords using index addressing, in which a 5-bit offset is sign-extended and added to a general-purpose register to derive the effective address, which is used to access memory.

3. * The M·CORE processor's controller and data operator have 96 bits of register storage, where general purpose registers r0 to r15 are generally interchangeable except that r1 is used to point to the stack, r15 is an implied register for some operations, and r0 is the return address. Add with carry is used in multiple precision arithmetic. The M·CORE architecture has an instruction to multiply one signed number by another signed number. The bkpt instruction is particularly useful as a subroutine call to a fast, short subroutine, because it is a fast instruction.

4. Identify which applications would be concerned about storage density and which about speed. Give reasons for your decisions.

  a. Pinball machine game
  b. Microwave oven control
  c. Home security monitor
  d. Fast Fourier transform (FFT) module for a large computer
  e. Satellite communications controller

5. Write the (hexadecimal) code for a BR L instruction, where the instruction code is at locations 0x12A and 0x12B and

  a. L is location 0x12E
  b. L is location 0x190
  c. L is location 0x12A

6. Write the op code for the following instructions, assuming r0 is 0x1000 and r1 is 0x8000, and explain in words what happens when it is executed (including the effects on the condition code C):

    a. `ld.b`   `r2,(r1,0)`
    b. `st.w`   `r0,(r1,4)`
    c. `addu`   `r0,r1`
    d. `andn`   `r0,r1`
    e. `cmphs` `r0,r1`

7. Write the op code for the following instructions, assuming r0 is 5, r1 is 0x80, r2 is 0x1000, r3 is 0x8000, and C is 1. Give the register values that are changed when these instructions are executed.

    a. `mult`   `r0,r1`
    b. `cmplt` `r0,r1`
    c. `movi`   `r2,50`
    d. `incf`   `r1`
    e. `xtrb2` `r1,r3`

8. Suppose a memory is filled except for the program that follows, like this: the word at address 0xUVSTWXYZ is 0xYZ (for example, location 0x12142538 has value 0x38). Assuming that an address in r1 never points to the program, what will the value of r1 be after each instruction is executed: (r1 is reinitialized to 0x12345678 before each instruction)

$$\text{ld.b r1,(r1,0)}$$
$$\text{ld.h r1,(r1,4)}$$
$$\text{ld.w r1,(r1,8)}$$

9. Suppose the condition code bit C is clear and the `addc r2, r1` instruction is executed. Give the value in the condition code bit C if

    a. Register r2 is 0x77, r1 is 0x77
    b. Register r2 is 0xFFFF80C8, r1 is 0xFFFF8077
    c. Register r2 is 0x1872338C, r1 is 0x000000C8
    d. Repeat part (c) for SUBC r2,r1

10. Give the shortest M·CORE instruction sequences that perform the same operation as the following nonexistent 16-bit M·CORE instructions. Make no assumptions about register contents before these instructions are executed, except that the address of ALPHA is in r6 and C is 0.

    a. `asr ALPHA,4` (shift 32-bit `ALPHA` right arithmetically four places)
    b. `neg r1` (two's complement negate r1)

c. muls (multiply, signed, 16-bit r2 times 16-bit r1 to get 32-bit result in r2.)
d. inc.h ALPHA (increment 16-bit global variable ALPHA)
e. dec.w ALPHA (decrement 32-bit global varialbe ALPHA)

11. How many times is the following loop repeated when the instruction COND is

```
a.  cmpnei    r2,0
b.  cmplti    r2,50
c.  cmpnei    r2,100
                  movi   r2,100
                  clc
          LOOP:   subi   r2,1
                  cond
                  bt     LOOP
```

Show calculations or explain your answers.

12. What is the value of r1 after the following program ends

```
                  movi    r2,100
                  movi    r1,10
                  clc
          LOOP:   addi    r2,1
                  loopt   r1,LOOP
          EXIT:   br      EXIT
```

13. Convert the following C language construct into the shortest M·CORE assembly language instructions, assuming A is a local variable in r5. As long as the expression in the while statement is true, the statements inside curly brackets are repeated.

```
          A = 10;
          while (A > 3)
          { statements;
            A = A -1; }
```

14. Repeat problem 13 with the following high-level programming language construct:

```
          A = 10;
            do
          { statements;
            A = A - 1; }
          while ( A > 3)
```

15. r1 initially contains the value 0x9C. What is the value of r1 after the following instructions:

```
a. bseti   r1,6
b. brev    r1
c. bmaski r1,5
```

16. What are the hexadecimal values of registers r2 and r1 after the DIVU r2, r1 instruction if they contain these values:

a. r2 = 0x40000000, r1 = 0x8000
b. r2 = 0x4000, r1 = 0x80000000

17. Repeat problem 16 with divs instruction and

a. r2 = 0x40000000, r1 = 0x8000
b. r2 = 0x80000000, r1 = 0x4000

18. What are the values of registers r2 and r1 after the mult instruction if they contain these values:

a. r2 = 0x80, r1 = 0x80
b. r2 = 0x8C, r1 = 0x45

19. Write a shortest assembly-language program that adds five 8-bit unsigned numbers stored in consecutive locations, the first of which is at location 0x800, and put the 8-bit sum in 0x814. If any unsigned number addition errors occur, branch to ERROR.

20. Repeat problem 19 where the numbers are 32-bit signed.

21. Repeat problem 20 where the numbers are 16-bit unsigned numbers.

22. Repeat problem 20 where the numbers are 16-bit signed numbers.

23. Write a shortest assembly-language program that adds a 5-byte unsigned number stored in consecutive locations, the first of which is at location 0x803, to a 5-byte unsigned number stored in consecutive locations, the first of which is at location 0x813. If any unsigned number addition errors occur, branch to ERROR.

24. Give the shortest M·CORE instruction sequences to implement 64-bit arithmetic operations for each case given below. In each case, the data arrive in register r1 (high 32 bits) and register r2 (low 32 bits), and are returned in the same way.

a. Complement
b. Increment
c. Decrement
d. Shift right logical

e. Shift right arithmetic
f. Shift left (arithmetic or logical)

25. Give the shortest M·CORE assembly language subroutines to implement 64-bit arithmetic operations for each case given below. In each case, one operand arrives in register r1 (high 32 bits) and register r2 (low 32 bits), and the result is returned in the same way, and the other operand is pushed on the stack, high-order byte at lowest address, just before the subroutine is entered, and is removed by the subroutine just before it returns. For subtract, the number being subtracted is on the stack. Compare subtracts the number on the stack, and returns a value in r1 where bits 3 and 2 have value 0 if less, 1 if equal, and 2 if greater, for signed comparisons, and bits 1 to 0 have value 0 if less, 1 if equal, and 2 if greater, for unsigned comparisons.

a. add
b. subtract (stack value from register)
c. multiply
d. compare, to set r1 if register value is greater than stack value

2

Programming In C and C++

We now consider programming techniques used in I/O interfacing. The interface designer must know a lot about them. As the industry matures, the problems of matching voltage levels and timing requirements, discussed in §4.2, are being solved by better-designed chips, but the chips are becoming more complex, requiring interface designers to write more software to control them.

The state-of-the-art MMC2001 clearly illustrates the need for programming I/O devices in a high-level language, and for programming them in object-oriented languages. The UART and ISPI ports may be a challenge to many assembly-language programmers. But the 32K-byte SRAM and 256K-byte ROM memory is large enough to support high-level language programs. Also, object-oriented features like modularity, information hiding, and inheritance will further simplify the task of controlling M·CORE systems.

This book develops C and C++ interfacing techniques. Chapter 1, describing the architecture of a microcomputer, has served well to introduce assembly language, although a bit more will be done in this chapter. We introduce C in this chapter. The simplest C programming constructs are introduced in the first section. Data structure handling is briefly covered in the next section. Programming styles including the writing of structured, modular, and object-oriented programming will be introduced in the last section. Subroutines will be further studied as an introduction to programming style. The use of classes in C++ will be introduced at the end of this chapter. While this introduction is very elementary and rather incomplete, it is adequate for the discussion of interfacing in this text. Clearly, these concepts must be well understood before we discuss and design those interfaces.

For this chapter, the reader should have programmed in some high-level language. From it, he or she should learn general fundamentals of programming in C or C++ to become capable of writing and debugging tens of statements with little difficulty, and should learn practices specifically applicable to an M·CORE embedded processor. If you have covered this material in other courses or absorbed it from experience, this chapter should bring it all together. You may pick up the material just by reading this condensed version. Others should get an idea of the amount of background needed to read the rest of the book.

2.1   Introduction to C

I/O interfacing has long been done in assembly language. However, experience has shown that the average programmer can write something like ten lines of (debugged and documented) code per day, whether the language is assembler or higher level. But a line of high-level language code produces about 6 to 20 useful lines of assembly-language code, so if the program is written in a high-level language, we might become 6 to 20 times more efficient. We can write the program in a high-level language like C or C++. However, assembly-language code produced by a high-level language is significantly less statically and dynamically efficient, and somewhat less precise, than the best code produced by writing in assembly language, because it generates unnecessary code. Thus in smaller microcontrollers using the 6805, after a high-level language program is written, it may be converted first to assembly language, where it is tidied up, after which the assembly-language program is assembled into machine code. As a bonus, the original high-level language can be used to provide comments to the assembly-language program. In microcontrollers using processors designed for efficient high-level language programming such as an M·CORE processor, C or C++ can control the device without being converted to assembly language. This has the advantage of being easier to maintain, because changes in a C program do not have to be manually translated into and optimized in assembly language. Or finally, a small amount of assembly language, the part actually accessing the I/O device, can be embedded in a larger C program. This approach is generally easier to maintain — ß because most of the program is implemented in C — and yet efficient and precise in small sections where the I/O device is accessed.

We will explain the basic form of a C *procedure*, the simple and the special numeric operators, conditional expression operators, and conditional loop statements, and functions. However, we do not intend to give all the rules of C that you need to write good programs. A C program consists of one or more procedures. The first procedure to be executed is called `main`. Other "subroutines" or "functions" called from `main` are also procedures. All the procedures, including `main`, are written as follows:

```
declaration of global variable;
declaration of global variable;
.
.
return type procedure_name(declaration of parameter_1, ...)
.
.
{
    declaration of local variable;
    declaration of local variable;
    ...
.
.
    statement;
    statement;
    ...
.
.
}
```

Each *declaration of a variable* and each statement ends in a semicolon (;), and more than one of these can be put on the same line. Carriage returns and tabs are equivalent to spaces in C programs (execpt in character strings), and can be used to improve readability. The periods (.) in the example do not appear in C programs, but are meant here to denote that one or more declarations or statements may appear.

Parameters and variables are usually 8-bit (`char`), 16-bit (`short`), or 32-bit (`long`) signed integer types. They can be declared unsigned by putting the word *unsigned* in front of `char`, `short`, or `long`. More than one variable can be put in a declaration; the variables are separated by commas (,). A vector having *n* elements is denoted by the name and square brackets around the number of elements *n*, and the elements are numbered 0 to $n - 1$. For example, the declaration `short a,b[10];` shows two variables, a scalar variable *a* and a vector *b* with ten elements. Variables declared outside the procedure (e.g., before the line with *procedure_name*) are global, and those declared within a procedure (e.g., between the curly brackets after *procedure_name* "{" and "}") are local. C++ permits declarations inside expressions, as in `for (short i = 0; i < 10; i++)`. Parameters will be discussed in §2.3.1. A *cast* redefines a value's type. A cast is put in parentheses before the value. If *i* is a `short`, `(char) i` is a `char`.

Statements may be algebraic expressions that generate assembly-language instructions to execute the procedure's activities. A statement may be replaced by a sequence of statements within a pair of curly brackets "{" and "}". This will be useful in conditional and loop statements to be discussed. Operators used in statements include addition, subtraction, multiplication, and division, and a number of very useful operators that convert efficiently to assembly-language instructions or program segments. Table 2.1 shows the conventional C operators that we will use in this book. Although not all of them are necessary, we use many parentheses so we will not have to learn the precedence rules of C grammar. The following simple C procedure *main* has (signed) 32-bit local variables *a* and *b*; it puts 10 into *a*, 1 into *b*, and the *a* + *b* th element of the 10-element unsigned global 8-bit vector *d* into 8-bit unsigned global *c*, and returns nothing (`void`) as is indicated by the data type to the left of the procedure name.

Table 2.1. Conventional C Operators Used in Expressions

=	make the left side equqal to the expression on its right
+	add
-	subtract
*	multiply
/	divide
%	modulus (remainder after division)
&	logical bit-by-bit AND
|	logical bit-by-bit OR
~	logical bit-by-bit negation
<<	shift left
>>	shift right

unsigned char c,d[10];
void main(void) { long a, b;
    a = 5; b = 1; c = d[a + b];
}

We use the HIWARE C++ compiler to generate machine code. The compiled and linked program is disassembled using the HIWARE Decoder program. We show in what follows the disassembled code. We hasten to note, however, that different code is produced by different compilers, or even by the same compiler when different optimization options are selected, or a different version of the compiler is used. The assembly language, as shown in what follows, is typical, however of what the compiler produces.

0000F090	2470	SUBI	R0,$08	allocate 8 bytes on stk.
0000F092	60A7	MOVI	R7, $05	constant 5 to reg.
0000F094	9700	ST.W	R7, (R0,0)	store reg in local var. a
0000F096	6016	MOVI	R6, 01	constant 1 to reg
0000F098	9610	ST.W	R6, (R0,4)	store in local var. b
0000F09A	1C67	ADDU	R7, R6	add a + b to get index
0000F09C	7505	LRW	R5, [*+20]	$0000F0B0 addr of d
0000F09E	1C75	ADDU	R5, R7	get addr. d[a + b]
0000F0A0	A705	LD.B	R7, (R5,0)	get value of d[a + b]
0000F0A2	7502	LRW	R5, [*+8]	$0000F0AC addr of c
0000F0A4	B705	ST.B	R7, (R5,0)	put into c
0000F0A6	2070	ADDI	R0, 08	deallocate 8 bytes
0000F0A8	00CF	JMP	R15	return to caller
0000F0AA	0000	BKPT		align to multiple of 4
0000F0AC	0000	BKPT		high bytes of addr. c
0000F0AE	1000	MFCR	R0, CR0	low bytes of addr. c
0000F0B0	0000	BKPT		high bytes of addr. d
0000F0B2	1004	MFCR	R4, CR0	low bytes of addr. d

After the procedure *main* is called, the procedure's first instruction SUBI R0, $08 makes room for *(allocates)* local variables on the stack. Note that the first instruction subtracts 8 from the stack pointer r0. Parameters and local variables will be obtained by adding offsets to r0. The last two instructions undo the effect of the initial instructions; ADDI R0, $08 removes the room for *(deallocates)* local variables on the stack and JMP R15 returns to the calling routine. Global variables are obtained by loading a register with the varaible's address and using it as an index register to access the global variable. This procedure *initializes* 32-bit local variables *a* and *b*, and then it uses these to index the vector d, to read an element from it into global variable c, as will be discussed in the next section.

The procedure's 32-bit constants are generated by LRW instructions. Observe that the machine code has an offset from the LRW instruction to the 32-bit constant. The Decode program calculates and displays the address of the constant. The constant is vainly disassembled, but the machine code gives the value of the constant. The reader should ignore the disassembled source code and merge the two 16-bit constant to get the 32-bit constant. An extra 16-bit filler is needed because the 32-bit constant must be aligned to a multiple of four.

Table 2.2 shows some of the very powerful special C operators used in this book. For each operator, an example is given together with its equivalent result using simpler operators. The assignment operator = assigns the value on its right to the variable named on its left and returns the value it assigns so that value can be used in an expression to the left of the assignment operation: the example shows 0 is assigned to *c*, and that value (0) is assigned to *b*, and then that value is assigned to *a*. The increment operator + + can be used without an assignment operator (e.g., *a*++ just increments *a*). It can also be used in an expression in which it increments its operand after the former operand value is returned to be used in the expression. For example, *b* = *a*[*i*++] will use the old value of *i* as an index to put *a*[*i*] into *b*, then it will increment *i*. Similarly, the decrement operator −− can be used in expressions. If the + + or −− appear in front of the variable, then the value returned by the expression is the updated value; *a*[++*i*] will first increment *i*, then use the incremented value as an index into *a*. The next row show the use of the + and = operators used together to represent adding to a variable. The following rows show -| *and* & appended in front of = to represent subtracting from, ORing to, or ANDing to a variable. Shift << and >> can be used in front of the = sign too. This form of a statement avoids the need to twice write the name of, and twice compute addresses for, the variable being added to or subtracted from. The last two rows of Table 2.2 show shift left and shift right operations and their equivalents in terms of simple shift operations.

A statement involving several operations may save intermediate values in other registers. The statement $i = (i << 3) + (i << 1) + c - '0';$ where *i* and *c* are local variables declared as: *unsigned char c; int i;* is compiled into the following assembly language:

```
0000F092   8710        LD.W    R7, (R0, 4)     get local variable i
0000F094   3C37        LSLI    R7, $03         get (i << 3)
0000F096   8610        LD.W    R6, (R0,4)      get local variable i
0000F098   3C16        LSLI    R6, 01          get (i << 1)
0000F09A   1C76        ADDU    R6, R7          get sum of terms so far
0000F09C   A700        LD.B    R7, (R0,0)      get c
0000F09E   1C67        ADDU    R7, R6          get sum of terms so far
0000F0A0   6306        MOVI    R6, 30          generate '0'
0000F0A2   0567        SUBU    R7, R6          get final expression
0000F0A4   9710        ST.W    R7, (R0, 4)     store it in i
```

A statement can be conditional, or it can involve looping to execute a sequence of statements which are written within it many times. We will discuss these control flow statements by giving the flow charts for them. See Figure 2.1 for conditional statements, Figure 2.2 for case statements, and Figure 2.3 for loop statements. These simple standard forms appear throughout the book, and we will refer to them and their figures.

Simple conditional expressions of the form *if then* (shown in Figure 2.1a), full conditionals of the form *if then else* (Figure 2.1b), and extended conditionals of the form *if then else if then else if then ... else* (shown in Figure 2.1c), use conditional

Table 2.2. Special C Operators

operator	example	equivalent to:
=	a=b=c=0;	c=0;b=c;a=b;
++	a++;	a=a+1;
— —	a— —;	a=a-1;
+=	a +=2;	a=a+2;
-=	a -= 2;	a=a-2;
|=	a |= 2;	a=a|2;
&=	a &= 2;	a=a&2;
<<=	a <<= 2;	a=a<<2;
>>=	a >>= 2;	a=a>>2;

Table 2.3.
Condition Expression Operators.

&&	AND
||	OR
!	NOT
>	Greater Than
<	Less Than
>=	Greater than or Equal
<=	Less Than or Equal
==	Equal to
!=	Not Equal To

expression operators (shown in Table 2.3). In the last expression, the *else if* part can be repeated as many times as needed, and the last part can be an optional *else*. Variables are compared using *relational operators* (e.g., > and <), and these are

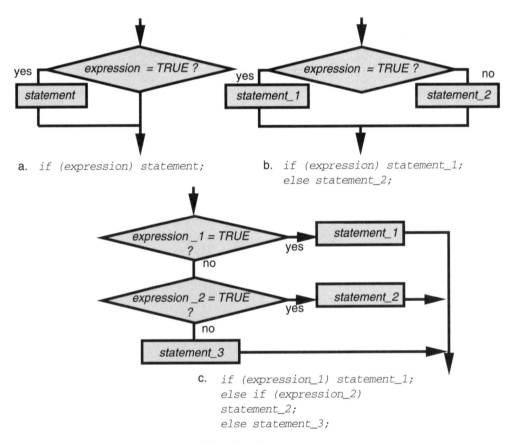

a. `if (expression) statement;`

b. `if (expression) statement_1;`
    `else statement_2;`

c. `if (expression_1) statement_1;`
    `else if (expression_2)`
    `statement_2;`
    `else statement_3;`

Figure 2.1. Conditional Statements.

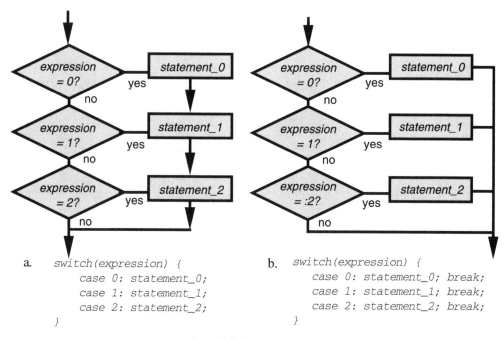

a.
```
switch(expression) {
    case 0: statement_0;
    case 1: statement_1;
    case 2: statement_2;
}
```

b.
```
switch(expression) {
    case 0: statement_0; break;
    case 1: statement_1; break;
    case 2: statement_2; break;
}
```

Figure 2.2. Case Statements.

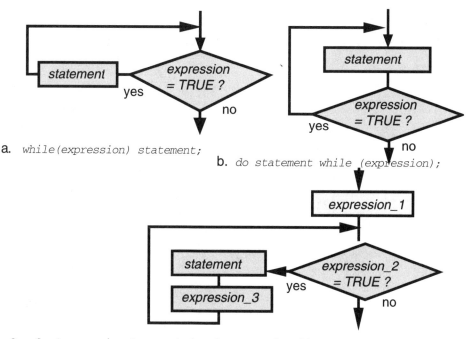

a. `while(expression) statement;`

b. `do statement while (expression);`

c. `for(expression_1;expression_2;expression_3)statement;`

Figure 2.3. Loop Statements.

combined using *logical operators* (e.g., `&&`). For example, `(a>5)&&(b<7)` is true if
`a>5` and `b<7`.

Consider a decision tree using conditional expressions, like `if(alpha! = 0)
beta = 10; else if(gamma == 0) delta++; else if((epsilon! =
0)&&(zeta == 1)) beta = beta<<3;` where each variable is local and of type
`char`. We use this example because it contains many operators just discussed. This
can be coded in assembly language as:

```
0000F092   A720        LD.B    R7, (R0, 2)   get alpha
0000F094   2A07        CMPNEI  R7, $00       see if nonzero
0000F096   E802        BF      *+6           $0000F09C
0000F098   60A7        MOVI    R7, 0A        if so, put 10 in r7
0000F09A   F00F        BR      *+32          $0000F0BA
0000F09C   A730        LD.B    R7,(R0, 3)    get gamma
0000F09E   2A07        CMPNEI  R7, $00       see if nonzero
0000F0A0   E004        BT      *+10          $0000F0AA
0000F0A2   A710        LD.B    R7, (R0,1)    if not, get delta
0000F0A4   2007        ADDI    R7,01         increment it
0000F0A6   B710        ST.B    R7, (R0,1)    put back delta
0000F0A8   F009        BR      *+20          $0000F0BC
0000F0AA   A740        LD.B    R7, (R0,4)    get epsilon
0000F0AC   2A07        CMPNEI  R7, 00        check if it is nonzero
0000F0AE   E807        BF      *+14          $0000F0BC
0000F0B0   A750        LD.B    R7, (R0,5)    get zeta
0000F0B2   2A17        CMPNEI  R7, 01        check if it is not 1
0000F0B4   E004        BT      *+8           $0000F0BC
0000F0B6   A700        LD.B    R7, (R0,0)    get beta
0000F0B8   3C37        LSLI    R7, 03        shift left 3 places
0000F0BA   B700        ST.B    R7, (R0,0)    store it in beta
```

The *case* statement is a useful alternative to the conditional statement. (See
Figure 2.2.) A numerical expression is compared to each of several possible values;
the match determines which statement will be executed next. The case statement
(such as the simple one in Figure 2.2a) jumps into the statements where the variable
matches the comparison value, and executes all the statements below it. The *break*
statement (shown in Figure 2.2b) exits the whole case statement, in lieu of executing
its remaining statements.

An expression `switch(n){ case 1:i = 1; break; case 3:i = 2; break;
case 6: i = 3;break; }` is coded in assembly language by comparing the same
number (the switch operand) against different case constants, conditionally
branching to the case statement's code

```
0000F092   8710        LD.W    R7, (R0, 4)   get n
0000F094   2A17        CMPNEI  R7, $01       check if it is not 1
0000F096   E805        BF      *+12          $0000F0A2
0000F098   2A37        CMPNEI  R7, $03       check if it is not 3
0000F09A   E805        BF      *+12          $0000F0A6
0000F09C   2A67        CMPNEI  R7, $R06      check if it is not 6
```

0000F09E	E805	BF	*+12	$0000F0AA
0000F0A0	F006	BR	*+14	$0000F0AE
0000F0A2	6017	MOVI	R7,$01	i is 1
0000F0A4	F003	BR	*+8	$0000F0AC
0000F0A6	6027	MOVI	R7, $02	i is 2
0000F0A8	F001	BR	*+4	$0000F0AC
0000F0AA	6037	MOVI	R7, $03	i is 3
0000F0AC	9700	ST.W	R7, (R0,0)	store i

An alternative implementation to conditional branch statements, generated in some compilers, is to call up a subroutine followed by a list of case values and relative branch offsets. When a sequence of consecutively numbered cases from 0 to N − 1 is presented, the compiler uses one of its several such subroutines, followed by a list of relative branch offsets.

Loop statements can be used to repeat a statement until a condition is met. A statement within the loop statement will be executed repeatedly. The expressions in both the following loop statements are exactly like the expressions of the conditional statements, using operators as shown in Table 2.3.

The *while* statement of Figure 2.3a tests the condition before the loop is executed and is useful if, for example, a loop may have to be done 0 times. Assume local variable i is initially cleared. Then the statement while(i<10) alpha[i++] = 0; is compiled into assembly language as is illustrated below to clear the local vector alpha[10]. The while statement branches to the loop test at the end, then performs the test, and then branches to the beginning of the loop if the expression is true.

0000F096	F009	BR	*+20	$0000F0AA
0000F098	A700	LD.B	R7, (R0,0)	loop begins: get i
0000F09A	1276	MOV	R6, R7	copy i
0000F09C	2007	ADDI	R7, $01	increment i
0000F09E	B700	ST.B	R7, (R0, 0)	save i
0000F0A0	1207	MOV	R7, R0	get location of locals
0000F0A2	2037	ADDI	R7, $04	get address of alpha
0000F0A4	1C76	ADDU	R6, R7	get address of alpha[i]
0000F0A6	6007	MOVI	R7, $00	clear r7
0000F0A8	B706	ST.B	R7, (R6, 0)	clear alpha[i]
0000F0AA	A700	LD.B	R7, (R0, 0)	test loop cond.: get, i
0000F0AC	2297	CMPLTI	R7, $0A	if 10 greater than i, loop
0000F0AE	E7F4	BT	*−22	$0000F098

The *do while statement* (shown in Figure 2.3b) tests the condition after the loop is executed at least once, but it tests the result of the loop's activities. It is useful in I/O software. For instance, do alpha[i++] = 0; while(i < 10); clears alpha; it compiles into the assembly language code that follows. This code is almost the same as the code for the while statement. However, this do while statement always executes the loop once before checking the loop test at the end, whether or not the compiler can ascertain that the loop test will cause the loop to execute again.

Generally the *do while( )* construct is more efficient than the *while( )* construct, because the latter has an extra branch instruction to jump to its end. But the HI-WARE compiler optimizes *while( )* statements, by removing this initial branch, if the initial value of the condition is able to be determined at compile time to be true.

0000F096	A700	LD.B	R7, (R0, 0)	loop begins: get *i*
0000F098	1275	MOV	R5, R7	get *i*
0000F09A	2007	ADDI	R7, $01	increment *i*
0000F09C	B700	ST.B	R7, (R0, 0)	save *i*
0000F09E	1206	MOV	R6, R0	get location of locals
0000F0A0	2036	ADDI	R6, $04	get location of *alpha*
0000F0A2	1C65	ADDU	R5, R6	get loc. of, *alpha[i]*
0000F0A4	6006	MOVI	R6, $00	constant 0
0000F0A6	B605	ST.B	R6, (R5, 0)	save in *alpha[i]*
0000F0A8	0147	ZEXTB	R7	extend unsigned
0000F0AA	2297	CMPLTI	R7, $0A	if less than 10, loop
0000F0AC	E7F4	BT	*−22	$0000F096

The more general *for statement* (shown in Figure 2.3c) has three expressions separated by semicolons (;). The first expression initializes variables used in the loop; the second tests for completion in the same style as the while statement; and the third updates the variables each time after the loop is executed. Any of the expressions in the for statement may be omitted. For example, *for(i = 0;i<10;i++) alpha[i] = 0;* will clear the vector *alpha* as the above loops did it. It is compiled into assembly language as shown below. As with the while loop, the loop test at the end should be done first, but since the compiler determines that the loop is done at least once, this step is skipped.

0000F096	6007	MOVI	R7, $00	constant 0
0000F098	B600	ST.B	R7, (R0, 0)	loop: get *i*
0000F09A	1207	MOV	R7, R0	get address of locals
0000F09C	2037	ADDI	R7, $04	address of *alpha*
0000F09E	A600	LD.B	R6, (R0, 0)	get *i*
0000F0A0	1C76	ADDU	R6, R7	address of *alpha[i]*
0000F0A2	6007	MOVI	R7, $00	constant 0
0000F0A4	B706	ST.B	R7, (R6, 0)	save in *alpha[i]*
0000F0A6	A600	LD.B	R6, (R0, 0)	get *i*
0000F0A8	2006	ADDI	R6, $01	increment *i*
0000F0AA	B600	ST.B	R6, (R0, 0)	save *i*
0000F0AC	0146	ZEXTB	R6	unsigned extension of *i*
0000F0AE	2296	CMPLTI	R6, $0A	if less than 10, loop
0000F0B0	E7F3	BT	*−24	$0000F098

This program segment is not particularly efficient. An assembly-language program that is equivalent to C statements *p = alpha; do *p++ = 0; while(p ! = alpha+10);* is:

```
        lrw      r8, 0x30001000
        mov      r7, r8            address of local variables
        addi     r7, 8             address of alpha
        movi     r6, 10            start with 10
        add      r6, r7            address of alpha[10]
        movi     r5, 0             constant 0 to be stored
L:      st.b     r5, (r7, 0)       write zero into alpha[i]
        addi     r7, 1             increment pointer
        cmpne    r7, r6            compare to end
        bt       L                 if not, loop again
```

However, the C compiler may not actually generate the assembly language shown here. If you need tight code, you may have to insert the assembly-language code into a C procedure, in a manner to be shown in §2.3.5. Note also that, as in this example, if it were efficiently compiled, the clearest C program does not always lead to the most efficient assembly-language program. A less clear program may generate better code.

The *break* statement will cause the *for, while,* or *do while* loop to terminate just as in the case statement, and may be used in a conditional statement. For instance, *for(;;) {i++; if(i == 30) break;}* executes the statement *{i++; if(i == 30) break;}* indefinitely, but the loop is terminated when *i* is 30.

An important feature of C, extensively used to access I/O devices, is its ability to describe variables and addresses of variables. If *a* is a variable, then &*a* is the address of *a*. If *a* is a variable that contains an address of another variable *b*, then **a* is the contents of the word pointed to by *a*, which is the contents of *b*. (Note that *a*b* is *a* times *b* but **b* is the contents of the word pointed to by *b*.) Whenever you see &, read it as "address of," and whenever you see *, read it as "contents of thing pointed to by." In a declaration statement, the statement *char *p;* means that the thing pointed to by *p* is a character, and *p* points to (contains the address of) a character. In an assignment statement, **p = 1;* means that 1 is assigned to the value of the thing pointed to by *p*, whereas *p = 1;* means that the pointer *p* is given the value 1. Similarly, *a = *p;* means that *a* is given the value of the thing pointed to by *p*, while *a = p;* means *a* gets the value of the pointer *p*. Some C compilers will give an error message when you assign an integer to a pointer. If that occurs, you have to use a cast. Write *p = (int *)0x4000;* to tell the compiler 0x4000 is really a pointer value to an integer and not an integer itself.

Finally a comment is anything enclosed by /* and */, or anything after//. Comments can be put anywhere in your program, except within quotation marks. We strongly encourage you to supply comments to document your code.

2.2   Data Structures

Data structures are at least as important as programming techniques, for if the program is one-half of the software, the data and their structure are the other half. When we discuss storage density as an architecture characteristic, we discuss only the

amount of memory needed to store the program. We are also concerned about data
storage and its impact on static and dynamic efficiency, as well as the size of memory
needed to store the data. Prudent selection of the data structures a program uses can
shorten or speed up the program. These considerations of data structures are critical
in microcontrollers.

A data structure is one among three views of data. The *information structure* is
the view of data the end user sees. For instance, the end user may think of his of her
data as a table, such as Table 2.1 in this book. The programmer sees the same data as
the *data structure:* strongly related to the way the data are accessed but independent
of details such as size of words and position of bits. It is rather like a painter's
template, which can be filled in with different colors. So the data structure may be an
array of characters that spell out the words in Table 2.1. The *storage structure* is the
way the information is actually stored in memory, right down to the bit positions.
Therefore the table may appear as an array of 8-bit words in the storage structure.

Practical engineers often find data structures hard to accept. These provide a
level of abstraction that facilitates some overall observations that can be applied to
similar storage techniques. For instance, if we can develop a concept of how to access
an array, we can use similar ideas to access arrays of 8-bit or 24-bit data, even
though the programs could be quite different. But here we must stress that a data
structure is simply a kind of template that tells us how data are stored and is also a
menu of possible ways the data can be written or read. Two data structures are
different if they have different templates that describe their general structure or if the
menus of possible access techniques are different.

Constants are often used with data structures, for instance to declare a size of a
vector and to use that same number in *for* loops. They can be defined by *define* or
enum statements, put before any declarations or statements, to equate names to
values. The define statement begins with characters *#define* and does not end with
a semicolon.

#define ALPHA 100

Thenceforth, we can use the label ALPHA throughout the program, and 100 will
effectively be put in place of ALPHA just before the program is actually compiled.
This permits the program to be better documented, using meaningful labels, and
easier to maintain, so that if the value of a label is changed it is changed everywhere
it occurs.

A number of constants can be created using the enum statement. Unless reinitialized with an = sign, each member has a value one greater than the value of the
previous member. The first member has value 0. Hexadecimal values are prefixed
with zero ex (0x):

enum { BETA, GAMMA, DELTA = 0x5};

indicates that *BETA* has value 0, *GAMMA* has value 1, and *DELTA* has value 5.

In a declaration, any scalar variable can be initialized by use of a "=" and a
value. For instance, if integers *i, j,* and *k* have values 1, 2 and 3, we write a global
declaration:

short i = 1, j = 2, k = 3;

C procedures access global variables using direct addressing, and such global variables may be initialized in a procedure _startup that is executed just before main is started. Initialized local variables of a procedure should generate machine code to initialize them just after they are allocated each time the procedure is called. The procedure

```
void fun(void){
        short i, j, k; /* allocate local variables */
        i = 1; j = 2; k = 3; /* initialize local variables */
}
```

is equivalent to the procedure

```
void fun(void){
        short i = 1, j = 2, k = 3; /* allocate and initialize local variables */
}
```

Data structures divide into three main categories: indexable, sequential, and linked. Indexable and sequential, discussed here, are more important. Linked structures are very powerful, but are not as easy to discuss in abstract terms. They will be discussed later.

2.2.1   Indexable Data Structures

Indexable structures include vectors, lists, arrays, and tables. A *vector* is a sequence of elements, where each element is associated with an index *i* used to access it. To make address calculations easy, C associates the first element with the index 0, and each successive element with the next integer (*zero-origin indexing*). Also, the elements in a vector are considered numbers of the same *precision* (number of bits or bytes needed to store an element). We will normally consider 1-word precision vectors, although we soon show an example of how the ideas can be extended to *n*-word vectors. Finally the *cardinality* of a vector is the number of elements in it. A vector is fully specified if its origin, precision and cardinality are given. A zero-origin, 16-bit, 3-element vector 31, 17, and 10 is generated by a declaration short v[3] and stored in memory as (hexadecimal):

                                        001F
                                        0011
                                        000A

and we can refer to the first element as v[0], which happens to be 31. However, the same sequence of values could be put in a zero-origin vector of three 8-bit elements, generated by a declaration char u[3] and stored in memory as:

                                        1F
                                        11
                                        0A

The declaration of a global vector variable can be initialized by use of a " = " and a list of values, in angle brackets. For instance, the three-element global integer vector *v* can be allocated and initialized by

$$short \; v[3] = \{31,17,10\};$$

The vector *u* can be similarly allocated and initialized by the declaration

$$char \; u[3] = \{31,17,10\};$$

The procedure `fun(void)` in §2.1 illustrated the accessing of elements of vectors in expressions. The expression `c = d[a+b];` accessed the *a+b* th element of the 8-bit 10-element vector *d*. The term "vector" is a general term, and similar declarations and statements can be used for 16-bit and 32-bit or other precision, and for other cardinality vectors. The concept of "data structure" is to generalize the storage and access used in one instance to cover other instances of the same kind of data handling technique. When reading the assembly code generated by C, be wary of the implicit multiplication of the vector's precision (in bytes) when calculating offset addresses of elements of the vector. And because C does not check that indexes are within the cardinality of a vector, your C program must be able to implicitly or explicitly assure this to avoid nasty bugs — when a vector's data is inadvertently stored outside the memory allocated to a vector.

    A *list* is like a vector, being accessed by means of an index, but the elements of a list can be any combination of different precision words, code words, and so on. For example, the list can have three elements: the 1-byte number 5, the 2-byte number 7, and the 1-byte number 9. This list is stored in machine code in some computers as follows:

05

0007

09

The powerful *structure* mechanism is used in C to implement lists. The mechanism is implemented by a declaration that begins with the word *struct* and has a definition of the structure within angle brackets, and a list of variables of that structure type after the brackets, as in

$$struct \; \{ \; char \; 11; \; short \; 12; \; char \; 13; \} \; list;$$

A globally defined list can be initialized as we did with vectors, as in

$$struct \; \{ \; char \; 11; \; short \; 12; \; char \; 13; \} \; list = \{5, \; 7, \; 9\};$$

The data in a list are identified by "dot" notation, where a dot "." means "element." For instance, `list.11` is the `11` element of the list `list`. If *P* is a pointer to a `struct`, then arrow notation, such as `P->11`, can access the element `11` of the list. The `typedef` statement, although it can be used to create a new data type in terms of existing data types, is often used with `structs`. If `typedef a struct { char 11; short 12; char 13; } list;` is written, then `list` is a data type, like `short` or `char`, and can be used in declarations such as `list b;` that declares *b* to be an instance of type *list*. We will find the *typedef* statement to be quite useful when a *struct* has to be declared many times and pointers to it need to be declared too. A structure can have *bit fields* which are unsigned integer elements having fewer than 32 bits. Such a structure as

$$struct \; \{unsigned \; a{:}1, \; b{:}2, \; c{:}3{;}\}1{;}$$

has a 1-bit field *1.a*, 2-bit field *1.b* and 3-bit field *1.c*. A *linked list* structure, a list wherein some elements are addresses of (the first word in) other lists, is flexible and powerful. It is widely used in advanced software, including interfacing.

An *array* is a vector whose elements are vectors of the same length. We normally think of an array as a two-dimensional pattern, as in

1	2	3
4	5	6
7	8	9
10	11	12

An array is considered a vector whose elements are themselves vectors. The array is stored in *row major order*: in this arrangement a row is stored with its elements in consecutive memory locations. (In *column major order* a column is stored with its elements in consecutive memory locations.) For instance, the global declaration

$$short \; ar1[4][3] \; = \; \{\{1,2,3\},\{4,5,6\},\{7,8,9\},\{10,11,12\}\};$$

allocates and initializes a row major ordered array *ar1*, and *a = ar1[i][j];* puts the row-*i* column-*j* element of *ar1* into *a*, as shown in the previous example of an array.

A *table* is to a list as an array is to a vector. It is a vector of identically structured lists (rows). Tables often store characters, where either a single character or a collection of n consecutive characters are considered elements of the lists in the table. Index addressing is useful for accessing elements in a row of a table, especially if the table is stored in row major order. If the address register points to the first word of any row, then the displacement can be used to access words in any desired column. Also, autoincrement addressing can be used to select consecutive words from a row of the table.

In C, a table *tbl* is considered a vector whose elements are structures. For instance, the declaration

$$struct \; \{char \; 11; \; short \; 12; \; char \; 13;\} \; tbl[3];$$

allocates a table whose rows are similar to the list *list* above. The "dot" notation with indexes can be used to access it, as in

$$a{=}tbl[2].11;$$

In simple compilers, multidimensional arrays and *structs* are not implemented. They can be reasonably simulated using one dimensional vectors. The user becomes responsible for generating vector index values to access row-column elements or *struct* elements.

2.2.2 Sequential Data Structures

The other important class of data structures is sequential structures, which are accessed by relative position. Rather than having an index i to get to any element of the structure, only the "next" element to the last one accessed may be accessed in a sequential structure. Strings, stacks, queues, and deques are sequential structures important in microcontrollers.

A *string* is a sequence of elements such that after the ith element has been accessed, only the $(i + 1)$st element, the $(i - 1)$th, or both, can be accessed. In particular, a string of characters, stored using the *ASCII code* shown in Table 2.4, is an ASCII *character string* and is used to store text. The ASCII code of a character is stored as a 7-bit code in a char variable. Character constants are enclosed by single quotes around the character, as 'A' is the character A. Special characters are *null* '\0', line feed '\n', tab '\t', form feed '\f' (begin on new page), *cr* carriage return '\r', and ' ' space. Strings are allocated and used in C as if they were char vectors, initialized by putting the characters in double quotes, and end in the null character '\0'. (Allow an extra byte for it.)

One can initialize a character c to be the code for the letter a and a string s to be *ABCD* with the declaration:

```
char c='a', s[5]=''ABCD'';
```

Strings are also very useful for input and output when debugging C programs; we will discuss the use of strings when we describe the *printf* function later. However, a source-level debugger for a C compiler provides better debugging tools. Even so, some discussion of string-oriented input and output is generally desirable for human interfacing. This discussion of *IoStreams* will be done in §4.4.5 and §5.3.5.

Table 2.4. ASCII Codes

	00	10	20	30	40	50	60	70
0	'\0'		"	0	@	P	'	p
1			!	1	A	Q	a	q
2			"	2	B	R	b	r
3			#	3	C	S	c	s
4			$	4	D	T	d	t
5			%	5	E	U	e	u
6			&	6	F	V	f	v
7			'	7	G	W	g	w
8			(	8	H	X	h	x
9			)	9	I	Y	i	y
A	'\n'		*	:	J	Z	j	z
B			+	;	K	[	k	{
C	'\f'		,	<	L	\	l	|
D	'\r'		-	=	M	]	m	}
E			.	>	N	^	n	~
F			/	?	O	_	o	

   Characters in strings can be accessed by indexing or by pointers. An index can be incremented to access each character, one after another from first to last character. Alternatively, a pointer to a character such as *p* can be used; *p* is the character that it points to and * (p++) returns the character pointed to and then increments the pointer *p* to point to the next character in the string.

   The characters you type on a terminal are usually stored in memory in a character string. You can use a typed word as a command to execute a routine, with unique words for executing each routine. A C program to compare a string that has just been stored in memory, and pointed to by *p*, to a string stored in *start_word*, is shown as *main(void)* next:

```
char *p, start_word[6] = ''START''; /* assume p points to a string stored elsewhere */
    void main (void) { int i, nomatch;
      for(i = nomatch = 0; i<5; i++) {
            if(*(p++) ! = start_word[i]) { nomatch = 1; break; }
      }
      if (nomatch == 0) strt(); /* if string is START then execute the strt proc */
    }
```

Inside the loop, we compare a character at a time of the input string against the string *start_word*. If we detect any difference, we set local variable *nomatch* because the user did not type the string *start_word*. However, if all five characters match up — the user did type the word "START" — the program calls *strt*, presumably to start something. The assembly language for this C program is listed here:

```
0000F094   25F0   SUBI     R0,$20              allocate local variables
0000F096   007B   STM      R11-R15,(R0)        save registers used here
0000F098   600E   MOVI     R14,$00             clear r14
0000F09A   DEB0   ST.H     R14,(R0,22)         clear nomatch
0000F09C   F013   BR       *+40                $0000F0C4 to loop end
0000F09E   7D10   LRW      R13,[*+64]          $0000F0E0 address of p
0000F0A0   8E0D   LD.W     R14,(R13,0)         get p
0000F0A2   12EB   MOV      R11,R14             copy p
0000F0A4   200E   ADDI     R14,$01             increment p
0000F0A6   9E0D   ST.W     R14,(R13,0)         replace p
0000F0A8   CEA0   LD.H     R14,(R0,20)         get i
0000F0AA   9E60   ST.W     R14,(R0,24)         save copy of i
0000F0AC   017E   SEXTH    R14                 signed extend i
0000F0AE   7C0B   LRW      R12,[*+44]          $0000F0DCstart_word
0000F0B0   1CEC   ADDU     R12,R14             addr. start_word[i]
0000F0B2   AE0B   LD.B     R14,(R11,0)         get *p
0000F0B4   AB0C   LD.B     R11,(R12,0)         get start_word[i]
0000F0B6   0FEB   CMPNE    R11,R14             if equal
0000F0B8   E803   BF       *+8                 $0000F0C0 skip next 3
0000F0BA   601E   MOVI     R14,$01             else set to 1
0000F0BC   DEB0   ST.H     R14,(R0,22)         save nomatch
0000F0BE   F006   BR       *+14                $0000F0CC exit
```

```
0000F0C0  8E60  LD.W    R14,(R0,24)     get saved i
0000F0C2  200E  ADDI    R14,$01         increment i
0000F0C4  DEA0  ST.H    R14,(R0,20)     save as i
0000F0C6  017E  SEXTH   R14             sign extend i
0000F0C8  224E  CMPLTI  R14,$05         see if 5 characters
0000F0CA  E7E9  BT      *-44            $0000F09E if so, loop
0000F0CC  CEB0  LD.H    R14,(R0,22)     get nomatch
0000F0CE  017E  SEXTH   R14             sign extend nomatch
0000F0D0  2A0E  CMPNEI  R14,$00         check nomatch
0000F0D2  E001  BT      *+4             $0000F0D6
0000F0D4  FFDD  BSR     *-68            if 1, execute strt
0000F0D6  006B  LDM     R11-R15,(R0)    restore registers used here
0000F0D8  21F0  ADDI    R0,$20          deallocate local variables
0000F0DA  00CF  JMP     R15             return to caller
0000F0DC  0000  BKPT                    address of start_word
0000F0DE  1000  MFCR    R0,CR0
0000F0E0  0000  BKPT                    address of p
0000F0E2  1008  MFCR    R8,CR0
```

Besides character strings, bit strings are important in microcontrollers. In particular, a very nice coding scheme called the *Huffman* code can pack characters into a bit stream and achieve about a 75% reduction in storage space when compared to storing the characters directly in an ASCII character string. It can be used to store characters more compactly and can also be used to transmit them through a communications link more efficiently. As a bonus, the encoded characters are very hard to decode without a code description, so you get a more secure communications link using a Huffman code. Further, we need to handle data structures and bit shifting using the << and >> operators and bit masking using the & operator in many I/O procedures. Procedures for Huffman coding and decoding provide a rich set of examples of these techniques.

We recommend that you test your ability to read C by studying the procedures below. We also suggest that you compile these procedures and step through them using a high-level debugger. In this example, we are particularly interested in pointing out that strings may have elements other than characters (here they are bits). Further, the elements of strings can be themselves strings or other data structures, provided such data are decipherable.

The code is rather like Morse code, in that frequently used characters are coded as short strings of bits, just as the often-used letter "e" is a single dot in Morse code. To ensure that code words are unique and to suggest a decoding strategy, the code is defined by a tree having two branches at each branching point (*binary tree*), as shown in Figure 2.4. The letters at each end (leaf) are represented by the pattern of 1s and 0s along the branches from the left end (root) to the leaf. Thus, the character string MISSISSIPPI can be represented by the bit string 111100010001011011010. Note that the ASCII string would take 88 bits of memory while the Huffman string would take 21 bits. When you decode the bit string, start at the root and use each successive bit of the bit string to guide you up (if 0) or down (if 1) the next branch until you get

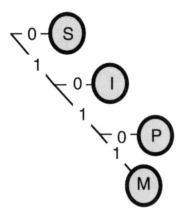

Figure 2.4. A Huffman Coding Tree.

to a leaf. Then copy the letter and start over at the root of the tree with the next bit of the bit string. The bit string has equal probabilities of 1s and 0s, so techniques used to decipher the code based on probabilities will not work. It is particularly hard to break a Huffman code.

A C program for Huffman coding is shown below. The original ASCII character string is stored in the *char* vector *strng*. We will initialize it to the string *MISSISSIPPI* for convenience, although any string of M, I, S, and P letters could be used. The procedure converts this string into a Huffman coded 48-bit bit string stored in the vector *code[3]*. It uses the procedure *shift(void)* to shift a bit into *code*. This procedure, shown at the end, is also used by the decoding procedure shown between these procedures.

```
extern short shift(void);
short code[3], bitlength; /* output code and its length */
char strng[12] = ''MISSISSIPPI''; /*input code, terminated in a NULL character */
struct table{ char letter; char charcode[4]; } codetable[4]
        = { 'S','XX0'', 'I',''X10'', 'P',''110'','M',''111'' };
void main(void){ int row, i; char *point, letter;
    for (point = strng; *point ; point++ ){
        for (row = 0; row < 4; row ++){
            if (((*point) & 0x7f) == codetable[row].letter){
                for (i = 0; i < 3; i++){
                    letter = codetable[row].charcode[i];
                    if (letter ! = 'X')
                        { shift();code[2] = (letter&1);
                        bitlength++;}
                }
            }
        }
    }
    i = bitlength; while((i++)<48)shift(); /* shift out unchanged bits */
}
```

Huffman decoding, using the same *shift(void)*, is done as follows:

```
short code[3] = {0xf116, 0xd000, 0}, bitlength = 21; /* input string */
    char tbl[3][2] = {{'S',1},{'I',2},{'P','M'}};
    char strng[20]; /* output string */
    void main(void){ int row,entry; char *point;
        point = strng; row = 0;
        while((bitlength-)>0){
            if((entry = tbl[row][shift()]) < 0x20) row = entry;
            else {row = 0; *(point++) = entry }
        }
        *point = '\0'; /* terminate C string with NULL character */
    }
    shorft shift(void) { int i;
        i = 0; if(code[0] & 0x8000) i = 1; code[0] = code[0] <<1;
        if (code[1] & 0x8000) code[0]  = 1; code[1] = code[1] <<1;
        if (code[2] & 0x8000) code[1]  = 1; code[2] = code[2] <<1;
        return(i);
    }
```

Now that we have shown how nice the Huffman code is, we must admit a few problems with it. To efficiently store some text, the text must be statistically analyzed to determine which letters are most frequent, in order to assign these the shortest codes. Note that S is most common, so we gave it a short code word. There is a procedure for generating the best Huffman code, which is presented in many information theory books, but you have to get the statistics of each letter's occurrences to get that code. Nevertheless, although less than perfect, one can use a fixed code that is based on other statistics if the statistics are reasonably similar. Finally, although the code is almost unbreakable without the decoding tree, if any bit in the bit string is erroneous, your decoding routine can get completely lost. This may be a risk you decide to avoid because the code has to be sent through a communications link that is as error-free as possible.

A *deque* is a generalized data structure that includes two special cases: the stack and the queue. A deque (pronounced deck) is a sequence of elements with two ends we call the top and the bottom. You can only access the top or bottom elements on the deque. You can *push an element on top* by placing it on top of the top element, which makes it the new top element, or you can *push an element on the bottom*, making it the new bottom element. Or you can *pull (or pop) the top element*, deleting the top element from the deque, making the next to top element the top element and putting the old top element somewhere else, or *pull (or pop) the bottom element* in like manner.

Deques are theoretically infinite, so you can push as many elements as you want on either the top or bottom: But practical deques have a maximum capacity. If this capacity is exceeded, we have an *overflow* error. Further, if you pull more elements than you push an *underflow error* exists.

In the procedure that follows, note that you cannot really associate the ith word from either end of a deque with a particular location in memory. In fact, in a pure sense, you can only access the deque's top and bottom words and cannot read or write any other word in the deque. In practice, we sometimes access the ith element from the deque's top or bottom by using a displacement with the indexes that point to the top and bottom words — but this is not a pure deque. We call it an *indexable deque* to give it some name.

C declarations and programs for initializing, pushing and pulling words from a deque are shown below. The *buffer* is an area of memory set aside for use as the deque expands, and cannot be used by any other data or program code. The programmer allocates as much room for the buffer as appears necessary for the worst case (largest) expansion of the deque. Two indexes are used to read or write on the top or bottom, and a counter is used to detect overflow or underflow. The deque buffer is implemented as a 10-element global vector deque, and the indexes as global *unsigned chars top* and *bottom* initialized to the first element of the deque, as in the C declaration

```
unsigned char deque [10], size, error, top, bottom;
```

As words are pulled from top or bottom, more space is made available to push words on either the top or bottom. To take advantage of this, we think of the buffer as a ring or loop of words, so that the next word below the bottom of the buffer is the word on the top of the buffer. That way, as words are pulled from the top, the memory locations can become available to store words pushed on the bottom as well as words pushed on the top, and vice versa. Then to push or pop data into or from the top or bottom of it, we can execute the following procedures:

```
void pstop(int item_to_push) {
      if ((++size) > 10) error = 1;
      if (top == 10) top = 0; deque[top++] = item_to_push;
}
int pltop(void) {
      if ((- size ) < 0) error = 1;
      if (top == 0) top = 10;
      return(deque[- - top]);
}
void psbot(int item_to_push) {
      if ((++size) > 10) error = 1;
      if (bottom == 0) bottom = 10;
      deque[- bottom] = item_to_push;
}
int plbot(void) {
      if ((- size ) < 0) error = 1;
      if (bottom == 10) bottom = 0;
      return( deque[bottom++] );
}
```

A *stack* is a deque in which you can push or pull on only one end. We have discussed the stack accessed by the stack pointer S, which permits the machine to push or pull words from the top of the stack to save registers for procedure calls, as well as `trap`, `bkpt`, and hardware interrupts. Now we consider the stack as a special case of a deque. (Actually, the M·CORE processor stack can be made a special case of indexable deque using r0 with index addressing.) It is an example of a stack that pushes or pulls elements from the top only. Another equally good stack can be created that pushes or pulls elements only from the bottom of the deque. In fact, if you want two different stacks in your memory, have one that pushes and pulls from the top and another that pushes and pulls from the bottom. Then both stacks can share the same buffer, as one starts at the top of this buffer (lowest address) and builds downward, while the other starts at the bottom (highest address) and builds upward. A stack overflow exists when the top pointer of the stack that builds upward is equal to the bottom pointer of the stack that builds downward. Note that if one stack is shorter, then the other stack can grow longer before an overflow exists, and vice versa. You only have to allocate enough words in the buffer for the maximum number of words that will be in both at the same time.

Programs to push or pull on the two stacks are simpler than the general program that operates on the general deque because the pointers do not roll around the top or bottom of the buffer.

The final structure that is important in microcomputer systems is the *queue*. This is a deque in which we can push data on one end and pull data from the other end. In some sense, it is like a shift register, but it can expand if more words are pushed than are pulled (up to the size of the buffer). In fact, it has been called an elastic shift register. Or conversely, a shift register is a special case of a queue, a fixed-length queue. Queues are used to store data temporarily, such that the data can be used in the same order in which they were stored. We will find them very useful in interrupt handlers and procedures that interact with them.

One of the rather satisfying results of the notion of data structures is that the stack and queue, actually quite different concepts, are found to be special cases of the deque. The two structures are, in fact, handled with similar programs. Other data structures — such as multidimensional arrays, trees, partially ordered sets, and graphs such as lattices and banyans — are important in general programming. You are invited to pursue the study of data structures to improve your programming skills. However, this section has covered the data structures we have found most useful in microcomputer interface software.

2.3   Writing Clear C Programs

M·CORE systems are almost always large enough to consider the advantages of writing clear assembly-language programs over the expense of writing short programs. There may be reason for concern about static efficiency in smaller microcontrollers. An implementation of the Motorola 6805 has so little memory — 2K words of ROM — and the programs are so short that static efficiency is paramount and readability is less important to a good programmer, who can comprehend even poorly written programs. Readability is significant for programs larger than 16K — even for a good programmer — because programs may have to be written by several programmers who read each other's code, and may have to be maintained long after the original programmers have gone.

A significant technique for writing clear programs is good documentation, such as using comments and flow charts. Of course, these do not take up memory in the machine code, so they can be used when static or dynamic efficiency must be optimized. Another technique is the use of consistent programming styles that constrain the programmer, thereby reducing the chance of errors and increasing the reader's ease of understanding. A major idea in clear programming methodology is modular top-down design. We also need to develop the concept of object-oriented programming. In order to discuss these ideas, we first need to further refine our understanding of procedures and arguments.

2.3.1   C Procedures and Their Arguments

Conceptually, *arguments* or *parameters* are data passed from or to the calling routine to or from a procedure, like the x in $sin(x)$. Inside the procedure, the parameter is declared as a variable, such as y, as in *short sin(short y) { ... }*. It is called the *formal parameter*. Each time the procedure is called the calling routine uses different variables for the parameter. The variable in the calling routine is called the *actual parameter*. For example, at one place in the program we put *sin(alpha)*, in another, *sin(beta)*, and in another, *sin(gamma)*. *alpha, beta*, and *gamma* are actual parameters.

At the conceptual level, arguments are called by value, result, reference, or name. In *call by value*, the actual parameters themselves are passed into the procedure. In *call by result*, a formal parameter inside the procedure is usually left in a register (r1). This value is usually then used in an expression or stored into its actual parameter after the procedure is finished.

In *call by reference*, or *call by name*, the data remain stored in the calling routine and are not actually moved to another location, but an address of the data is given to the procedure and the procedure uses this address to get the data whenever necessary. In C or C++, call by name puts & in front of an actual parameter but the formal parameter has an asterisk * in front; see actual parameter *a* and corresponding formal parameter *b* below:

```
void main(void) { char a;     void f(char *b) {
        f(&a);                        *b = '1';
    }                             }
```

C++ uses call by reference; & is put in front of a formal parameter but the actual parameter has no operator in front; see actual parameter *a* and corresponding formal parameter *b* below:

```
void main(void) { char a;     void f(char &b) {
        f(a);                         b = '1';
    }                             }
```

A procedure in C may be called by another procedure in C as a function. The arguments may be the data themselves, which is call by value, or the address of the data, which is call by name.

HIWARE's C compiler passes a function's single input and output argument in a register r2. Consider the function: *short square(short i) {return i*i;}* coded as:

```
0000F090   0322   MULT   R2,R2                        multiply arg. by itself
0000F092   00CF   JMP    R15                          return to caller
```

The input in r2 is multiplied by itself. It leaves the result in r2. This convention is followed by all single argument functions, which return a value. Note how an efficient compiler such as HIWARE's C++ compiler optimizes this subroutine.

Local variables can be held in registers, and if these overflow, on the stack. The following example, the function: *unsigned short swap(unsigned short i) { return (i >> 8) (i << 8); }*, shows a slightly more complex subroutine.

```
0000F090   1227   MOV    R7,R2                        duplicate sub's argument
0000F092   0162   ZEXTH  R2                           unsigned number extend
0000F094   3C82   LSLI   R2,$08                       shift left 8 bits
0000F096   0167   ZEXTH  R7                           unsigned number extend
0000F098   3E87   LSRI   R7,$08                       shift right 8 bits
0000F09A   1E72   OR     R2,R7                        combine
0000F09C   00CF   JMP    R15                          return
```

The registers can be used for up to six arguments, which use registers r2 to r7. Consider *RaisePower(&k, &j, i)* which returns in *k* the value *j* to the power *i*, where *i*, *j*, and *k* are integers; *i* is passed by value, and *j* and *k* are passed by name.

```
void RaisePower (short *k, short *j, short i) {
      for(*k = 1; i-; ) *k = *k * *j;
   }
```

The called procedure is implemented as

```
0000F090   24F0 SUBI    R0,$10                        allocate 16 bytes
0000F092   007C STM     R12-R15,(R0)                  save used registers there
0000F094   0174 SEXTH   R4                            extend argument i
0000F096   6017 MOVI    R7,$01                        put 1
0000F098   D702 ST.H    R7,(R2,0)                     into *k
0000F09A   122E MOV     R14,R2                        get k
0000F09C   123D MOV     R13,R3                        get j
0000F09E   124C MOV     R12,R4                        get i
0000F0A0   F004 BR      *+10                          $0000F0AA
0000F0A2   C70E LD.H    R7,(R14,0)                    get *k
0000F0A4   C60D LD.H    R6,(R13,0)                    get *j
0000F0A6   0376 MULT    R6,R7                         calculate new product
0000F0A8   D60E ST.H    R6,(R14,0)                    put into *k
0000F0AA   12C7 MOV     R7,R12                        get i
```

```
0000F0AC  240C SUBI     R12,$01        decrement i
0000F0AE  0177 SEXTH    R7             sign extend i
0000F0B0  2A07 CMPNEI   R7,$00         if nonzero, loop
0000F0B2  E7F7 BT       *-16           $0000F0A2
0000F0B4  006C LDM      R12-R15,(R0)   restore used registers
0000F0B6  20F0 ADDI     R0,$10         deallocate 16 bytes
0000F0B8  00CF JMP      R15            return
```

Call by value, as i is passed, does not allow data to be output from a procedure, but any number of call by value input parameters can be used in a procedure. Actual parameters passed by name in the calling procedure have an ampersand "$\&$" prefixed to them to designate that the address is put in the parameter. In the called procedure, the formal parameters generally have an asterisk "$*$" prefixed to them to designate that the data at the address are accessed. Observe that call by name formal parameters j or k used inside the called procedure all have a prefix asterisk "$*$". A call by name parameter can pass data into or out of a procedure, or both. Data can be input to a procedure using call by name, because the address of the result is passed into the procedure and the procedure can read data at the given address. A result can be returned from a procedure using call by name, because the address of the result is passed into the procedure and the procedure can write new data at the given address to pass data out of the procedure. Any number of call by name input/output parameters can be used in a procedure.

A procedure may be used as a function that returns exactly one value and can be used in the middle of algebraic expressions. The value returned by the function is put in a *return statement*. For instance, the function *power* can be written

```
short power(short i, short j) { short k,n;
     for(n = 1, k = 0;k<j;k++) n = n*i; return n;
}
```

This function can be called within an algebraic expression by a statement a = *power(b,2)*. The output of the function named in the `return` statement is passed by call by result. Its assembly-language code is given below:

```
0000F090  24F0  SUBI     R0,$10         allocate 16 bytes
0000F092  007C  STM      R12-R15,(R0)   store regs used here
0000F094  0173  SEXTH    R3             expand j
0000F096  601C  MOVI     R12,$01        put 1 into n
0000F098  122E  MOV      R14,R2         move i to r14
0000F09A  123D  MOV      R13,R3         move j to r13
0000F09C  F001  BR       *+4            $0000F0A0 to loop end
0000F09E  03EC  MULT     R12,R14        recompute n
0000F0A0  12D7  MOV      R7,R13         move j to r7
0000F0A2  240D  SUBI     R13,$01        decrement j
0000F0A4  0177  SEXTH    R7             sign extend j
0000F0A6  2A07  CMPNEI   R7,$00         if nonzero, loop
0000F0A8  E7FA  BT       *-10           $0000F09E
0000F0AA  006C  LDM      R12-R15,(R0)   restore regs used here
```

```
0000F0AC  20F0  ADDI          R0,$10          deallocate 16 bytes
0000F0AE  00CF  JMP           R15             return to caller
```

In C, the address of a character string can be passed into a procedure, which uses a pointer inside it to read the characters. For example, the string *s* is passed to a procedure *puts* that outputs a string by outputting to the user's display screen one character at a time using a procedure *putchar*. The procedure puts is written

```
void puts(s) char *s; {
    while(*s! = 0) putchar(*(s++));
}
```

It can be called in either of three ways, as shown side by side:

```
void main(void){      void main(void){         void main(void){
  char s[6] = ''ALPHA'';  char s[6] = ''ALPHA'';    puts(''ALPHA'');
  puts(&s[0]);            puts(s);}                }
}                     }
```

The first calling sequence, though permissible, is clumsy. The second is often used to pass different strings to the procedure, while the third is better when the same constant string is passed to the procedure in the statement of the calling program. The third calling sequence is often used to write prompt messages out to the user and to pass a format string to a formatted input or output procedure like *printf*, to be described shortly.

A *prototype* for a procedure can be used to tell the compiler how arguments are passed to and from it. At the beginning of a program we write all prototypes, such as

```
extern void puts(char *);
```

The word *extern* indicates that the procedure *puts(char *)* is not actually here but is elsewhere. The procedure itself can be later in the same file or in another file. The argument *char* * indicates that the procedure uses only one argument and it will be a pointer to a character (i.e., the argument is called by name) . In front of the procedure name a type indicates the procedure's result. The type *void* indicates that the procedure does not return a result. After the prototype has been declared, any calls to the procedure will be checked to see if the types match. For instance, a call *puts('A')* will cause an error message because we have to send the address of a character (string), not a value of a character to this procedure.

The prototype for *power* is:

```
extern short power(short, short);
```

to indicate that it requires two arguments and returns one result, all of which are call-by-value-and-result 16-bit signed numbers. The compiler will use the prototype to convert arguments of other types if possible. For instance, if *x* and *y* are 8-bit signed numbers (of type *char*) then a call *power (x,y)* will automatically extend these 8-bit to 16-bit signed numbers before passing them to the procedure. If a procedure has a *return n* statement that returns a result, then the type statement in front of the procedure name indicates the type of the result. If that type is declared to be *void* as in the `puts` procedure, there must not be a `return n` statement that returns a result.

At the beginning of each file, prototypes for all procedures in that file should be declared. While writing a procedure name and its arguments twice, once in a prototype and later in the procedure itself, may appear clumsy, it lets the compiler check for improper arguments and, where possible, instructs it to convert types used in the calling routine to the types expected in the called routine. We recommend the use of prototypes.

The *macro* is similar to a procedure, but is either evaluated at compile time or is inserted into the program wherever it is used, rather than being stored in one place and jumped to whenever it is called. The macro in C is implemented as a *#define* construct. As *#defines* were earlier used to define constants, macros are also "expanded" just before the program is compiled. The macro has a name and arguments rather like a procedure, and the rest of the line is the body of the macro. For instance

$$\textit{\#define f (a,b,c) a = b*2+c}$$

is a macro with name `f` and arguments `a`, `b`, and `c`. Wherever the name appears in the program, the macro is expanded and its arguments are substituted. For instance if `f ( x, y, 3 )` appeared, then `x = y * 2 + 3` is inserted into the program. Macros with constant arguments are entirely evaluated at compile time, and generate a constant used at run time.

Procedures `getchar` and `gets` input characters and character strings, `InDec` and `InHex` input decimal numbers and hexadecimal numbers, and `putchar`, `puts`, and `printf` output characters, character strings, and formatted character strings. These very powerful functions are actually executed in a host computer on whose keyboard the user is typing and on whose screen the user is reading the results, rather than the target M·CORE processor, to avoid the loading of a lot of machine code along with the program, which may be a serious problem if the target computer's memory is limited. They are not available when the M·CORE processor is used without a host computer in a real standalone system. The target M·CORE processor being debugged is designed to execute predetermined illegal instructions to actually execute these procedures in the host computer. When the monitor gets an illegal instruction "interrupt," the host computer reads the instruction from the target computer's memory. If it is one of the predetermined illegal instructions selected to call these input-output procedures, the host will execute the procedures, examining the target memory, and writing data into the target memory as needed, and then resume the program after the illegal instruction. C procedures `strcpy` and `strcat` are also very useful but can be easily loaded into the target computer to manipulate strings being input or output.

The procedure `printf` requires a character string format as its first parameter and may have any number of additional parameters as required by the format string. The format string uses a percent sign "%" to designate the input of a parameter, and the characters following the "%" sign establish the format for the output of the parameter value. While there are a large number of formats, we generally use only a few. The string "%d" will output the value in decimal. For instance, if *i* has the value 123, then

```
printf(''The number is %d'', i);
```

will print on the terminal

The number is 123

Similarly, "%X" will output the value in hexadecimal. If *i* has the value 0x1A, then

```
printf(''The number is 0x%X'', i);
```

will print on the terminal

0x1A

If a number is put between the % and d or X letters, that number gives the maximum number of characters that will be printed.

Similarly, "%s" will output a string of characters passed as a parameter. For instance,

```
char st[6] = ''ALPHA'';
printf(''%s'', st);
```

will print on the terminal

ALPHA

Observe that the integers for decimal or hexadecimal output are passed by value, but the string is passed by name as we discussed at the end of §2.3.1.

```
printf(''Hi There\nHow are you?'');
```

will print on the terminal:

Hi There
How are you?

Decimal numbers can be input using *inDec*, and hexadecimal numbers are input using *inHex*. They stop inputting when a nondigit character is typed in. Character strings input using *gets* can be analyzed and disassembled using indexes in or pointers to the strings. Character strings can be assembled for output using *puts* by the procedures `strcpy` and `strcat`. The procedure `strcpy (s1,s2);` will copy string *s2* (up to the null character at its end) into string *s1*. The procedure `strcat (s1,s2);` will concatenate string *s2* (up to the null character at its end) onto the end of string *s1*. These simple procedures are shown below.

```
void strcpy(char *s1, char s2) { while(*s2) *s1++ = *s2++; }

void strcat(char *s1, char s2) {while(*s1)s1++; while(*s2) *s1++ = *s2++;}
```

We have examined techniques for calling subroutines and passing arguments. We have also learned to use some simple tools for input and output in C. We should now be prepared to write subroutines for interface software.

2.3.2 Programming Style

We conform our programming techniques to some style to make the program easier to read, debug, and maintain. The use of a consistent style is recommended, especially in longer programs where static efficiency is not paramount. For instance we can rigidly enforce reentrancy and use some conventions to make this rather automatic. Another programming style — structured programming — uses only simple conditional and loop operations and avoids GOTO statements. After this we discuss top-down and bottom-up programming. This will lead to an introduction to object-oriented programming.

An element of *structured programming* is the use of single entry point, single exit point program segments. This style makes the program much more readable because, to get into a program segment, there are no circuitous routes which are hard to debug and test. The use of C for specification and documentation can force the use of this style. The conditional and loop statements described in §2.1 are single entry point, single exit point program segments, and they are sufficient for almost all programs. The *while* loop technique is especially attractive because it tests the termination condition before the loop is done even once, so programs can be written that accommodate all possibilities, including doing the loop no times. And the *for* loop is essentially a beefed-up *while* loop. You can use just these constructs. That means avoiding the use of GOTO statements. Several years ago, Professor Edsger Dijkstra made the then controversial remark, "GOTOs Considered Harmful!" Now, most good programmers agree with him. We heard a story (from Harold Stone) that Professor Goto in Japan has a sign on his desk that says "Dijkstras Considered Harmful!" (Professor Goto denies this.) The only significant exception is the reporting of errors. We sometimes GOTO an error reporting or correcting routine if an error is detected — an abnormal exit point for the program segment. Errors can alternatively be reported by a convention such as using the carry bit to indicate the error status: You exit the segment with carry clear if no errors are found and exit with carry set if an error is found. Thus, all segments can have single exit points.

Top-down design produces programs more quickly than ad hoc and haphazard writing. You write a main program that calls subroutines (or just program segments) without yet writing the subroutines (or segments). A procedure is used if a part of the program is called many times, and a program segment — not a procedure — is used if a part of the program is used only once. The abstract specification is translated into a main program, which is executed to check that the subroutines and segments are called up in the proper order under all conditions. Then the subroutines (or segments) are written in lower-level subroutines (or segments) and tested. This is continued until the lowest-level subroutines (or segments) are written and tested. Superior documentation is needed in this methodology to describe the procedure and program segments so they can be fully tested before being written. Also, subroutine inputs and outputs have to be carefully specified.

The inverse of top-down design is bottom-up design, in which the lowest-level subroutines or program segments are written first and then fully debugged. These are built up, bottom to top, to form the main program. To test the procedure, you write a short program to call the procedure, expecting to discard this program when the next higher level program is written. Bottom-up design is especially useful in interface design. The lowest-level procedure which actually interfaces to the hardware is usually the trickiest to debug. This methodology lets you debug that part of the program with less interference from other parts. Bottom-up design is like solving an algebra problem with three separate equations, each equation in one unknown. Arbitrarily putting all the software and hardware together before testing any part of it is like simultaneously solving three equations in three unknowns. As the first algebraic problem is much easier, the use of bottom-up design is also a much easier way to debug interfacing software. In a senior level interfacing course at the University of Texas, students who tried to get everything working at once spent 30 hours a week in the lab, while those who used bottom-up design spent less than 10 hours a week on the same experiments.

Combinations of top-down and bottom-up design can be used. Top-down design works well with parts of the program that do not involve interfacing to hardware, and bottom-up design works better with parts that do involve interfacing.

2.3.3   Object-Oriented Programming

The concept of object-oriented programming was developed to program symbolic processes, database storage and retrieval systems, and user-friendly graphic interfaces. However, it is ideally suited to the design of I/O devices and systems that center on them. It provides a programming and design methodology that simplifies interface design.

Object-oriented programming began with the language SMALLTALK. Programmers using C wanted to use object-oriented techniques. Standard C cannot be used, but a derivative of C, called C++, has been developed to utilize objects with a syntax similar to that of C. A M·CORE C++ compiler written by HIWARE was used to generate code for M·CORE-based microcontrollers to check out the ideas described below.

An object's data are *data members* and its procedures are *member functions;* data and member functions are *encapsulated* together in an *object*. Combining them is a good idea because the programmer becomes aware of both together and logically separates them from other objects. As you get the data, you automatically get the member functions used on it. In the class for a queue shown below, observe that data members $QSize$, $QLen$, $Error$, QIn, $QOut$ and $QEnd$ are declared much as in a C struct, and member functions $push$, $pull$, and $error$ are declared pretty much like prototypes are declared in C. The class member protection terms, $protected$, $public$, and $virtual$, will be soon explained.

```
class Queue {
    protected: char Error; short *QIn, *QOut, *QEnd;
    public: char QSize, QLen; Queue(short); char error(void);
    virtual void push(short); virtual short pull(void);
};
```

A class' member functions are written rather like C procedures with the return type and class name in front of two colons and the member function name.

```
void Queue:: push (short i)
    {if((QLen + = 2)>QSize)Error = 1;if(QEnd == QIn)QIn- = QSize;
        *(QIn++) = i;}
short Queue:: pull (void)
    {if((QLen- = 2)<0)Error = 1;if(QEnd == QOut)QOut- = QSize; return
        *(QOut++);}
char Queue:: error(void){ char i; i = Error; Error = 0; return i; }
```

Any data member, such as *QSize*, may be accessed inside any member function of class *queue*, such as *push*. Inside a member function when a name appears in an expression, the name is first searched against local variables and function formal parameters. If the name matches, the variable is local or an argument. Then the variable is matched against the object data members, and finally against the global variables. In a sense, object data members are global among the member functions because each of them can get to these same variables. However, it is possible that a data member and a local variable or argument have the same name such as *size*. The data member can be identified as *this->size*, using the key word *this* as a pointer to the object that called the member function, while the local variable or argument is just *size*.

C++ uses constructors, allocators, destructors, and deallocators. An *allocator* allocates data member storage. A *constructor* initialize these variables, it has the same function name as the class name, a *destructor* terminates the use of an object, and a *deallocator* recovers storage for data members for later allocation. We do not use a deallocator in our experiments; it is easier to reset the M·CORE processor to deallocate storage. A destructor has the same function name as the class name, but has a tilde ($\sim$) in front of the member function name. We will use destructors later. Here is *Queue*'s constructor:

```
Queue::Queue(short i)
    {QEnd=(QIn=QOut=(short*)allocate(i))+(QSize=i);QLen=Error=0;}
```

Throughout this text, a conventional C procedure *allocate* provides buffer storage for an object's data members, as its allocator, and for an object's additional storage such as its queues. The contents of global variable *free* are initialized to the address just above the last global; storage between *free* and the stack pointer is subdivided into buffers for each object by the *allocate* routine. The stack used for return addresses and local variables builds from one end and the allocator builds from the other end of a common RAM buffer area. *allocate*'s return type *void* * means a pointer to anything.

```
char *free = (char*)0x30001000;
void *allocate(short i) { void *p = free; free + = i; return p; }
```

A global object of a class is declared and then used as shown below:

```
Queue Q(10);
void main(void) { short i;
    Q.push(1); i = Q.pull();
}
```

The object's data members, *QSize*, *QLen*, *Error*, *QIn*, *QOut*, and *QEnd*, are stored in global memory just the way a global *struct* is stored. If a data member could be accessed in *main*, as in *i = Q.Error* or *i = Qptr->Error* (we see later that it can not be accessed from *main*), the data member is accessed by using a predetermined offset from the base of the object exactly as a member of a C *struct* is accessed. Member functions can be called using notation similar to that used to access data in a *struct*; *Q.push(1)* calls the *push* member function of *Q* to push 1 onto *Q*'s queue. The "*Q.*" in front of the member function is rather like a first actual parameter, as in *push(Q, 1)*, but can be used to select the member function to be run, as we will see later, so it appears before the function.

The class' constructor is executed before the main procedure is executed, to initialize the values of data members of the object. This declaration *Q(10)* passes actual parameter *10* to the constructor, which uses it, as formal parameter *i*, to allocate 10 bytes for the queue. The queue is stored in a buffer assigned by the *allocate* routine.

Similarly a local object of a class can be declared and then used as shown below:

```
void main(void) { short i; Queue Q(10);
    Q.push(1); i = Q.pull();
}
```

The data members *QSize*, *QLen*, *Error*, *QIn*, *QOut*, and *QEnd*, are stored on the stack, and the constructor is called just after *main* is entered to initialize these data members; it then calls *allocate* to find room for the queue. The member functions are called the same way as in the first example when the object was declared globally.

Alternatively, a pointer *Qptr* to an object can be declared globally or locally; then an object is set up and then used as shown below.

```
void main(void) { Queue * Qptr; short i;
    Qptr = new Queue (20);
    Qptr ->push(1); i = Qptr ->pull();
}
```

In the first line, *Qptr*, a pointer to an object of class *queue*, is declared here as a local variable. (Alternatively it could have been declared as a global variable pointer.) The expression *Qptr = new Queue (20);* is put anywhere before the object is used. This is called *blessing* the object. The allocator and then the constructor are both called by the operator *new*. *allocate* automatically provides room for the data members *QSize*, *QLen*, *Error*, *QIn*, *QOut*, and *QEnd*. The constructor explicitly calls up the allocate procedure to obtain room for the queue itself, and then initializes all the object's data members. After it is thus blessed, the object can be used in the program. Alternatively, use a pointer to an object with a *#define* statement to insert the asterisk, as follows:

```
#define Q (*Qptr)
void main(void) { Queue *Qptr = new Queue(20); short i;
    Q.push(1); i = Q.pull();
}
```

Wherever the symbolic name Q appears, the compiler substitutes *(*Qptr)* in its place. Note that **ptr.member* is the same as *ptr->member*. So this makes the syntax of the use of pointers to objects match the syntax of the use of objects most of the time. However, the blessing of the object explicitly uses the pointer name.

A hierarchy of derived and base classes, inheritance, overriding, and factoring are all related ideas. These are described below, in that order.

A class can be a *derived class* (also called *subclass*) of another class, and a hierarchy of classes can be built up. We create derived classes to use some of the data or member functions of the base class, but we can add members to, or replace some of the members of, the base class in the derived class. For instance the aforementioned class *Queue* can have a derived class *CharQueue* for *char* variables; it declares a potentially modifiable constructor and different member functions *pull* and *push* for its queue. When defining the class *CharQueue* the *base class* (also called *superclass*) of *CharQueue* is written after its name and a colon as :*Queue*. A class such as *Queue*, with no base class, is called a *root class;* it has no colon and base class shown in its declaration.

```
class CharQueue : public Queue {public:
    CharQueue(char);virtual void push(short);virtual short pull
        (void);
};
CharQueue::CharQueue(char i) : Queue(i) {}
void CharQueue:: push (short i) {
    if((QLen++)>(QSize))Error = 1; if(QEnd == QIn)QIn- = QSize;
    *(((char*)QIn)++) = i;
}
short CharQueue:: pull (void) {
    if((QLen-) == 0) Error = 1; if(QEnd == QOut)QOut- = QSize;
    return *(((char*)QOut)++);
}
```

The notion of *inheritance* is that an object will have data and member functions defined in the base class(es) of its class as well as those defined in its own class. The derived class inherits the data members or member functions of the parent that are not redefined in the derived class. If we execute *Qptr->error();* then the member function *Queue:: error()* is executed because *CharQueue* does not declare a different *error* member function. If a member function cannot be found in the class which the object was declared or blessed for, then its base class is examined to find the member function to be executed. In a hierarchy of derived classes, if the search fails in the class' base class, the base class' base class is searched, and so on, up to the root class. *Overriding* is the opposite of inheritance. If we execute *Qptr->push(1);* the member function *CharQueue:: push* is executed rather than *Queue::push* because the class defines an overriding member function. Although we did not need additional variables in the derived class, the same rules of inheritance and overriding would apply to data members as to member functions.

Most programmers face the frustration of several times rewriting a procedure, such as one that outputs characters to a terminal, wishing they had saved a copy of it and used the earlier copy in later programs. Commonly reused procedures can be kept in a library. However when we collect such common routines, we will notice some common parts in different routines. Common parts of these library procedures can be put in one place by *factoring*. Factoring is common to many disciplines — for instance, to algebra. If you have *ab* + *ac* you can factor out the common term *a* and write *a (b + c)* which has fewer multiplies. Similarly, if a large number of classes use the same member function, such a member function could be reproduced for each. Declaring such a member function in one place in a base class would be more statically efficient, where all derived classes would inherit it. Also, if an error were discovered and corrected in a base class' member function, it is automatically corrected for use in all the derived classes that use the common member function. We will use this idea of factoring, and inheritance, to develop a library of classes for M·CORE I/O interfacing. *CharQueue*'s constructor, using the notation : *Queue(i)* just after the constructor's name *CharQueue::CharQueue(char i)* before the constructor's body in *{ }*, calls the base class' constructor before its own constructor is executed. In fact, *CharQueue*'s constructor does nothing else, as is denoted by the empty procedure *{ }*. All derived classes need to declare their constructor, even if that constructor does nothing but call its base class' constructor. Other member functions can call their base's member functions by the key word *inherited* as in *inherited::push(i);* or by explicitly naming the class, in front of the call to the function, as in *Queue::push(i);*

Consider the hypothetical situation where a program can declare classes *Queue* and *CharQueue*. Inside *main*, are a number of statements *Qptr->push(1);* and *i = Qptr->pull();* At compile time, either of the objects can be declared for either *Queue* or *CharQueue*, using conditional compilation; for instance, see the program below on the left: this program declares *Q* a class *Queue* object if *mode* is *#declared*, otherwise it is a class *CharQueue* object. Then the remainder of the program is written unchanged. Alternatively, at compile time, a pointer to objects can be blessed for either the *Queue* or the *CharQueue* class. The program below right shows this technique.

```
void main(void){ short i;          void main(void){ short i; Queue *Qptr;
#ifdef mode                        #ifdef mode
    Queue Q(10);                       Qptr = new Queue(10);
#else                              #else
    CharQueue Q(10);                   Qptr = new CharQueue (10);
#endif                             #endif
    Q.push(1); i = Q.pull();           Qptr->push(1); i = Qptr->pull();
}                                  }
```

Moreover, a pointer can be blessed to be objects of different classes at run time. At the very beginning of *main*, assume a variable called *range* denotes the actual maximum data size saved in the queue:

```
void main(void){ short i, range; Queue *Qptr;
    if(range < 128) Qptr = new CharQueue(10); else Qptr = new Queue(10);
    Qptr->push(1); i = Qptr->pull();
}
```

Qptr->push(1); and *i = Qptr->pull();* will use the queue of 8-bit members if the range is small enough to save space, otherwise it will use a queue that has enough room for each element to hold the larger data, as will be explained shortly.

Polymorphism means that any two classes can declare the same member function name and argument, especially a class and its inherited classes. It means that simple intuitive names like push can be used for interchangeable member functions of different classes. Polymorphism will be used later when we substitute one object for another object; the member function names and arguments do not have to be changed. You do not have to generate obscure names for functions to keep them separate from each other. Moreover, in C++, the number and types of operands, called the function's *signature*, are part of the name when determining if two functions have the same name. For instance, *push(char a)* is a different function than *push(short a)*.

When a pointer to an object is able to be blessed at run-time as a pointer to different classes, the *virtual* function becomes very useful. If we do not insert the word *virtual* in front of a member function in the class declaration, then the function is directly called by means of a JSR or BSR instruction, just like a normal C procedure. If a member function is declared virtual, then to call it, we look its address up in a table shown on the right side of Figure 2.5. This table is used because, generally, many objects of the same class might be declared or blessed, and they might have many virtual member functions. For instance there could be queues for input and for output, and queues holding temporary results in the program. As Figure 2.5 shows, to avoid storing the pointers to virtual member functions with every object that uses them, their pointers are collected together and put in a common table for the class. This is accomplished by the *new* operator at run-time putting the address of a different table into a hidden pointer (Figure 2.5) that points to a table of virtual member function addresses, depending on the run-time value of *range* in the last example. Then, data members are easily accessed by the pointer, and virtual member functions are almost as easily accessed by means of a pointer to a pointer. If a member function is executed, as in *Q.push(i)* or *Qptr->push(i)*, the object's hidden

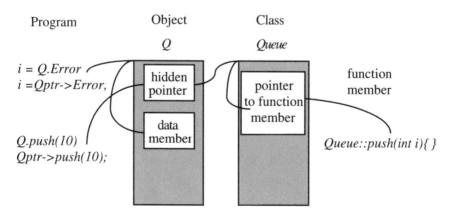

Figure 2.5. An Object and its Pointers

pointer has the address of a table of member functions; the specific member function is jumped to by using a predetermined offset from the hidden pointer. Different objects of the same class point to the same table. Note that data members of different objects of a class are different data items, but member functions of different objects of a class are common to all the objects of the same class via this table.

The user may want to enable or disable virtual functions throughout an application. We suggest writing *VIRTUAL* in place of `virtual` in the class definition. If we then put #*define VIRTUAL* `virtual` at the beginning of the file, all designated class member functions are virtual (permitting run time redirection, §5.1.5). But if we write #*define VIRTUAL* /**/ then designated member functions are not virtual, to improve efficiency.

When object pointers are blessed at run time and have virtual member functions, if a virtual member function appears for a class and is overridden by member functions with the same name in its derived classes, the sizes and types of all the arguments should be the same, due to the fact that the compiler does not know how an object will be blessed at run time. If they were not, the compiler would not know how to pass arguments to the member functions. For this reason, we defined the arguments of *CharQueue*'s *push* and *pull* member functions to be *short* rather than *char*, so that the same member function name can be used for the *short* version, or a *char* version, of the queue. This run-time selection of which class to assign to an object is not needed with declarations of objects, but only with blessing of object pointers, since the run-time program can not select at compile time which of several declarations might be used. Also the pointer to the object must be declared an object of a common base class if it is to be used for several classes.

Information hiding limits access to data or member functions. A member can be declared *public*, making it available everywhere, *protected*, making it available only to member functions of the same class or a derived class of it, or *private*, making it available only to the same class' member functions, and hiding it from other functions. These words appearing in a class declaration apply to all members listed after them until another such word appears; the default if no such words appear is *private*. The data member *Error* in the class *Queue* cannot be accessed by a pointer in *main* as in *i = Q.Error* or *i = Q->Error* because *Error* is not *public*, but only through the public member function *error*. This way, the procedure *main* can read (and automatically clear) the *Error* variable, but cannot accidentally or maliciously set *Error*, nor can it read *Error*, forgetting to clear it. You should protect your data members to make your program much more bug-proof. Declare all data and member functions as *private* if they are only to be used by the class' own member functions, declare them *protected* if they might be used by derived classes, and declare them *public* if they are used outside the class and its member functions.

Templates generalize object-oriented programming. A *template class* is a class that is defined for an arbitrary data type, which is selected when the object is blessed or declared. The class declaration and the member functions have a prefix like `template <class T>` to allow the user to bless or declare the object for a specific class having a particular data type, as in *Q = new Queue<char>(10)*. The generalized class definition is given below; you can substitute the word *char* for the letter *T* everywhere in the declarations or the class member functions.

The following class also exhibits another feature of C++, which is the ability to write the member function inside the declaration of the class. The function is written in place of the prototype for the function. This is especially useful when templates are used with short member functions, because otherwise the notation *template <class T>* and the class name *Queue<T>::* would have to be repeated before each member function.

```
template <class T> class Queue {
    private: T *QIn, *QOut, *QEnd, QSize, QLen; char Error;
    public : Queue<T>::Queue(short i)
      {QEnd = (QIn = QOut = (T*)allocate(i)) + (QSize = i); QLen = Error = 0;}
    virtual void Queue<T>:: push (T i)
      {if((++QLen)> = QSize)Error = 1; if(QEnd == QIn) QIn- = QSize;
        *(QIn++) = i; }
    virtual T pull (void)
      {if((-QLen)<0)Error = 1;if(QEnd == QOut)QOut- = QSize;return
        *(QOut ++);}
    virtual char error(void) { char i; i = Error; Error = 0; return i; }
};
```

If you declare *Queue<char> Q(10);* or bless *Qptr = new Queue <char>(10);* then a queue is implemented that stores 8-bit data, but if you declare *Queue<short> Q(10)* or bless *Qptr = new Queue<short>(10);* then a queue is implemented that stores 16-bit data. It is clear that templates permit us to define one generalized class which can be declared or blessed to handle 8-bit, 16-bit, or 32-bit signed or unsigned data when the program is compiled. This selection must be made at compile time, since it generates different calls.

Operator overloading means that the same operator symbol generates different effects depending on the type of the data it operates on. C++ can overload operators so they do different things when an operand is an object, which depends on the object's definition. In effect, the programmer can provide a new part of the compiler that generates the code for symbols, depending on the types of data used with the symbols. For instance, the << operator used for shift can be used for input or output if an operand is an I/O device. The expression *Q >> a* can be defined to output the character *a* to the object *Q* and *Q << a* can be defined to input a character from the object *Q* and put it into *a*. This type of operator overloading is used in I/O streams for inputting or outputting formatted character strings. Without this feature, we simply have to write our function calls as *a = Q.Input()* and *Q.Output(a)* rather than *Q<<a* or *Q>>a*. However, with overloading we write a simpler program; for instance we can write an I/O stream *Q << a << '' is the value of '' << b;* Overloading can also be used to create arithmetic-looking expressions to evaluate them. Besides operators like + and -, C++ considers the cast to be an operator, as well as the assignment = . In the following example, the cast operator is overloaded by *operator T ();* and the assignment operator by *T operator = (T); T* will be a cast, like *char*, so *operator T ();* will become *operator char ();* whenever the compiler has an explicit cast like *(char)i*, where *i* is an object, or an implicit cast where object *i* appears in an expression needing a *char*, the compiler calls *operator T ();* to cast the data. Similarly, wherever the compiler has calculated an expression having a *char* value but the assignment statement has an object *i* on its left, the compiler uses the overloaded = operator specified by *T operator = (T);*

```
template <class T> class Queue {
    private: T *QIn, *QOut, *QEnd, QSize, QLen; char errors;

    public: Queue(short i)
        {QEnd= (QIn=QOut= (T*)allocate(i) + (QSize=i));QLen=errors=0;}

    virtual void push(T i)
        {if((++QLen)>QSize)errors = 1;  if(QEnd == QIn)QIn- = QSize;
        *(QIn++) = i;}

    virtual T pull(void)
        {if((-QLen)>0)errors = 1;if(QEnd == QOut)QOut- = QSize;
        return*(QOut++);}

    virtual char error(void){ char i; i = errors; errors = 0; return i; }

    operator T () { return pull(); }; /* cast */

    T operator = (T data) { push(data); return data; }; /* assignment */
};
```

Whenever the compiler sees an object on the left side of an equal sign when it has evaluated a number for the expression on the right side and it would otherwise be unable to do anything correctly, the compiler looks at your declaration of the overloaded assignment operator, to determine that the number will be pushed onto the queue. The expression *Q = 1;* will do the same thing as *Q.push(1);* and **Qptr = 1;* will do the same thing as *Qptr->push(1);* Similarly, whenever the compiler sees an object anywhere on the right side of an equal sign when it is trying to get a number and it would otherwise be unable to do anything correctly, the compiler looks at your declaration of the overloaded cast operator, to determine that the number will be pulled from the queue. The expression *i = Q;* will do the same thing as *i = Q.pull();* and *i = *Qptr;* will do the same thing as *i = Qptr->pull();* Now if a queue *Q* returns a temperature in degrees Centigrade, you can write an expression like *degreeF = (Q * 9)/5 + 32;* or *degreeF = (*Qptr * 9)/5 + 32;* and the compiler will pull an item from the queue each time it runs into the *Q* symbolic name. While overloading of operators is not necessary, it provides a mechanism for simplifying expressions to look like common algebraic formulas.

A derived class usually defines an overloaded assignment operator even if its base class has defined an overloaded assignment operator in exactly the same way, because some C++ compilers can get confused with the "=" sign. If *Q1* and *Q2* are objects of class *Queue<char>*, then *Q1 = Q2;* will not pop an item from *Q2* and push it onto *Q1*, as we would wish when we use overloaded assignment and cast operators, but "clones" the object, copying *Q2*'s contents into *Q1* as if the object were a *struct*. That is, if *Q1*'s class' base class overrides "=" but *Q1*'s class itself does not override " = ", *Q1* = *Q2;* causes *Q2* to be copied into *Q1*. However, if " = " is overridden in *Q1*'s class definition, the compiler treats " = " as an overridden assignment operator, and *Q1* = *Q2;* pops an item from *Q2* and pushes it onto *Q1*. The derived class has to override " = " to push data. The " = " operator, though useful, needs to be carefully handled. All our derived classes explicitly define *operator* = if " = " is to be overridden.

C++ object-oriented programming offers many useful features. Encapsulation associates variables with procedures that use them in classes, inheritance permits factoring out of procedures that are common to several classes, overriding permits the redefinition of procedures, polymorphism allows common names to be used for procedures, virtual functions permit different procedures to be selected at run time, information hiding protects data, template classes generalize classes to use different data types, and operator overloading permits a program to be written in the format of algebraic expressions. If the programmer does not have C++, but has a minimal C compiler, many of the features of object-oriented programming can be simulated by adhering to a set of conventions. For instance, in place of a C++ call *Queue.push()*, one can write instead *QueuePush()*. Information hiding can be enforced by only accessing variables like *QptrQSize* in procedures like *QueuePush()*. C++ gives us a good model for useful C conventions.

2.3.4   Optimizing C Programs Using Declarations

We will discuss some C++ techniques that can be used to improve your interface software, and some techniques you can use to get around its limitations for this application. While the techniques discussed here are specific to a C++ compiler, if you are using another compiler or cross-compiler, similar ideas can be implemented.

C and C++ have some additional declaration keywords. If the word *register* is put in front of a local variables, that variable should be stored in a register. Putting often-used local variable in registers instead of on the stack obviously speeds up procedures. It also puts them in known places that can be used in embedded assembly language which is discussed in the next subsection. You can check your understanding of the use of these registers by writing a C procedure with embedded assembly language, then disassembling your program.

If the word *static* is put in front of a local variable, that variable will be initialized, stored, and accessed as a global variable is, but it will only be "known" to the procedure like a local variable is. If the word *static* is put in front of a global variable or a procedure name, that variable or procedure will only be known within the file and not linked to other files. For instance if a C++ project is composed of several files of source code such as file1.c, file2.c and so on, then if a procedure *fun()* in file1.c is declared static, it cannot be called from a procedure in file2.c. However both file1.c and file2.c can have procedures *fun()* in them without creating duplicate procedure names when they are linked together to run (or download) the procedures. If the word *static* is put in front of a class data member, that variable is common to all the objects of the class just like member functions are common to all objects of the class.

2.3.5   Optimizing C Programs with Assembler Language

Assembly language can be embedded in a C program. It is the only way to insert some privileged instructions like MTCR used to enable interrupts. It can be used to implement better procedures than are produced from C source code by the compiler. For instance, the LOOPT instruction can be put in assembly language embedded in C to get a faster do-while-loop. Finally, the .byte or .short directives can be used to build the machine code of instructions that are in the M·CORE instruction set, such as the use of illegal instructions as calls to debug routines, and are thus not generated by C.

Many C++ compilers restrict insertion of assembly language into its C++ procedures. Having implemented protection using *private, protected* and *public* declarations these compilers do not want the programmer to get around this protection using embedded assembly language. In some C++ compilers, to embed assembly language in them, the procedure body is completely written in machine code in *inline procedures*, as in `inline f() = 0x1234;` or a list, as in `short inline f(short) = {0x1234, 0x5678, 0x9abcd};` (parameters are optional). However, HIWARE's C++ compiler permits embedded assembly language.

A single embedded assembly language instruction is put after an expression *asm*. For instance, the line asm MTCR r3,cr7 will insert the MTCR r3,cr7 instruction in the C procedure. However, no other statements may appear after this construction. Also, several assembly language instructions can be put on consecutive lines; the first line is preceded by *asm{* and the last is followed by a matching *}*. These first and last lines should have no assembly-language statements on them. Each intervening line will have a different assembly-language statement on it.

Parameters can be used in assembly-language instructions. For instance, a procedure *clr* to quickly clear a block of *N* bytes starting at location *A* can be written

```
void clr(char *A, short N) { asm {
            movi    r7,0
     11:    st.b    r7,(r2,0)
            addi    r2,1
            loopt   r3,11
} }
```

The leftmost argument, a *char* pointer variable **A* is passed in GPR *r2*, and the rightmost argument, label *N*, is passed in GPR *r3*.

Consult Chapter 1 and the M·CORE Reference Manual to understand the machine coding of M·CORE instructions and the meaning of instructions. Using these resources, you should be able to insert assembly language and machine code into your C++ programs.

2.4   Conclusions

In this chapter, we have surveyed some software background needed for microcomputer interfacing. The first section introduced C. It was followed by the description and handling of data structures. C constructs were introduced in order to make the implementation of storage structures concrete. Indexed and sequential structures were surveyed. We then covered programming style and procedure calling and argument passing techniques. We then covered structured and top-down, and bottom-up programming, introduced object-oriented C. Finally, we showed some techniques used to improve C procedures or insert necessary corrections to the code produced by the compiler.

If you found any section difficult, we can recommend additional readings. *The C Programming Language* by Kernighan and Richie, who were the original developers of C, remains the bible and fundamental reference for C. Tutorials on object-oriented programming are available from the IEEE Computer Society Press. We recommend that you read it as well as subsequent articles on object-oriented design. Other fine books are available on these topics, and more are appearing daily. We might suggest contacting a local college or university instructor who teaches architecture, microprocessors, or C programming for the most recent books on these topics.

Do You Know These Terms?

See page 30 for instructions.

procedure	data structure	overflow error	constructor
declaration of a	storage structure	indexable deque	destructor
parameter or a	vector	buffer	deallocator
variable	zero-origin indexing	deque	derived class
char	precision	stack	subclass
short	cardinality	queue	base class
long	list	argument	superclass
unsigned	array	parameter	enum
statement	row major order	formal parameter	overriding
void	column major order	actual parameter	inheritance
allocate	root class	call by value	factoring
deallocate	table	call by result	polymorphism
initialize	string	call by reference	signature
if then	ASCII code	call by name	virtual
if then else	character string	return statement	information hiding
relational operators	Huffman code	prototype	template class
logical operators	binary tree	extern	operator over-
case	deque	macro	loading
break	push	structured	register
while statement	pull	programming	static
do while statement	pop	member function	
for statement	underflow error	encapsulate	
information	top-down design	object	
structure	data member	allocator	

Problems

Problem 4 is a paragraph correction problem (see guidelines at the end of Chapter 1). Other problems in this chapter and many in later chapters are C and C++ language programming problems. We recommend the following guidelines for problems answered in C: In main( ) or "self-initializing procedures" each statement must be limited to C operators and statements described in this chapter, should include all initialization operations, should have comments as noted at the end of §2.1, and C subroutines should follow the C++ style for C procedures recommended in §2.3.1. Unless otherwise noted, you should write programs with the greatest static efficiency.

1. Write a shortest C or C++ procedure `void main` that will find x and y if $ax + by = c$ and $dx + ey = f$. Assume that a, b, c, d, e, f are global integers that somehow are initialized with the correct parameters, and your answers, x and y, are to be stored in local variables in `main`. (You might verify your program with a source-level debugger.)

2. Write a shortest C or C++ procedure `void main(void)` that will sort five numbers in global integer vector `a[5]` using an algorithm that executes five passes, where each pass compares each `a[i]` with all `a[j]`, $j < i$, and puts the smaller element in `a[i]` and larger in `a[j]` for all i running from 0 to 4.

3. Write a C or C++ procedure `void main(void)` to generate the first five Fibonacci numbers F(i), (F(0) = F(1) = 1 and for i > 1, F(i) = F(i-1) + F(i-2)) in global integers `a0, a1, a2, a3`, and `a4` so that `ai` is F(i). Compute F(2), F(3) and F(4) from the previous two numbers.

4.* The information structure is the way the programmer sees the data, and is dependent on such details as the size of words and positions of bits. The data structure is the way the information is actually stored in memory, right down to the bit positions. A queue is a sequence of elements with two ends, in which an element can be pushed or pulled from either the top or bottom. A stack is a special case of queue, where an element can only be pushed from one end and pulled from the other. An important element in constricted programming is the use of single entry and single exit point in a program segment. A calling routine passes the address of the arguments, called formal parameters, to the procedure. In call by value and result, the data are not actually moved to another location, but the address of the data is given to the procedure. Large vectors, lists, and arrays can be more effectively called by reference than by value.

5. A two-dimensional array can be simulated using one-dimensional vectors. Write a shortest C or C++ procedure `void main(void)` to multiply two 3 × 3 integer matrices, **A** and **B**, putting the result in **C**, all stored as one-dimensional vectors in row major order. Show the storage declarations/directives of the matrices, so that A and B are initialized as

$$A = \begin{matrix} 1 & 2 & 3 \\ 4 & 5 & 6 \\ 7 & 8 & 9 \end{matrix} \qquad B = \begin{matrix} 10 & 13 & 16 \\ 11 & 14 & 17 \\ 12 & 15 & 18 \end{matrix}$$

6. A *long* can be simulated using one-dimensional *char* vectors. Suppose A is a zero-origin 5-by-7 array of 32-bit numbers, each number stored in consecutive bytes most significant byte first, and the matrix stored in row major order, in a 140-byte *char* vector. Write a C or C++ procedure *short get (char *a, unsigned char i, unsigned char j, char *v)* where a is the storage array, i and j are row and column, and v is the vector result. If $0 \le i < 5$ and $0 \le i < 7$, this procedure puts the *i*th row, *j*th column 32-bit value into locations *v, v+1, v+2, and v+3*, most significant byte first, and returns 1; otherwise it returns a 0 and does not write into *v*.

7. A *struct* can be simulated using one-dimensional arrays *char* vectors. The *struct {long v1; unsigned short v2:4, v3:8, v4:2, v5:1};* has, tightly packed, a 32-bit element *v1*, a 4-bit element *v2*, an 8-bit element *v3*, a two-bit element *v4*, a one-bit element *v5*, and an unused bit to fill out a 16-bit *unsigned short*. Write shortest C or C++ procedures *void getV1 (char *s, char *v), void getV2 (char *s, char *v), void getV3 (char *s, char *v), void getV4 (char *s, char *v), void getV5 (char *s, *v), void putV1 (char *s, char *v), void putV2 (char *s, char *v), void putV3 (char *s, char *v), void putV4 (char *s, char *v), void putV5 (char *s, *v),* in which *get...* will copy the element from the *struct* to the vector and *put...* will copy the vector into the *struct*. e.g *getV2 (s, v)* copies element V2 into *v*, and *putV5 (s, v)* copies *v* into element V5.

8. Write a shortest C or C++ procedure *void main (void)* and procedures it calls, without any assembly language, which will first input up to 32 characters from the keyboard to the M·CORE processor (using *getchar (void)*), and will then jump to one of the procedures, given below, whose name is typed in (the names can be entered in either upper or lower case, or a combination of both, but a space is represented as an underbar). The procedures: *void start (void), void step_up (void), void step_down (void), void recalibrate (void),* and *void shut_down (void),* just type out a message; for instance, *start (void)* will type out "Start Entered" on the host computer monitor. The *main (void)* procedure should generate the least number of bytes of object code possible and should run on HIWAVE. Although you do not have to use HIWAVE to answer this problem, you can use it without penalty, and it may help you get error-free results faster.

9. Suppose a strings such as "SEE THE MEAT", "MEET A MAN", or "THESE NEAT TEAS MEET MATES" is stored in *char string[40];* Using one dimensional vector rather than linked list data structures to store the coding/decoding information:

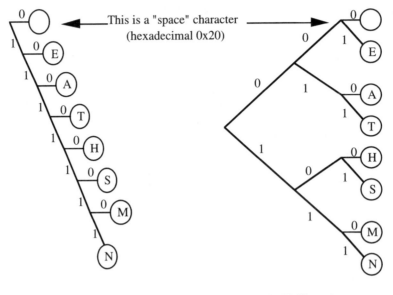

a. Chain   b. Uniform tree

Figure 2.6. Other Huffman Codes.

a. Write a C or C++ procedure *encode ()* to convert the ASCII *string* to Huffman code, as defined by the coding tree in Figure 2.6a, storing the code as a bit string, first bit as most significant bit of first element of *short code[16];*

b. Write a C or C++ procedure *decode ()* that decodes such a code in *short code[16]*, using the coding tree in Figure 2.6a, putting the ASCII string back as it was in *char string[40]*.

10. Repeat problem 9 for the Huffman coding tree in Figure 2.6b.

11. Write an initialization and four shortest C or C++ procedures *void pstop(short)* push to top, *short pltop(void)* pull from top, *psbot (-short)* push to bottom, *short plbot(void)* pull from bottom, of a 10-element 16-bit word deque. The deque's buffer is *short deque[10]*. Use global *short* pointers, *top* and *bottom*. (Note: the deque in §2.2.2 used indexes where this uses pointers. See also §2.3.3 for a queue that uses pointers.) Use global *char* variables for the size of the deque, *size*, and error flag *errors*, which is to remain cleared if there are no errors, and to be 1 if there are underflow or overflow errors. Note that C or C++ always initializes global variables to zero if not otherwise initialized. The procedures should manage the deque correctly as long as *errors* is zero. Procedures *pstop(short)* and *psbot(short)* pass by value, and procedures *pltop(void)* and *plbot(void)* pass by result.

12. Write embedded assembly language in a C or C++ procedure *get(char *a, short i)* which moves *i* bytes following address *a* into a *char* global vector *v*, assuming *v* has a dimension larger than or equal to *i*. To achieve speed, use a LOOPT instruction. The call to this procedure, get(s, n), is implemented:

```
ld.w r2,s
ld.w r3,i
jsr get
```

13. Write a shortest C or C++ procedure *hexString(unsigned short n, char *s)* that runs in HIWAVE to convert an unsigned integer *n* into printable characters in *s* that represent it in hexadecimal so that *s[0]* is the ASCII code for the 1000's hex digit, *s[1]* is the code for the 100's hex digit, and so on. Suppress leading 0s by replacing them with blanks.

14. Write a shortest procedure *short inhex(void)* in C or C++ to input a 4-digit hexadecimal number from the keyboard (the letters A through F may be upper or lower case, typing any character other than '0'..'9', 'a'..'f', 'A'..'F', or entering more than 4 hexadecimal digits, terminates the input, and starts the conversion) and convert it to a binary number, returning the converted binary number as an unsigned short. Although you do not have to use HIWAVE to answer this problem, you can use it without penalty, and it may help you get error-free results faster.

15. Write a shortest C or C++ program *short check(short base, short size, short range)* to write a checkerboard pattern in a vector of *size* s $= 2^n$ elements beginning at *base*, and then check to see that it is still there after it is completely written. It returns 1 if the vector is written and read back correctly, otherwise it returns 0. A checkerboard pattern is *range* r $= 2^k$ elements of 0s, followed by 2^k element of \$FF, followed by 2^k elements of 0s, ... for $k < n$, repeated throughout the vector. (This pattern is used to check dynamic memories for pattern sensitivity errors.)

16. Write a class *BitQueue* which is fully equivalent to the class *Queue* in §2.3.3, but pushes, stores, and pulls 1-bit values, and all sizes are in bits rather than 16-bit words. The bits are stored in 16-bit *short* vector allocated by the *allocate(int)* procedure.

17. Write a class *ShiftInt* which is fully equivalent to the class *Queue* in §2.3.3, but the constructor has an argument *n*, and member function *j = obj.shift(i);* shifts an *short* value *i* into a shift register of *n* ints, and shifts out an *short* value to *j*.

18. Write a class *ShiftChar* which is a derived class of the class *ShiftInt* in problem 17, where member function *j = shift(i);* shifts a *char* value *i* into a shift register of *n* chars, and shifts out a *char* value to *j*. *ShiftChar* uses *ShiftInt*'s constructor.

19. Write a class *ShiftBit* which is fully equivalent to the class *ShiftInt* in problem 19, but shifts 1-bit values, and all sizes are in bits rather than 16-bit words. The bits are stored in 16-bit *short* vector allocated by the *allocate(int)* procedure.

20. Write a templated class *Deque* which is a derived class of templated class *Queue*, and which implements a deque that can be pushed into and pulled from either end. The member functions *pstop(void)* push to top, *pltop(void)* pull from top, *psbot(void)* push to bottom, and *plbot(void)* pull from bottom. Use inherited data and member functions wherever possible.

21. Write a templated class *IndexDeque* which is a derived class of templated class *Queue*, which implements an indexable deque that can be pushed into and pulled from either end, and in which the *i*th member from the top or bottom can be read. The member functions *pstop(void)* push to top, *pltop(void)* pull from top, *psbot(void)* push to bottom, *plbot(void)* pull from bottom, *rdtop(i)* reads the *i*th element from the top, and *rdbot(i)* reads the *i*th element from the bottom of the deque. Member functions *rdtop(i)* and *rdbot(i)* do not move the pointers. Use inherited data and member functions wherever possible.

22. Write a templated class *Matrix* which implements matrix addition and multiplication for square matrices (number of rows = number of columns). Overloaded operator + adds two intervals resulting in an interval, overloaded operator * multiplies two matrices resulting in a matrix, and overloaded operators = and *cast* with overloaded operator *[ ]* writes or reads elements; for instance if *M* is an object of class *Matrix* then *M[i][j] = 5;* will write 5 into row *i*, column *j*, of matrix *M*, and *k = M[i][j];* will read row *i*, column *j*, of matrix *M* into *k*. *Matrix*'s constructor has an argument *size* which is stored as a data member *size*, and allocates enough memory to hold a *size* by *size* matrix of elements of the template's data width, using a call to the procedure *allocate*.

23. Intervals can be used to calculate worst-case possibilities, for instance, in determining if an I/O device's setup and hold times are satisfied. An interval <a,b>, a b, is a range of real numbers between a and b. If <a,b> and <c,d> are intervals A and B, then the sum of A and B is the interval <a+c, b+d>, and the negative of A is <-b, -a>. Interval A contains interval B if every point in B is also in A. Write a templated class *Interval* having public overloaded operators + for adding two intervals resulting in an interval, - for negating an interval resulting in an interval, and an overloaded operator > returning a *char value* 1 if the left interval contains the right interval, otherwise it returns 0. If *A*, *B*, and *C* are of class *Interval*, the expression *A = B + C;* will add intervals *B* and *C* and put the result in *A*, *A = - B;* will put the negative of *B* into *A*, and the expression *if (A > B) i = 0;* will clear *i* if *A* contains *B*. The template allows for the values such as *a* or *b* to be *char*, *short*, or *long*. The class has a public variable *error* that is initially cleared, and set if an operation cannot be done or results in an overflow.

24. Write a templated class *Interval*, having the operators of problem 25 and additional public overloaded operators * for multiplying two intervals to get an interval, and/for dividing two intervals to get an interval, and a procedure *sqrt(Interval)*, which is a friend of *Interval*, for taking the square root. Use the naive rule for multiplication, where all four terms are multiplied, and the lowest and highest of these terms are returned as the product interval, and assume there is already a procedure *long sqrt(long)* that you can use for obtaining the square root (do not write this procedure). If A, B, and C are of class *Interval*, the expression A = B * C; will multiply intervals B and C and put the result in A, A = B/C will divide B by C, putting the result in A, and A = *sqrt(B)*; will put the square root of B into A. Note that a/b is a * (1/b), so the multiply operator can be used to implement the divide operator, a - b is a + (-b), so the add and negate operators can be used to implement the subtract operator, and 4 * a is a + a + a + a, so scalar multiplication can be done by addition. Also *Interval* has a public data member *error* that can be set if we invert an interval containing 0 or get the square root of an interval containing a negative value. Finally, write a *main(void)* procedure that will initialize intervals a to <1,2>, b to <3,4>, and c to <5,6>, and then evaluate the result of the expression (-b + sqrt (b * b - 4 * a * c))/(a + a).

3

Operating Systems

This chapter shows how a multitasking operating system, such as Microware Systems Corporation's Ariel operating system, provides an infrastructure for a powerful embedded processor such as Mototola's MMC2001. The user will make an operating system call (or execute a *service call*) to manage time, memory, synchronization, communications, and I/O for many tasks that appear to be executed simultaneously by the embedded processor. These call on the Ariel kernel, which contains the code needed to manage these resources.

We first introduce the operating system to understand what a service call is and what an operating system is supposed to do. We then introduce service calls to manage time, memory, and intertask communication. This section can be scanned over at a high level, in order to understand the idea of a service call and appreciate its complexity and its power. This section also provides some detailed information for writing Ariel service calls. The file Ariel.h, which is on your CD-ROM, can be searched for procedure names or constant names used in the examples in this section, to locate their definitions in Ariel.h, so as to identify alternative constants you may need to answer the questions at the end of the chapter. We suggest that you skip over these details, such as the procedure arguments, to get the fundamental ideas of what an operating system can do in a microcontroller. Upon a later reading, the details can help you do these exercises at the end of the chapter.

Upon completing this chapter, the reader should understand what an operating system is and how it controls the processor. He or she should be able to use operating system service calls and should be prepared to use them to write an Ariel device driver for an M·CORE microcontroller system.

3.1 What Is an Operating System?

In this section we discuss what an operating system is. We identify the main functions and features of an operating system, then the features of the microcontroller-oriented operating system Ariel. We show why we use such an operating system in this chapter. We then survey the component parts of such a system. This example serves to introduce the nature of multitasking operating systems in general. It provides sufficient background for the understanding of how a device driver works, which is then treated in later chapters.

3.1.1   Three Views of an Operating System

In this introductory section, we present three different views of the operating system. It can be thought of as a main program, for which all user's main programs are subroutines, as a collection of frequently used subroutines, and as an extension of the hardware.

The notion of an operating system began with the need to assist users to more efficiently use large machines. Well before operating systems became common, users signed on and sat at the computer while it ran, just as you may run I/O experiments on a computer in the laboratory for later sections of this book. After many users began using these machines, the large machines began to be run by human operators who fed the machine with data on punch cards and thus controlled it. Later, a program called the *operating system* replaced the human operator and scheduled the loading of punched cards and the use of the machine. Later still, remote CRT terminals were used, and the operating system quickly sequenced the programs in the computer, controlling computer time and resources for most efficient use of the machine. One large machine appeared to be many small machines, one for each user. In this sense, the operating system was the main program, and each user's program was a subroutine of that program.

The phenomenon we called factoring (§2.3.3) began to be useful in operating systems. If a large number of users use the same program segment, such a segment would be reproduced in each program. It would be more efficient to put such a program in one place, where all the user's programs would be able to use it. The omnipresent operating system became home to most of these program segments that were originally in most user's programs, and the operating system became essentially a collection of commonly used subroutines.

A third view of the operating system is as an extension to the hardware. A computer includes I/O devices, which are quite difficult for the typical programmer to control. You have to read a book like this to know what they are doing. If they were controlled by subroutines, written by someone who read this book, subroutine users would be isolated from needing to know all that detail. They would only need to know the conventions for calling the subroutine. The subroutine call instruction would appear rather like a machine instruction, except that it would execute more slowly. The subroutine call is thought of as atomic and indivisible; the programmer learns to use it without knowing how it works, much as he or she learns about a machine instruction. In this sense, I/O subroutines augment the instruction set of the computer, especially to handle I/O.

The program segments should be used in the same way whenever possible. Rather than calling them with a JSR instruction and ending them with an RTS instruction (a subroutine convention), Ariel services are called by a TRAP instruction, that behaves like an "instruction" described above, that is emulated using a vectored interrupt. This puts the processor in the supervisor state, so it can access I/O devices and protected memory;

TRAP #0

When it is executed, register r1 has a number corresponding to the program used to implement the operation. This service call acts as a virtual instruction. All such Ariel service calls will be handled this way. We discuss the specific technique we use to call an Ariel service in §3.2.1.

High-level languages were then designed to use these service calls. It is difficult to find a widely used high-level language that does not depend on these service calls. For instance, C generally utilizes the UNIX service calls. The operating system can become essential to the environment for the running of programs written in such high-level languages. It is more as a support for high-level languages, whether or not it is used, as originally used, to efficiently schedule many users on an expensive mainframe computer.

The personal computer generally uses an operating system to support high-level languages. Generally, the operating system must be run when the computer is started, and will call the user's application, and the user's application will call various program segments in the operating system using a service call. The user's application can handle I/O, including I/O to a special-purpose I/O device, using such service calls. The application can also access an I/O device directly, or an assembly language program segment can be embedded in a high-level language to access I/O devices. However, if a program segment called a *device driver* is part of the operating system to access the device, then any and all programs can use the same service call to use this device driver. In principle, the I/O device is available to any and all high-level languages with equal ease, when it is accessed through a device driver. A programmer who hacks away in BASIC to get an understanding of a process, and then rewrites the process in a C program, can use the same driver without modification, to make it easier to switch from one language to another, or use the I/O with programs written in different languages.

Before we leave this motivational discussion, we observe that the environment created for the high-level languages includes the architecture, which is essentially the instruction set of the computer, together with the service calls. This creates a *virtual architecture*, which is the appearance of the machine as seen by the programmer. A virtual architecture is a view seen by some programmer other than the system programmer (who sees the real architecture), and there may be a hierarchy of virtual architectures. The virtual architecture seen by the compiler and compiler writer is the set of machine language instructions in the machine and the service calls to the operating system. The virtual architecture seen by the applications programmer is the high-level language supported by the compiler. Ultimately, the application has some capabilities and procedures for its use, and this is the final "architecture" of the system, as it appears to the user of that application.

3.1.2   Real-Time Operating Systems

The remainder of this chapter is largely derived from various Motorola documents, and Microware Systems Corporation's well-written manual, *Using Microware Ariel.* A *real-time operating system,* or *kernel,* or *executive,* is software that manages system resources, and provides an architectural framework, or infrastructure, for applications. Real-time applications operate on a time scale imposed by the external world. For example, a real-time application must respond to inputs as soon as they are presented, even if these inputs are frequent and asynchronous. Furthermore, the outputs generated by a real-time system cannot be delayed, even though they often bear a complex relation to the past history of the input system. Real-time service calls should be dynamically efficient, but should also be predictably able to work within physical process timing constraints.

A real-time operating system contains a *scheduler*, which schedules the use of the CPU, and a *library* of service calls, which are invoked by the application programs. Objects and the service calls that use them are patterned after objects introduced in §2.3.3, but are programmed in C without C++'s class syntax, in order to be available to a larger range of applications. Ariel's service calls are equivalent to class member functions.

Because the operating system provides infrastructure, an application developer can focus all efforts on the application, reducing software development cost and time-to-market delays, and enhancing product reliability, maintainability, and quality. However, the operating system requires careful adherence to design policies and operational rules. This requires more care in designing and writing software, and this can cause more overhead than writing all the infrastructure as classes, in whatever nonstandard way the developer chooses to adopt, possibly using object-oriented C as discussed in §2.3.3.

Most operating systems are capable of using a floppy disk, or of using ROMs without a disk at all. An operating system that does not use floppy or hard disks, but uses ROMs to store it, is called *ROM-based*. An operating system, where an operating system and applications programs are permanently built into the hardware and the whole is considered as a single indivisible system, is called *embedded*. Ariel is ROM-based in many of Motorola's development boards, and is designed for embedded systems.

Other operating systems, widely used in larger computers, are large and sophisticated, and having been developed over many years, they have some anomalous features making them difficult to teach. By comparison, Ariel was written for several microcontrollers. As it was developed for small but powerful microcontrollers, it can be economically implemented. Ariel was developed as a *multitasking* operating system, in which a number of different essentially independent tasks, or threads, are run at the same time, each taking different turns using the machine in *time-sharing* mode. Ariel is one of the most compact and simple operating systems in which the student can "poke around" in a multitasking operating system. The requirements of a multitasking operating system make the device driver somewhat more complex. Studying Ariel device drivers prepares you for other multitasking operating system device drivers.

3.2   Functions and Features of Ariel

In this section, we will discuss the Ariel kernel's features, start-task request, support for both common and exclusive use memory pools, wait and cancel-wait operations on a task, event flags, message buffers, mailboxes, semaphores, controlled shared variables, and signals. However, we only overview these features here; see the *Microware Ariel Programming Reference* for details. There, physical and logical input and output (I/O) are discussed. In physical I/O, the task selects a particular console, printer, disk or other unit, and indicates the hardware parameters of the operation to be performed. For example, for a disk write, the starting logical block address and the number of blocks are indicated, as well as the source data address. For logical I/O, the task deals in terms of a generic byte stream. Such a stream is a source of, or destination for, an ordered sequence of bytes. The last subsection covers the techniques for calling physical I/O sevices. The writing of I/O drivers is considered in sections in Chapter 5 and Chapter 6. Finally, time management, covered in Chapter 7 rather than here, facilitates the synchronization of tasks to real time.

3.2.1 Common Ariel Conventions

Before we look at Ariel's mechanisms, we will discuss some of its common conventions. These include the data type convention, a "class mechanism," the use of an id, which also indicates an error, and a 4-character external name to specify tasks.

Ariel redefines data types to simplify its maintenance for different compiler and microcontroller environments. For instance, the type *uxid_t* is defined to be *unsigned long* for M·CORE. Ariel uses the data type *uxid_t* in place of some *unsigned long* variables. (Ariel has a preference for names that include many underscore letters "_".)

Generally, some Ariel mechanisms, such as multitasking, will be provided by an object. This object is similar to a C++ class object, but a C *struct* is used in place of the C++ data members, and C procedures used in place of the C++ member functions have a "class name" followed by a "member function name"; for instance Ariel's C procedure *_tsk_create()* is rather like C++ *tsk.create()*. Of course, this makes Ariel available to the large number of users of C compilers. These objects are often created by a procedure such as *_tsk_create()* and are deleted by procedure *_tsk_delete()*, very much like a C++ object is created by a constructor, and deleted by a destructor. Before creation, they are generally identified by a four-letter *external name*, which is how the object is known externally. External names can be packed into long variables by a macro *_EXTNAME;* for instance the name 'TSK4' can be packed by *_EXTNAME ('T','S','K','4');*. Creation returns an identifier (*id*) which can be saved in a local or global variable to be used internally. The object is generally used by procedures, such as *_tsk_start()*, by passing this id as one of that procedure's arguments. Most procedures return an unsigned long result which may indicate an error. This result should generally be checked for errors. When either an id or an error is returnable from a procedure, the macro *ISSVCERR* can be used to identify if it is an error message; for instance *ISSVCERR(id)* is nonzero if id indicates an error.

Many service calls face the possibility of waiting for their completion. A maximum wait time parameter is provided, where a number such as 10 can be ORed with a constant indicating the unit of time. For instance _TIME_MS is 0x100 and _TIME_MS_TENS is 0x200. To wait 4 ms, the parameter is written 4 | _TIME_MS.

Service calls pass arguments as C and C++ procedures do, in registers r2 to r7, but also put a service call number in r1, which is normally the scratch register. For instance, to create a task using the service call _tsk_create(), the number 68 is put into r1 and a TRAP #0 instruction is executed. However, as the service call's arguments are the procedure's arguments, they are simply passed through the procedure to the service call. This is conveniently done in a *glue-code procedure* such as is shown here:

```
uxid_t _tsk_create(struct tcd *i) { asm { // create task
        movi r1,68
        trap #0
}}
```

However, Ariel is designed to be an embedded operating system, in which only the procedures used by mechanisms needed by the application are loaded into the memory. If fewer service calls are loaded into a system, fewer numbers are needed, so the number in r1 for _tsk_create() may be different than 68, as already shown here.

3.2.2   Task management

Since the CPU is generally much faster than the machinery it controls, it can be shared among several tasks so it can be constantly engaged in doing productive work. A task is an application's execution of an instruction sequence, such as the operations that control a traffic light, or operations to collect statistics on the traffic. Tasks operate independently of each other, but may communicate or coordinate with each other. But tasks controlling different traffic lights might actually execute the same interpreter program, or C, or C++ procedure. Tasks are not the same as procedures. These tasks and another task, that of collecting statistics, might occasionally interact to coordinate their operations or communicate results. For instance, collected statistics might be used to modify each of the traffic light sequences. Component tasks and their support structures may be created and deleted as the application runs. Task facilities within Ariel are fully dynamic.

A task is always in one of three *states* — dormant, blocked, or ready — and a ready task may be active. A *dormant* task has not yet started, or has executed and terminated. A *blocked* task is waiting for a requested service call to be completed, or has been halted by a blocking (error) signal. A *ready* task can execute, and is neither dormant nor blocked. An *active* task is executing, and is generally the highest priority ready task.

Task management encompasses: creating tasks, starting tasks, setting task priority, determining task properties, such as identifier and name, terminating tasks, and deleting tasks. See Table 3.1.

Table 3.1. Task Control Services

_tsk_create()	Create task.
_tsk_start()	Start given task.
_tsk_delete()	Delete requesting task.
_tsk_exit()	Terminate requesting task.
_tsk_exitrestart()	Terminate task with restart after given interval of time.
_tsk_startcoord()	Start given task and transfer coordination to new task.
_tsk_connect()	Connect given task to interrupt or exception.
_tsk_setpri()	Set current priority of given task.
_tsk_execproc()	Set/reset processor on which task can execute.
_tsk_getextname()	Get external name (key) of given task.
_tsk_getid()	Get identifier of task with given key.
_tsk_getdseg()	Get address of data segments of requesting task.

Before it can execute, a task must be created. Application tasks can be created as part of the normal Ariel initialization, as specified as part of a source file *userintf.c*, or dynamically via a task-level request, where *tcdptr* points to a structure, *TSKT_tcd* (Task Control Data) containing fixed task properties. This structure is defined in header file, *os.h* which is described in the Microware Ariel Programming Reference, Appendix D. The following shows how to dynamically create a task. Its *struct TSKT_tcd* appears following it.

```
void main (void) /* creating task */ { uxid_t TSKT_id; /* id. of created task */
    TSKT_id = _tsk_create ((struct tcd *)&TSKT_tcd);
}
```

The *struct* used in the creation of the previous task is shown here.

```
static const struct tcd TSKT_tcd = {
    _EXTNAME('T', 'S', 'K', '4'),    /* a unique, 4-character external name */
    _ATTR_ABS | _ATTR_DURABLE | _ATTR_APPL,    /*attributes */
    -1,                /* use -1 for single processor systems such as M·CORE*/
    250,               /* default priority: relative task importance, 0 is lowest */
    {0,0,0},           /* reserved */
    0,                 /* usage flags: do not uses floating point coprocessor */
    (tep_t)Task_T;     /* entry point, defined by typedef void(*tep_t)(uptr_t);*/
    0x300;             /* length of stack segment, in multiples of 0x100 */
    0,NULL,            /* length and address of uninitialized data segment */
    0,0,0,             /* time interval, increment, and limit for auto priority change */
    NULL               /* reserved for pointer to file name */
};
```

When a task is created, it is dormant. A start request wakes and makes it ready. When one task starts another via the service call *_tsk_start()*, the requester may supply *arg* which is the address of an application-dependent structure containing a variety of param⸍⸍⸍⸍rs. However, a task can have no argument (*void*) if it needs no information fr⸍⸍⸍⸍ ⸍ requester. An example of a procedure for a task is:

```
void Task_T(IOARG *arg /* user-defined type */ ) { /* task code here */ }
```

For example, to start the previously created task at priority 123, pass it an argument and wait until the task starts, as shown here:

```
void main (void) {uxid_t TSKT_id;/* id. of task */rv_t result;/* request result */
    TARG T_data; T_data.T_arg1 = 1; T_data.T_arg2 = 0; /* data for task ... */
    _tsk_start (
    TSKT_id,                    /* selects task to be started (from previous example) */
    123 | _COOR_CURPRI,         /* priority , ORed with _COOR_STDPRI */
    (uptr_t) T_data,            /* run-time argument presented to the task, or NULL*/
    result,                     /* addr. of buffer that receives result of start request */
    _CONT_WAIT | _COOR_START);
        /* coordination qualifier: what to do if ca not start */
    }
}
```

In multitasking, each task runs until: (1) it completes; (2) it must wait for an event to occur; (3) it needs a resource that is unavailable; or (4) it is interrupted by something more important. A task should not monopolize a system resource if a task of higher priority requires the same resource, so a task of low priority may have its execution preempted by a task of higher priority. To accomplish this goal, Ariel keeps track of each task's resources, priority, and state.

In time-sliced scheduling, a periodic real-time interrupt permits tasks to be switched each time it occurs. A task has an *age*, which is essentially the number of real-time interrupts that have occurred since the task ran last, and, after each real-time interrupt, the task with the oldest age is run. Although Ariel does not have this type of scheduling, object-oriented time-slice scheduling will be discussed in §7.2.

When an executing task is suspended, the current hardware register values are stored in a dedicated memory area. Later, when the task resumes, the register images from memory are loaded into the hardware registers and execution continues as though there had been no break. This process, known as a *context switch*, is invisible to the task.

If scheduling is preemptive and task 2 has higher priority than task 1, then task 1 creates task 2 and task 2 will run to completion, before task 1 resumes. In effect, task 2 behaves just like a subroutine. But if task 2 has lower priority that task 1, then it runs when task 1 becomes blocked. If task 1 is never blocked, task 2 never runs. If scheduling is time-sliced and both tasks have the same priority, then each task runs in every other time slice until task 2 terminates.

The service call, `_tsk_exit()` or `_tsk_delete()`, is called to terminate a task. `_tsk_exit()` does not delete the task, so it can be started again. `_tsk_delete()` exits and deallocates the task's data structure. These pass a parameter, `retcode,` which is not used in this implementation, but future releases may pass it to the task that created it. No error is indicated by *SUCCESS*, which is zero. When a task terminates, it can be restarted by executing `_tsk_exitrestart()`. For instance `_tsk_exitrestart(SUCCESS, 4 | _TIME_MS)` restarts the terminated task in 4 ms.

3.2.3   Memory Management

Commonly, the sum of the maximum memory requirements far exceeds its in-
stantaneous needs. For example, each of 20 tasks might need 10K bytes of work
space at some time (for a theoretical worst case total of 200K). Yet, because of the
way the memory needs are phased among the tasks, only 50K maximum is ever
needed at any given time.

Memory management assigns memory to users as they request memory. Primary
read-write memory is initially "owned" by the operating system and is "rented" to
applications programs as they are loaded into memory, or as they request memory
for data.

A *pool* is a contiguous chunk of memory that Ariel allocates to tasks upon
demand. Ariel supports a flexible *common memory pool* (CMP) and a more efficient
fixed block pool (FBP). Memory within each pool is divided into fixed-size blocks. In
an FBP, each allocation delivers exactly one block; in a CMP, a task can receive a
contiguous area of any desired integral number of blocks. Memory pool services are
listed in Table 3.2.

A global declaration to let the compiler and linker provide space for the pool,
and a service call to create a common memory pool, is shown here.

```
char s[1000];            /* global variable, where the pool will be located */
cmp0 = _cmp_create(      /* returns common pool id */
  -1,                    /* use -1 for single processor systems, M.CORE */
  _EXTNAME('C', 'M', 'P', 'O'),/* a unique, 4-character external name */
  s,                     /* address of buffer */
  4000,                  /* number of bytes in pool */
  7                      /* log base 2 of block size: a block is 2**7 bytes */
);
```

A common memory pool can also be created during startup. One of them is
".GTA." After it is created, *n* bytes can be allocated and the address returned to
`char *ptr` by:

```
_cmp_alloc(
  cmp0,                  /* use common pool id - returned by _cmp_create */
  n,                     /* number n of bytes requested */
```

Table 3.2. Shared memory control services

`_cmp_create()`	Create common memory pool
`_cmp_getid()`	Get identifier of common memory pool
`_cmp_alloc()`	Allocate contiguous area from common memory pool
`_cmp_free()`	Deallocate area taken from common memory pool
`_cmp_delete()`	Delete common memory pool
`_fbp_create()`	Create fixed block memory pool
`_fbp_getid()`	Get identifier of fixed block memory pool
`_fbp_alloc()`	Allocate one block from fixed block memory pool
`_fbp_free()`	Deallocate block taken from fixed block memory pool
`_fbp_delete()`	Delete fixed block memory pool

```
  (uptr *) &ptr,            /* pointer where address of allocated memory is put */
  _CONT_WAIT                /* coordination qualifier: wait for service to complete */
);
```

_cmp_alloc() can be coordinated, such as limiting the wait time, in its last parameter. After some allocated bytes are no longer needed, they can be deallocated by

```
_cmp_free(
  cmp0,                     /*use common pool id - returned by _cmp_create */
  n,                        /* number n of bytes in allocated block */
  ptr,                      /* pointer to block */
);
```

When the pool is no longer needed, it can be deleted by _cmp_delete(cmp0); where cmp0 is the id returned by _cmp_create. Fixed block pools are handled in similar manner, but with fbp in place of cmp in the name, and with different parameters, since the size of a block is fixed rather than variable.

3.2.4   Synchronization

In many applications, tasks must share a common set of data or section of code. Typical of shared data is a table that is read by one task and updated by another. While the data is being read, the update task must be blocked; while the data is being updated, the reader task must be blocked. Examples of shared code are any non-reentrant subprograms that could be called by different tasks. Ariel provides three different facilities to achieve such mutual exclusion: semaphores (SF), controlled shared variables (CSV), and event flags (EFG). Since the semaphore is the simplest, its use will be described in some detail. The other mechanisms will be mentioned at the end of the section. Table 3.3 lists the service calls for these synchronization mechanisms.

Table 3.3. Synchronization Services

_sem_create()	Create counting semaphore
_sem_wait()	Wait for given counting semaphore to be free
_sem_release()	Release semaphore
_sem_delete()	Delete semaphore
_csv_create()	Create group of controlled shared variables
_csv_wait()	Wait for control over whole group of controlled shared variables
_csv_release()	Release group of controlled shared variables
_csv_delete()	Delete group of controlled shared variables
_efg_create()	Create group of global event flags
_efg_global()	Set or reset event flags immediately
_efg_local()	Set or reset local event flags of given task immediately
_efg_timesend()	Set event flags after given interval of time
_efg_wait()	Wait until event flags are set
_efg_delete()	Delete a group of global or local event flags

Shared data and code are sometimes called *critical regions*. Not every shared resource is critical, however. A fixed data table would not be critical. Nor would a reentrant subprogram. A region is critical if it is shared and is (or can be) altered as part of its normal use. Critical regions are protected by guaranteeing one-task-at-a-time access.

A separate SF is assigned to each critical region. (One could use one SF to protect several regions, but this usually leads to excessive contention delays.) Prior to using the critical region, every task waits for the corresponding SF to be free. If another task already has taken that SF, the new task waits. Queuing is based on the current priority of the waiting tasks. Upon exit, the current user releases control to the next waiting task.

Ariel's *semaphore* is an internal counting semaphore. When a SF is created, its count is initialized to zero, which indicates a resource is available; a nonzero count indicates the resource is in use. When a task wishes to use the resource, it executes a *_sem_wait()* service call, which, if the count is zero, makes the count nonzero and returns to permit the task to continue; otherwise it blocks the task and lets another task run instead. As a task discontinues use of the resource, it executes a *_sem_release()* service call, which reduces the count, and if the resulting count becomes zero, it lets a blocked task run. Only one task, that the semaphore blocked, will be allowed to run.

Semaphores are initialized using a service call *_sem_create()*. When no longer needed, they are deleted using *_sem_delete()*.

```
extern int32 (void)
void task1 (void) { uxid_t sema0;
 sema0 = _sem_create
    {_EXTNAME('C', 'M', 'P', 'O')/* a unique, 4-character external name */);
    _sem_delete(sema0); /* use id returned by _sem_create */
}
```

A simple way to protect critical subprograms is to wait immediately as it is entered and to release just before returning. This places all control aspects completely within the subprogram. As a result, a caller does not have to know that the subprogram is critical. We illustrate this technique with the *dining philosopher's problem*, which is a classic example of a large class of synchronization problems. Five philosophers sit at a round table. They each have a chopstick on their right, but need two chopsticks to eat. The chopsticks are represented by semaphores. Assume that five semaphores have been created, and have ids *sema[0]* to *sema[4]*. Each task is given a unique index i as its argument and executes exactly the same procedure *philosopher()*.

The program below exhibits a *deadlock* or *deadly embrace*; it is possible that no task will get both the resources it needs. This can occur if the fifth philosopher grabs the chopstick that the first philosopher needs to eat before the first philosopher gets to its second *_sem_wait* service call. An alternative scheme, implementing *deadlock avoidance*, is to have every even-numbered philosopher pick up their left chopstick while the odd-numbered philosophers pick up their right chopstick. Another deadlock avoidance scheme is to always pick up the lower-numbered chopstick first, before picking up the higher-numbered chopstick. Philosopher 5 violates this rule, which causes this deadlock. Coding these schemes is straightforward, and is left as an exercise at the end of the chapter.

```
typedef IOARG struct{int i};

void philosopher(IOARG *arg) {int i = arg->i; rv_t st; /* place for status */
    do{
        _sem_wait(sema[i], &st, _CONT_WAIT);/* wait on one of the semaphores */
        _sem_wait(sema[(i + 1) % 5], &st, _CONT_WAIT););/* wait on next sem*/
        // eat
        _sem_release(sema[i]); /* release a semaphore */
        _sem_release(sema[(i+1) % 5]); /* release the other semaphore */
        // do something other than eat
    }while(1);
    }
```

The problem with waiting until a service is completed is that there is no guarantee that the service will ever be completed (perhaps because of deadlock), or that the waiting time will not compromise other more urgent aspects of the task's work. Many critical real-time applications cannot risk any uncontrolled waits. Thus, _sem_wait() permits inclusion of a maximum wait time as part of its third parameter, which is the parameter _CONT_WAIT in the example above. Also, that parameter can specify setting an event flag or sending a signal. An example of setting flag 3, waiting for at most 10 s, is _CONT_LEF3 | _TIME_SECONDS | 10. Signals and time synchronization will be discussed in later sections.

The semaphore provided by Ariel is similar to, but not exactly the same, as the semaphore proposed by Dijkstra. When a Dijkstra semaphore is created, a non-negative number s is assigned to the semaphore. Thereafter, s can increase or decrease (down to 0) via P and V. A Dijkstra semaphore maintains only the tally, s and the list of blocked tasks; it does not record the current "owner" of the semaphore (as does Ariel). Thus, a Dijkstra semaphore permits s tasks to proceed into a critical region. These could be s different tasks, the same task s times, or any other combination that sums to s. Ariel permits the same task to proceed any number of times, but blocks all other tasks.

Controlled shared variables provide another coordination mechanism. A set of tasks may share a group of variables. While a semaphore grants each task exclusive access to the whole group, Ariel guarantees that only one task at a time is within the critical region. In some cases, a task must also be blocked until a certain relation exists among the group variables. Such a shared variable can be a global *struct* or array. For example, the task might have to be blocked until one element of the *struct* is greater than another of its elements. Thus, the task might leave the critical region to allow other tasks to change the variables, but then reenter it when the desired relation is true. A _csv_create() service call renders a shared variable controlled (see Table 3.3), a _csv_delete() service call removes this shared variable's control. _csv_wait() waits until it is not used, and _csv_release() releases control of the shared variable.

Another synchronization primitive is the *event flag*. A *flag group* is 16 flags. A task can set or clear a flag or a collection of flags in a group, either immediately or after a given time interval. A task may wait until all of a set of flags in a group are set, or else wait until one of a set of flags is set. The wait service call's last argument can specify a time limit for waiting for the flags to become set. Global event flag groups can be created and deleted, analogously to semaphores. Each task automatically is provided with a local event flag group which can neither be created nor deleted.

Some service calls for these controlled shared variables and event flags are similar to those of semaphores (see Table 3.2), but in the function name the letters csv or efg are used in place of sem and some arguments are different.

A major class of problems is *readers* and *writers* (of some data). Readers can share the data, permitting simultaneous access, but writers cannot share the data, requiring sole access to it while writing. These problems can be handled by the techniques given above.

Although there are three control primitives (and communication primitives can also be used for synchronization), each one has its advantages for some applications. However, in this book, we will develop just the semaphore mechanism. We leave the other mechanisms for problems at the end of this chapter, which you can work if you have a copy of Microware's Ariel Operating System documentation.

3.2.5 Communication

The operating system provides several means of communicating among tasks. See Table 3.4. These include mailboxes and message buffers, which we discuss in the next subsection, and signals, which are discussed in the following subsection.

Table 3.4. Communication services

_mbx_create()	Create mailbox
_mbx_send()	Send message to mailbox
_mbx_recv()	Receive first available message from mailbox
_mbx_close()	Close mailbox
_mbx_delete()	Delete mailbox
_msb_create()	Create message buffer
_msb_getid()	Get identifier of message buffer
_msb_get()	Get message from a buffer
_msb_put()	Post message to a buffer
_msb_delete()	Delete message buffer
_sig_get()	Get response to given signal
_sig_set()	Set response to one or more signals
_sig_send()	Send signal to one task or group of tasks
_sig_cancel()	Cancel pending signals of requesting task
_sig_wait()	Wait until signal arrives
_sig_timesend()	Send specified signal after given interval of time

3.2.5.1   Mailboxes and Message Buffers

Mailboxes are intuitively simple, as they are similar to email. In a mailbox, a message is a string of characters. The task sending the message assigns to it a priority and sends the message into a mailbox, and the receiver receives the message from the mailbox. The size or number of messages that are able to be put in a mailbox are essentially not limited. Messages are put into the mailbox in descending order of message priority, the highest priority message being retrieved first. Among messages with the same priority, the messages are retrieved in order of insertion; the earliest message is retrieved first.

A task receiving a message can specify an input buffer shorter than the incoming message. The message is truncated, with the excess text discarded, and is removed totally from the mailbox. A message shorter than the buffer leaves excess buffer locations unmodified. After a task sends a message, it has the option of continuing, or waiting until the message is received. Similarly, if no messages are in the mailbox, a task receiving a message can either continue, or wait for the next message to arrive.

Many problems in multitasking operating systems have *producers* that generate data and *consumers* that use the data. Suppose two producer tasks generate output to the printer, and two printer tasks consume the print data (printing it). A mailbox is created, and each producer puts its data into a mailbox. Each consumer attempts to get a message from the mailbox; if there is no message queued, it waits. When the printing is completed, the printer task seeks the next message, in an endless loop. Mailboxes provide an orderly way to coordinate producers and consumers, allowing the producer to send work to the next free consumer, without knowing which one that is.

Mailboxes are "created" by each task that sends or receives messages by the task's calling _mbx_create with the same unique 4-character external name. Only the first such call will actually create it, the others merely return the mailbox's id. As each task is done, it calls _mbx_close. Only the last task actually closes and deletes it.

A producer creates a mailbox, opens it, and sends data as follows:

```
producer(){ uxid_t mbx; rv_t result; char msg[100];
    mbx = _mbx_create{/* create and open a mailbox */
        _EXTNAME('M', 'B', 'X', 'O'),/* a unique, 4-character external name */
        _mbx_send /* mode - make a sender */
    };
    if(!ISSVCERR(mbx)) /* if no error */ do { /* do "forever" */
        ... /* generate values in msg. The first 32-bits of msg are the number of bytes */
        _mbx_send(mbx, (uptr_t)&msg, &result, _CONT_WAIT, 100L);/* send */
    } while(1); /* end of "forever" loop */
    _mbx_close(mbx, _mbx_send); /* when done with the mailbox, close it */
}
```

Consumers can execute the following procedure to possibly create, and open, a mailbox and get the mailbox data into a receive buffer. Before _mbx_recv is called, the first 32-bit word of the receive buffer is initialized to the maximum number of characters that will be transferred into it.

```
consumer () { uxid_t mbx; rv_t result; char datblk[100];
    mbx = _mbx_create{ _EXTNAME ('M', 'B', 'X', 'O'), _mbx_recv};
    if (!ISSVCERR(mbx)) /* if no error */ do { /* do "forever" */
        datblk[0]=datblk[1]=datblk[2]=0;datblk[3]=100;/* set maximum size */
        _mbx_recv(commbx, (uptr_t)&datblk,&result, _CONT_WAIT);/* get data */
        if (result != SUCCESS) break; /* messages no longer available */
        ... /* use values */
    } while(1); /* end of "forever" loop */
    _mbx_close(mbx, _MBX_ RECV); /* when done with the mailbox, close it */
}
```

As an alternative to a mailbox, a *message buffer* can also be used for communication. A pointer to a string or *struct* may be put into a message buffer, and such a pointer may be received from it. This kind of message does not have a priority, but the sender may place it at the end of the deque (which acts like a FIFO) or at the beginning of the deque (which acts like a LIFO). A message buffer has a maximum number of pointers, established when it is created. The sender cannot effect coordination, such as to send a signal when the message has been received. Message buffers are therefore quite limited compared to mailboxes, but are provided as an alternative to mailboxes because they are more efficiently stored and processed. The service calls for these message buffers are similar to those of mailboxes (see Table 3.4), but in the function name the letters msb are used in place of mbx and some arguments are different.

Client-server problems have *clients* that request services, and *servers* that provide these services. These are similar to producer-consumer problems already discussed here, but after consuming data the server can return data to the client.

3.2.5.2   Signals

A *signal* is a software interrupt that may be handled at the task level. Ariel automatically sends a signal to a task when a breakpoint is encountered, when tracing, or an error exception occurs, such as a divide by zero, or execution of an unimplemented instruction. A task may elect to have a signal sent when a requested service is completed, or to send a signal to another task, or to a group of tasks, as a means of communication.

Signals are numbered 0 to 31. As shown in Table 3.5, signal 31 is used to force termination (which is to "kill" the task). Signals 16 to 30 are for debugging and error recovery. Any task can send signals 0 to 15 (or 31) to any task, including itself, for an end-of-service indicator. A task can respond when it receives such a signal: it can ignore it (_SIG_IGNORE), it can terminate (_SIG_TERM), it can perform a subprogram, or it can become blocked (_SIG_BLOCK). The default response to each signal is indicated in the column "Default".

Table 3.5. Signal usage

Num	Lvl	Use Within Ariel	Default
0-15	1	available for coordination	_SIG_IGNORE
16	2	(unassigned)	_SIG_BLOCK
17	3	(unassigned)	_SIG_BLOCK
18	3	(unassigned)	_SIG_BLOCK
19	3	(unassigned)	_SIG_BLOCK
20	3	trace	_SIG_BLOCK
21	3	breakpoint reached	_SIG_BLOCK
22	3	divide-by-zero	_SIG_BLOCK
23	3	(unassigned)	_SIG_BLOCK
24	3	unimplemented software interrupt	_SIG_BLOCK
25	3	privilege violation	_SIG_BLOCK
26	3	bad parameter in service call	_SIG_BLOCK
27	4	illegal instruction	_SIG_BLOCK
28	5	memory access error	_SIG_BLOCK
29	5	memory alignment error	_SIG_BLOCK
30	6	(unassigned)	_SIG_BLOCK
31	7	terminate "kill task"	_SIG_TERM

The response to a signal does not occur until the task becomes READY. Normally, signals are held if they arrive while a task is BLOCKED or DORMANT. However, a Kill Signal (31) arriving while a task is DORMANT is discarded.

A task can send a signal by invoking `rv_t _sig_send(uxid_t tid, u_int32 sig);` where `tid` specifies the target task. The special value 0 sends the signal to all other application tasks. `sig` specifies the signal of interest (0 to 15, or 31). A task can have a specified signal sent, even to itself, after a given interval via the service call `_sig_timesend()`. Other services can send signals when their action is completed. For instance when memory has been allocated, the last parameter of `_cmp_alloc()` can be `_CON_SIG3`; this sends signal 3 when allocation is complete.

A task can wait until a signal arrives by calling the service `rv_t _sig_wait()`, which returns the signal number (0 to 31) or an error. The first signal to arrive cancels the wait. For instance, `_sig_wait(200)` waits for a divide-by-zero error signal.

Alternatively, a signal can cause a *signal response subprogram* to be executed. An example of such a subprogram to handle a divide-by-zero error is:

`void subprogram(struct scf *) {/* response to divide-by-zero error signal */}`

When `subprogram` is entered upon receipt of a divide-by-zero error signal, the argument is a `struct` shown below. The subprogram can be written to examine this `struct`'s variables such as `srv`, and change the task's global variables, or to print out error messages. To prepare for this reaction, a service call `_sig_set(_SIG_19, subprogram)` is executed before the signal is sent. The first parameter, `_SIG_19`, indicates which signal is to cause the second parameter, `subprogram` to be executed. Instead of a procedure, the second argument can be a response such as `_SIG_IGNORE`. `_sig_set(_SIG_19, _SIG_IGNORE)` cancels responding to divide-by-zero errors by calling `subprogram`.

```
struct scf {                    /* signal context frame */
    struct scf *flk;            /* forward link */
    u_int16 fsn;                /* frame type (= 0x00) // sig num */
    u_int8 rsvd0[2];            /* reserved */
    prcid_t pi;                 /* processor index (0 - 15) */
    uxid_t tti;                 /* tid of target task */
/* register values, CPU-dependent, including pc (program counter) */
    context_t rsa;              /* register save area */
    rv_t srv;                   /* SVC return value */
    bool_t et;                  /* except type: 0 = non-SVC, 1 = SVC */
    u_int8 lsn;                 /* last sig num (0xFF = none) */
    u_int8 lsl;                 /* last signal level */
    u_int8 rsvd1;               /* reserved */
    struct scf *lsf;            /* last cur scf */
};
```

At the entry point of a signal subprogram, registers are saved. A signal sub-program is part of the task with the task's privileges and restrictions. The sub-program can execute any service call; in particular, signals can be sent and _sig_set() can be invoked. Upon executing an explicit return statement, or by reaching the end of the subprogram, saved register values are restored.

Every signal has a level of severity in the range 1 to 7. (See the column Lvl in Table 3.5.) A task starts at signal level 0. When a signal arrives, its level is compared with the task's current signal level. If the new signal is of the same or lower level, its arrival is noted, but no action is taken. In this case, if the same signal is already pending, the new one is effectively ignored, no matter what the response is supposed to be. Furthermore, the response action is not yet examined, so that the signal remains pending even if the response is currently to ignore the signal.

If a signal arrives with a level greater than the current signal level of the task, then the new signal is processed immediately. The status of the task is examined. If the task is waiting for a signal, the wait is canceled making the task READY. If the task is DORMANT, ACTIVE or BLOCKED, the signal is noted in the task's internal pending-signals variable, and further action is deferred until the task be-comes READY.

For a READY task, the signal response is examined. For _SIG_IGNORE, no further action is taken. For _SIG_TERM, the task is terminated. For _SIG_BLOCK, if the Debugger is present, the task is blocked and the *debugger* is started (passing to it the same block that a signal subprogram would receive). For _SIG_BLOCK without a Debugger, the error/signal block is passed to an *error logger task* (if present) and then the errant task is terminated. Finally, for the remaining possibility, execute a signal subprogram, the task context (status register, program counter, general reg-isters and signal level) all are saved on the task stack, the task signal level is raised to that of the new signal, and the subprogram is reentered. When the signal processing is postponed because the level of the incoming signal is too low, the aforementioned steps are carried out when the processing returns to the required level.

The signal will be an important mechanism for (hardware) interrupt handlers to cause tasks to resume operation, after waiting for the interrupt to occur.

3.2.6   Calling a Device Driver

A *device driver* is part of the operating system used to access an I/O device. Ariel accesses I/O in a manner that is quite similar to a mailbox. This approach facilitates the use of the device by several tasks, which can even have priorities for their messages, so that higher priority messages are output first. I/O service calls for I/O are shown in Table 3, 6. Before an I/O device can be used, the driver must be created, or a copy of its id must be obtained. A task can get the id of a *console* I/O device, which can interface to the user, by `_pio_getcons()`. The console handles control keys such as control-D (0x4) to invoke a debugger and escape (0x1b) to start a special task such as to take care of screen-oriented cursor positioning escape sequences. Ariel provides C I/O procedures `getchar, gets, printf, putchar, puts,` and `scanf,` which use the console device.

In the example below, `inient` prints out the message "Hello World" on the console printer. The `_pio_cmd` service call, which writes the message, uses a `struct cdev`, which tells it what to print; `cdev` is initialized just before the service is called. The address and length of the output string are given, the optional address of an auxiliary buffer is given, and the coordination conditions are given, in this `struct`. The service call has arguments id `uid`, command `_PIO_WRITE` (to write), and priority `OL` before the `struct` address, and an address of a status variable and coordination parameter, after it. The status of the request is posted in the status variable after the service is completed.

```
char pattern[] = ''Hello World\r\n'';
struct cdev { chqr *aub;              /* address of data to be output */
     u_int32 lub              /* length of data to be output *
     char *axb                /* auxiliary buffer */
     u_int32 spl              /* time limit for service */
}
void main(void) { int i;      /* task entry */
     uxid_t uid;              /* id for device */
     cdev_t iop;              /* struct for _pio_cmd service call */
     rv_t sb[2]               /* service status buffer */
     uid = _pio_getcons();    /* get id of standard console */
     iop.aub = pattern;       /* initialize character parameter block */
     iop.lub = strlen(pattern); iop.axb = NULL; iop.spl = 0;
     _pio_cmd(uid,_PIO_WRITE,OL,&iop,&sb[0],(qual_t) _CONT_WAIT);
}
```

Table 3.6. I/O device services

`_pio_create()`	Create peripheral unit to run under a given driver
`_pio_getcons()`	Get identifier of standard console for requesting task
`_pio_setcons()`	Install given unit as standard console of requesting task
`_pio_getid()`	Get identifier of unit with given external name
`_pio_cmd()`	Perform I/O and related functions on given peripheral unit
`_pio_control()`	Perform control functions on given peripheral unit

3.3 Object-oriented Operating Systems Functions

An object-oriented alternative can be implemented for each operating system function. §7.2 shows the implementation of a time-sharing system that can support multiple tasks or threads. These tasks can use semaphores (§3.2.4) to synchronize them. The key problem in the implementation of a semaphore (or other operating system function) is to avoid the possibility that when one task is executing a semaphore procedure, an interrupt might occur, thereby permitting another task to execute the same semaphore, which can then result in two tasks getting permission to use a common resource. The semaphore is supposed to guarantee that only one task gets control of the resource. To avoid this possibility, we merely have to disable interrupts during the time that one task executes the semaphore procedure. Then the procedure is indivisible, and mutual exclusion can be guaranteed. §6.2.3.1 gives procedures *enableInt()* and *disableInt()* that enable and disable interrupts. It suffices for now, without seeing how they are written, for you to consider them to be the way we enable and disable interrupts. A test-and-set procedure, shown below, permits a task to read the former value of a bit, and set it:

```
char test_and_set(char *p) { char result;
        disableInt(); result = *p; *p = 1; enableInt(); return result;
}
```

In order to enable or disable interrupts, the program needs to be in the supervisor state (§1.2). If a program is in a user state, it can switch to supervisor state using a *trap #3* instruction whose handler merely sets the supervisor bit of the PSW (§6.2). The user mode may be reentered using a procedure *toUser()* (§6.2). If user programs are to be executed in user mode to implement the protection afforded by supervisor-user modes, then the procedures *disableInt()* and *enableInt()* can be expanded so that *disableInt()* has a *trap #3* instruction and *enableInt()* also clears the supervisor bit when it sets the EE and IE bits.

Other operating system procedures can be implemented using the procedures *enableInt()* and *disableInt()* to ensure mutual exclusion. Simplified memory management using *allocate()* in §2.3.3 can be made mutually exclusive with these two procedures. A counting semaphore can be implemented by expanding the *test_and_set()* procedure above. The queue (§2.3.2) or queue class (§2.3.3) can be enclosed between *disableInt()* and *enableInt()* procedure calls to prevent two tasks from interfering with each other, to push data into or to pull data from the queue. This can be used to implement a mailbox or a message buffer (§3.2.5.1) or a pipe (§7.2.3). And a signal (§3.2.5.2) can be implemented by temporarily increasing a task's priority (§7.2.1), so it will examine a global variable; it can be made mutually exclusive by enclosing the procedure in *disableInt()* and *enableInt()* procedure calls.

The alternative to using a time-sharing operating system is to build the procedures that you need in your application, which are made mutually exclusive by disabling interrupts. Alternatively, you do not have to build, debug, and test these procedures if you utilize an operating system such as Microware's Ariel operating system.

3.4   Conclusions

An operating system kernel is generally used to manage time, memory, and communication in a larger microcontroller such as the MMC2001. The Ariel kernel infrastructure enables user-level programmers to concentrate on the application, rather than on duplicating often-written code. However, a programmer must adhere to operating system conventions, which might poorly fit the application, and might take time to learn. You still may find it better to control a device directly in C++. However, to make this decision, you need to understand the capabilities and limitations of an operating system kernel like Ariel. Further information on operating systems can be obtained from many textbooks available for operating system courses. See *Operating System Concepts*, by Sibershatz and Galvin, Addison-Wesley, 1994 for a more traditional coverage of operating systems.

Further, device drivers will be written to control I/O devices in the operating system. Knowledge of how a device works and how to control it, taught in subsequent chapters of this book, is essential to writing device drivers. But service calls will be used in device drivers. Some part of the code has to implement the conventions of the operating system: setting up and using memory and moving data between tasks and interrupt handlers, and so on. We will cover these extra requirements in the coming chapters.

Do You Know These Terms?

See page 30 for instructions.

service call	embedded	memory	signal response
operating system	operating system	management	controlled shared
device driver	multitasking	pool	variable
virtual architecture	operating system	common memory	reader
real-time operating	process	pool (CMP)	writer
system	time-sharing	fixed block pool,	mailbox
kernel	task	(FBP)	producer
executive	state	critical regions.	consumers
scheduler	dormant	semaphore	message buffer
library	blocked	dining	signal
ROM-based	ready	philosopher's	debugger
operating system	active	problem	error logger task
glue-code	age	deadlock	console
procedure	event flag	deadly embrace	subprogram
flag group	context switch	deadlock avoidance	external name

Problems

Problems 1 to 5 are paragraph correction problems; see guidelines at the end of Chapter 1. For other problems, use Microware's Ariel documentation for detailed description of its service calls. General programming guidelines are given at the end of Chapter 2, and hardware design guidelines are at the end of Chapter 3.

1. *Once, users signed on and sat at the computer while it ran, to be its operator. Later, a program called the Device Driver replaced the human operator and scheduled the loading of punched cards and the use of the machine. Each user's program was effectively a 'subroutine' of that program. The former program became a collection of commonly used program segments used in the personal computer to support such capabilities as floating-point arithmetic. It is part of the end user's visual architecture, which is his or her view of the machine. That program can also be viewed as an extension of the computer's memory.

2. *The Ariel colonel uses a BKPT instruction for a service call, where register r1 points to a one-dimensional array corresponding to the program used to implement the operation. This instruction puts the M·CORE processor in its superior mode. A procedure called a shell is part of the operating system used to access an I/O device. A real-time executive manages system resources and provides infrastructure for time-critical applications.

3. *Ariel's service calls operate on objects. A program called MERLIN includes as much of Ariel as an application requires. Ariel is a multicomputer operating system, wherein user programs, called tasks, run on different computers in space-sharing mode. Politics establish general design guidelines: Ariel should be *(name 4)* application independent, should provide for a variety of applications, should be incomprehensible, and should be reliable. Ariel should be multicomputing and dynamically efficient. Ariel is capable of running on M·CORE microcontrollers, which may use a floppy disk, or else may be designed to use SRAMs instead of a disk to hold the programs; the latter type of operating system is called embedded.

4. *Memory management attempts to give idle time in one process to another process, but tries to be fair to all processes if a process does not have any idle time. Each procedure has an importance level, which is higher if it gets use of I/O after more CPU cycles. Telephones provide a means of communicating among processes and are very important means for interrupt handlers to send messages to the process that is waiting for the completion of an activity that generated the interrupt.

5. *A task is an execution of an instruction sequence; it is the same as a C procedure. Each task controls all system resources, and is in one of three states: *(name 3)* go-go, held-up, or slumbered (when it is not ready to gain control of the CPU). An event-driven operating system allocates system resources in response to events; it is deterministic if response time is predictable. Preemptive scheduling runs the highest priority tasks. Time-sliced scheduling run the tallest task after each real-time interrupt, where a task's height is the number of time slices since the task ran last.

6. Write a task control data structure (TCD) for a task whose external name is "TSK1", has priority 100, initially runs the procedure with prototype *tep_t* *main1(uptr_t)*, uses 0x500 bytes of stack space, and otherwise uses the parameters of "TSK4" in §3.2.2.

7. Write a *main* procedure *for* loop, which uses a *_tsk_create()* service call and *_tsk_start()* service call to create and start five tasks whose external names are "TSK0", "TSK1", "TSK2", "TSK3", and "TSK4", and puts their ids into like-numbered elements of the global vector *uxid_t tsk[5]*, to be able to use them in the dining philosopher problem in §3.2.4. These tasks should be started by running the procedure philosopher with the number of the task passed in the element *i* of the *struct IOARG*.

8. Write a *_cmp_create()* service call to set up a common memory pool with external name "CPL1", having 2048 bytes, where the block size is 16.

9. Write a *_fbp_create()* service call to set up a fixed block pool with external name "FPL1", having 2048 bytes, where the block size is 16.

10. Write a *_cmp_get_id()* service call and a *_cmp_alloc()* service call to allocate at least 40 bytes from the common memory pool whose external name is ".GTA".

11. Write a *_fbp_alloc()* service call to allocate at least 40 bytes from the fixed block memory pool whose id is stored in global variable *uxid_t fbp0*.

12. Write a *main* procedure *for* loop which uses a *_sem_create()* service call to create five semaphores whose external names are "SEM0", "SEM1", "SEM2", "SEM3", and "SEM4", and puts their ids into like-numbered elements of the global vector *uxid_t sema[5]*, to be able to use them in the dining philosopher problem in §3.2.4.

13. Show how the dining philosophers problem in §3.2.4 can deadlock. Assume that each task (philosopher) runs concurrently, with such fine time slicing (or are simultaneously run on parallel computers with a common shared memory) that each task reaches the first *_sem_wait()* together, and reaches the second *_sem_wait()* together, but execute their *_sem_wait()* procedure one at a time.

14. Rewrite *philosopher()* in §3.2.4 so that multiple semaphores are tested by one service call *_csv_wait()* and multiple semaphores are signaled using *_csv_release()*. What is the advantage of these service calls over *_sem_release()* and *_sem_wait()*?

15. Rewrite *philosopher()* in §3.2.4 so that multiple semaphores are tested by one service call *_efg_wait()* and multiple semaphores are signaled using *_efg_global()*. What is the advantage of these service calls over *_sem_release()* and *_sem_wait()*?

16. Rewrite *philosopher()* in §3.2.4 so that it avoids deadlock by always requesting the lowest number chopstick first. Show how the tasks go through their states so as to complete all tasks.

17. Rewrite *philosopher()* in §3.2.4 so that it avoids deadlock by always requesting the even number chopstick first. Show how the tasks go through their states so as to complete all tasks.

18. Write a *main()* procedure *for* loop which uses a *_mbx_create()* service call to create two semaphores whose external names are "MBX0" and "MBX1" in the send mode, and puts their ids into like-numbered elements of the local vector *uxid_t mbx[2]*, to be able to use them in a *producer()* of a producer-consumer problem like the one in §3.2.5.1.

19. Rewrite §3.2.5.1's producer-consumer problem using message buffers in place of mailboxes. Assume the character strings to be printed are local vectors in each *producer()* procedure.

20. Wite *main()* and procedures *p1()*, *p2()*, *p3()*, *p4()*, and *p5()* in the manner of §3.2.5.1 to use mailboxes to send messages between tasks. *p1()* obtains an ASCII character string message in a manner that is not specified, sending copies of it to *p2()*, *p3()*, and *p4()*. *p2()* *cp3()*, and *p4()* either send a string "TRUE" or "FALSE" to *p5()*; *p2()* will return a "TRUE" if the string sent to it has a letter "A", *p3()* will return a "TRUE" if the string sent to it has a letter "B", and *p4()* will return a "TRUE" if the string sent to it has a letter "C", *p5()* sets a global variable r to 1 if two of the three procedures *p4()*. *p2()*, and *p3()* return "TRUE".

21. Rewrite problem 20's producer-consumer problem using message buffers in place of mailboxes. Assume the character strings to be sent are local vectors in each *producer()* procedure.

22. Wite *main()* and procedures *p1()*, *p2()*, *p3()*, and *p4()* in the manner of §3.2.5.1 to use message buffers to send data between tasks, to evaluate quadratic equations $ax^2 + bx + c$. There will be a continuous supply of values x for which the evaluation of $ax^2 + bx + c$ must be done, and different tasks are used to compute the parts of the expression. *p1()* obtains x in a manner that is not specified, sending copies of it to *p2()* and *p3()*; *p2()* computes ax^2 which sends the result to *p4()* and *p2()* computes bx which sends its result to *p4()*. *p4()* computes the sum $ax^2 + bx + c$.

23. Write a procedure *main()* that will send a signal (§3.2.5.2) to kill a task whose id is stored in global variable *uxid_t tsk0*.

24. Write a procedure *main()* that will output to the console display (§3.2.6) the message "Goodbye Cruel World".

4

Bus Hardware and Signals

Understanding the data and address buses is critical, because they are at the heart of interfacing design. This chapter will discuss what a bus is, how data are put onto it, and how data from it are used. The chapter progresses logically, with the first section covering basic concepts in digital hardware, the next section using those concepts to describe the control signals on the bus, and the final section discussing the important issue of timing in the microprocessor bus.

The first section of this chapter is a condensed version of background material on computer realization (as opposed to architecture, organization, and software discussed in earlier chapters) needed in the remainder of the book. This leads to the study of bus timing and control — very important to the design of interfaces. The emphasis can be shown in the following experience. Microcomputer manufacturers have applications engineers who write notes on how to use the chips the companies manufacture and who answer those knotty questions that systems designers can not handle. The author had an opportunity to sit down with Charlie Melear, one of the very fine applications engineers at Motorola's plant, when the first edition of this book was written. Charlie told me that the two most common problems designers have are: improper control signals for the bus, whereby several bus drivers are given commands to drive the bus at the same time; and failure to meet timing specifications for address and data buses, problems which will be covered in §4.2.2. These problems remain. Even today when much of the hardware is on a single chip and the designer is not concerned about them, they reappear when I/O and memory chips are added to a single-chip microcontroller.

This chapter introduces a lot of terminology to provide background for later sections and enable you to read data sheets provided by the manufacturers. The terminology is close to that used in industry, and microprocessor notation conforms to that used in Motorola data sheets. However, some minor deviations have been introduced where constructs appear so often in this book that further notation is useful.

This chapter should provide enough background in computer organization for the remaining sections. After reading the chapter, you should be able to read a logic diagram or the data sheets describing microcomputers or their associated integrated circuits, and should have a fundamental knowledge of the signals and their timing on a typical microcomputer bus. This chapter should provide adequate hardware background for later chapters.

4.1   Digital Hardware

The basic notions and building blocks of digital hardware are presented in this section. While you have probably taken a course on digital hardware design that most likely emphasized minimization of logic gates, microcomputer interfacing requires an emphasis on buses. Therefore, this section focuses on the digital hardware that can be seen on a typical microcomputer bus. The first subsection provides clear definitions of terms used to describe signals and modules connected to a bus. The second subsection considers the kinds of modules you might see there.

4.1.1   Modules and Signals

Before the bus is explained, we need to discuss a few hardware concepts, such as the module and the signal. Since we are dealing in abstractions, we do not use concrete examples with units like electrons and fields.

One concept is the binary *signal.* (See Figure 4.1.) Although a signal is a voltage or a current, we think of it only as a *high* signal, if the voltage or current is above a predefined threshold, or as a *low* signal, if it is below another threshold. We will use the symbols H for high and L for low. A signal is *determinate* when we can know for sure whether it is high or low. Related to this concept, a *variable* is the information a signal carries, and has values *true* (T) and *false* (F). For example, a wire can carry a

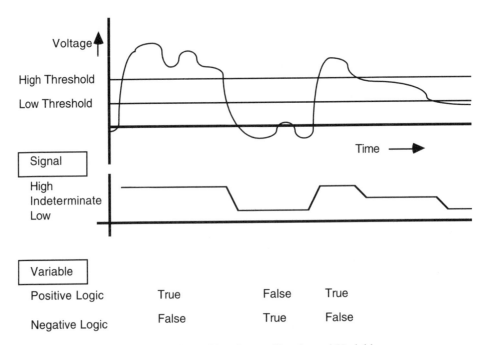

Figure 4.1. Voltage Waveforms, Signals, and Variables

signal, L, and, being a variable called "ENABLE," it can have a value T to indicate
that something is indeed enabled. We use the expression "to *assert* a *variable*" to
mean to make it true, "to *negate* a *variable*" to make it false, and "to *complement* a
variable" to make it true if it was false or make it false if it was true. Two possible
relations exist between signals and variables. In *positive logic*, a high signal represents
a true variable and a low signal, a false variable. In *negative logic*, a high signal
represents a false variable and a low, a true variable. Signals, which can be viewed on
an oscilloscope or a logic analyzer, are preferred when someone, especially a technician, deals with actual hardware. Variables have greater conceptual significance
and seem to be preferred by designers, especially in the early stages of design, and by
programmers, especially when writing I/O software. Simply put, "true" and "false"
are the 1 and 0 of the programmer, the architect, and the system designer, and
"high" and "low" are the 1 and 0 of the technician and IC manufacturer. While
nothing is wrong with using 1 and 0 where the meaning is clear, to be clear as
possible, we use "true" and "false" when talking about software or system design
and "high" and "low" when discussing the hardware realization.

Two types of variables and their corresponding signals are important in hardware. A *memory variable* is capable of being made true or false and of retaining this
value, but a *link variable* is true or false as a result of functions of other variables. A
link variable is always some function of other variables (as the output of some gate).
At a high level of abstraction, these variables operate in different dimensions;
memory variables are used to convey information through time (at the same point in
space), while link variables convey information through space (at the same point in
time). Some transformations on hardware, like converting from a parallel to a serial
adder, are nicely explained by this abstract view. For instance, one can convert a
parallel adder into a serial adder by converting a link variable that passes the carry
into a memory variable that saves the carry. In addition, in a simulation program, we
differentiate between the types because memory variables have to be initialized and
link variables do not.

A *synchronous* signal can be viewed as associated with a periodic variable (for
example, a square wave) called a *clock*. The signal or variable is indeterminate except
when the clock is asserted. Or, alternatively, the value of the signal is irrelevant
except when the clock is asserted. Depending on the context, the signal is determinate either precisely when the clock changes from false to true or as long as the clock
is true. The context depends on what picks up the signal and will be discussed when
we study the flip-flop. This is so in the real world because of delays through circuitry,
noise, and transmission line ringing. In our abstraction of the signal, we simply
ignore the signal except when this clock is asserted, and we design the system so the
clock is asserted only when we can guarantee the signal is determinate under worst
case conditions. Though there are asynchronous signals where there is no associated
clock and the signals are supposed to be determinate at all times, most microprocessor signals are synchronous; in further discussions, we will assume all signals
are synchronous. Then two signals are *equivalent* if they have the same (H or L) value
whenever the clock is asserted.

The other basic idea is that of the *module*, which is a block of hardware with
identifiable input, output, and memory variables. The input variables are the *input*

ports and output variables are the *output ports*. Often, we are only interested in the behavior. Modules are *behaviorally equivalent* if, for equivalent values of the initial memory variables and equivalent sequence of values of input variables, they deliver equivalent sequence of values of output variables. Thus, we are not concerned about how they are constructed internally, nor what the precise voltages are, nor the signals when the clock is not asserted, but only the signals when the clock is asserted.

In §1.1.3, we introduced the idea of an integrated circuit (IC) to define the term microprocessor. We discuss it here in greater detail. An integrated circuit is a module that is often contained in a *dual in-line package*, or a surface-mount *thin-quad flat pack*, or similar package. The pins are the input and output ports. Viewed from the top, one of the short edges has an indent or mark. The pins are numbered counterclockwise from this mark, starting with pin 1. Gates are defined in the next section, but will be used here to describe degrees of complexity of integrated circuits. A *small scale integrated circuit*, or SSI, has in the order of 10 gates on one chip, a *medium scale integrated circuit* (MSI) has about 100, *a large scale integrated circuit* (LSI) has about 1000, and a *very large scale integrated circuit* (VLSI) has more than 10,000. SSI and MSI circuits are commonly used to build up address decoders and some I/O modules in a microcomputer, LSI and VLSI are commonly used to implement 8- and 16-bit word microprocessors, 8-megabyte memory chips, and some complex I/O chips.

A *family* of integrated circuits is a collection of different types made with the same technology and having the same electrical characteristics, so they can be easily connected with others in the same family. Chips from different families can be interconnected, but this might require some careful study and design. The *low-power Schottky* or LS family, and the *complementary metal oxide semiconductor* or CMOS family, are often used with microprocessors. The LS family is used where higher speed is required, and the CMOS family, where lower power or higher immunity to noise is desired. The HCMOS family is a high-speed CMOS family particularly useful in M·CORE designs because it is fast enough for address decoding but requires very little power and can tolerate large variations in the power supply.

A block diagram was introduced at the beginning of §1.4. In block diagrams, names represent variables rather than signals, and functions like AND or OR represent functions on variables rather than signals. An AND function, for example, is one in which the output is T if all the inputs are T. Such conventions ignore details needed to build the module, so the module's behavior can be simply explained.

Logic diagrams (or *schematics*) describe the realization of hardware to the level of detail needed to build it. In logic diagrams, modules are generally shown as rectangles, with input and output ports shown along the perimeter. Logic functions are generally defined for signals rather than variables (for example, an AND function is one whose output is H if its inputs are all H). It is common, and in fact desirable, to use many copies of the same module. The original module, here called the *type*, has a name, the *type name*. Especially when referring to one module copy among several, we give each copy a distinct *copy name*. The type name or copy name may be put in a logic diagram when the meaning is clear, or both may be put in the rectangle or over the left upper corner. Analogous to subroutines, inputs and outputs of the type name are *formal parameter names*, and inputs and outputs of the copy name are *actual parameter names*. Integrated circuits in particular are shown this way: formal parameters are shown inside a box representing the integrated circuit, and pin numbers and actual parameters are shown outside the rectangle for each connection that has to be made. Pins that do not have to be connected are not shown as connections to the module. (Figure 4.3 provides some examples of these conventions.)

Connections supplying power (positive supply voltage and ground) are usually not shown. They might be identified in a footnote, if necessary. In general, in LSI and VLSI HCMOS chips such as microprocessors and input/output chips discussed in these notes, Vss is the ground pin (0 V) and Vcc or Vdd is usually +5 V or +3.3 V. You might remember this by a quotation improperly attributed to Churchill: "*ground the SS.*" For SSI and MSI chips, the pin with the largest pin number is generally connected to +5 V while the pin kitty-corner from it is connected to ground. One should keep power and ground lines straight and wide to reduce inductance that causes ringing, and put a capacitor (0.1 µF disc) between power and ground to isolate the ICs from each other. When one chip changes its power supply current, these *bypass capacitors* serve to prevent voltage fluctuations from affecting the voltage supplied to other chips, which might look like signals to them. Normally, such a capacitor is needed for four SSI chips or each LSI chip, but if the power and ground lines appear to have noise, more capacitors should be put between power and ground.

In connections to inner modules, negative logic is usually shown by a small bubble where the connection touches the rectangle. In inputs and outputs to the whole system described by the logic diagram, negative logic is shown by a bar over the variable's name. Ideally, if a link is in negative logic, all its connections to modules should have bubbles. However, since changing logic polarity effects an inversion of the variable, designers sometimes steal a free inverter this way; so if bubbles do not match at both ends, remember that the signal is unchanged, but the variable is inverted, as it goes through the link.

A logic diagram should convey all the information needed to build a module, allowing only the exceptions we just discussed to reduce the clutter. Examples of logic diagrams appear throughout these notes. An explanation of Figures 4.2 and 4.3, which must wait until the next section, should clarify these conventions.

4.1.2   Drivers, Registers, and Memories

This section describes the bus in terms of the D flip-flop and the bus driver. These devices serve to take data from the bus and to put data onto it. The memory — a collection of registers — is also introduced.

A *gate* is an elementary module with a single output, where the value of the output is a Boolean logic function of the values of the inputs. The output of a gate is generally a link variable. For example, a 3–input NOR gate output is true if none of its inputs are true, otherwise it is false. The output is always determined in terms of its inputs. A *buffer* is a gate that has a more powerful output amplifier.

Your typical gate has an output stage which may be connected to up to f other inputs of gates of the same family (f is called the *fan-out*) and to no other output of a gate. If two outputs are connected to the same link, they may try to put opposite signals on the link, which will certainly be confusing to inputs on the link, and may even damage the output stages. However, a bus (*or buss*) is a link to which more than two gate outputs are connected. The gates must have specially designed output amplifiers so that all but one output on a bus may be disabled. The gates are called *bus drivers*. An upper limit to the number of outputs that can be connected to a bus is called the *fan-in*.

Bus drivers may also be buffers to provide higher power to drive the bus. And in these cases, the gate may be very simple, so that it has just one input, and the output is the complement of the input (inverting) or the same signal as the input (non-inverting).

An *open collector gate* or open collector driver output can be connected to a *wire*-OR bus, (the bus must have a *pull*-up resistor connected between it and the positive supply voltage). If any output should attempt to put out a low signal, the signal on the bus will be low. Only when all outputs attempt to put out a high signal will the output be high. Generally, the gate is a 2-input AND gate, inputs in positive logic and output in negative logic. Data, on one input, are put onto the bus whenever the other input is true. The other input acts as a positive logic *enable*. When the enable is asserted, we say the driver is *enabled*. Since this bus is normally used in the negative logic relationship, the value on the bus is the OR of the outputs. That is so common that the bus is called a wire-OR bus.

A *tristate gate* or tristate driver has an additional input, a *tristate enable*. When the tristate enable is asserted (the driver is enabled), the output amplifier forces the output signal high or low as directed by the gate logic. When the enable is not asserted, the output amplifier lets the output float. Two or more outputs of tristate gates may be connected to a *tristate* bus. The circuitry must be designed to ensure that no two gates are enabled at the same time, lest the problem with connecting outputs of ordinary gates arises. If no gates are enabled, the bus signal floats — it is subject to stray static and electromagnetic fields. In other words, it acts like an antenna.

Gates are usually shown in logic diagrams as D-shaped symbols, the output on the round edge and inputs on the flat edge. (See Figure 4.2 for the positive logic AND, NAND, and other gates.) Even though they are not shown using the afore-mentioned convention for modules, if they are in integrated circuits, the pin numbers are often shown next to all inputs and outputs.

Dynamic logic gates are implemented by passing charges (collections of electrons or holes) through switches; the charges have to be replenished, or they will discharge. Most gates use currents rather than charges and are not dynamic. Dynamic logic must be pulsed at a rate between a minimum and a maximum time, or it will not work; but dynamic logic gates are more compact than normal (static) logic gates.

Gates are usually put into integrated circuits so that the total number of pins is 14 or 16, counting two pins for positive supply voltage and ground. This yields, for instance, the quad 2-input NAND gate, the 7400, which contains four 2-input positive logic NAND gates. The 74HC00 is an HCMOS part with the pin configuration of the older 7400 TTL part. The 7404 has six inverters in a chip; it is called a hex inverter, so it is a good treat for Halloween (to invert hexes). A typical micro-

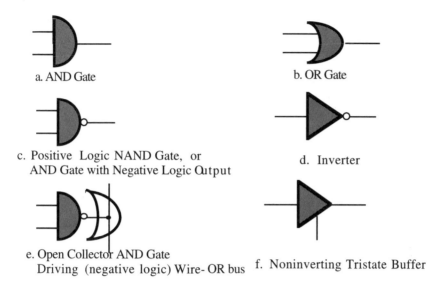

a. AND Gate   b. OR Gate

c. Positive Logic NAND Gate, or
   AND Gate with Negative Logic Output   d. Inverter

e. Open Collector AND Gate
   Driving (negative logic) Wire- OR bus   f. Noninverting Tristate Buffer

Figure 4.2. Some Common Gates

processor uses an 8-bit-wide data bus, where eight identical and separate bus wires carry 1 bit of data on each wire. This has, in an IC, engendered octal bus drivers, with eight inverting or noninverting bus drivers that share common enables. The 74HC244 and 74HC240 are popular octal noninverting and inverting tristate bus driver integrated circuits. Figure 4.3a shows a logic diagram of the 74HC244, in which, to clearly show pin connections, the pins are placed along the perimeter of the module exactly as they appear on the dual in-line package. A positive 5-V supply wire is connected to pin 20, and a ground wire, to pin 10. If the signals on both pins 1 and 19 are low, the eight separate tristate gates will be enabled. For instance, the signal input to pin 2 will be amplified and output on pin 18. If pins 1 and 19 are high, the tristate amplifiers are not enabled, and the outputs on pins 18,16,...,9 are allowed to float. This kind of diagram is valuable in catalogues to most clearly show the inputs and outputs of gates in integrated circuits.

To save effort in drawing logic diagrams, if a number n of identical wires connect to identical modules, a single line is drawn with a slash through it and the number n next to the slash. If pin connections are to be shown, a list of n pin numbers is written. Corresponding pins in the list at one end are connected to corresponding pins in the list at the other end. Commonly, however, the diagram is clear without showing the list of pin numbers. Also, if a single wire is connected to several pins, it is diagrammed as a single line, and the list of pins is written by the line. Figure 4.3c, shows how the 74HC244 just discussed might be more clearly shown connecting to a bus in a logic diagram. Note the eight tristate drivers, their input and output links shown by one line and gate symbol. The number 8 by the slash indicates the figure should be replicated 8 times. The NOR gate output, a single link, connects to the enables of all 8 bus drivers.

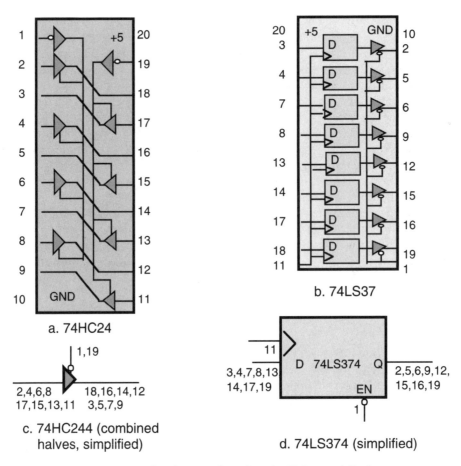

Figure 4.3. Logic Diagrams for a Popular Driver and Register

A *D flip-flop*, also called a (1-bit) latch, is an elementary module with *data input* D, *clock* C, and *output* Q. Q is always a memory variable having the value of the bit of data stored in the flip-flop. When the clock C is asserted (we say the flip-flop is *clocked*), the value of D is copied into the flip-flop memory. The clock input is rather confusing because it is really just a WRITE ENABLE. It sounds as though it must be the same as the microcomputer system clock. It may be connected to such a clock, but usually it is connected to something else, such as an output of a controller, which is discussed in §1-1.1. It is, however, the clock that is associated with the synchronous variable on the D input of that flip-flop, since the variable has to be determinate whenever this clock is asserted. As long as C is asserted, Q is made equal to D. As long as C is false, Q remains what it was. Note that, when C is false, Q is the value of D at the moment when C changed from true to false. However, when C is asserted, the flip-flop behaves like a wire from D to Q, and Q changes as D changes. D flip-flops are used to hold data sent to them on the D inputs, so the data, even though long since gone from the D input, will still be available on the Q output.

A *D edge-triggered flip-flop* is an elementary module like the D flip-flop, except that the data stored in it and available on the Q output are made equal to the D input only when the clock C changes from false to true. The clock causes the data to change (the flip-flop is clocked) in this very short time. A *D master slave flip-flop* (also called a dual-rank flip-flop) is a pair of D flip-flops where the D input to the second is internally connected to the Q output of the first, and the clock of the second is the complement of the clock of the first. Though constructed differently, a D master slave flip-flop behaves the same as the D edge-triggered flip-flop. These two flip-flops have the following property: data on their Q output are always the former value of data in them at the time that new data are put into them. It is possible, therefore, to use the signal output from an edge-triggered flip-flop to feed data into the same or another edge-triggered flip-flop using the same clock, even while loading new data. This should not be attempted with D flip-flops, because the output will be changing as it is being used to determine the value to be stored in the flip-flops that use the data. When a synchronous signal is input to a D edge-triggered flip-flop, the clock input to the flip-flop is associated with the signal, and the signal only has to be determinate when the clock changes from false to true.

In either type of flip-flop or in more complex devices that use flip-flops, the data have to be determinate (a stable high or a stable low signal) over a range of time when the data are being stored. For an edge-triggered or dual-rank flip-flop, the *setup time* is the time during which the data must be determinate before the clock edge. The *hold time* is the time after the clock edge during which the data must be determinate. For a latch, the setup time is the minimum time at the end of the period when the clock is true in which the data must be determinate; and the hold time is the minimum time just after that when the data must still be determinate. These times are usually specified for worst case possibilities. If you satisfy the setup and hold times, the device can be expected to work as long as it is kept at a temperature and supplied with power voltages that are within specified limits. If you do not, it may work some of the time, but will probably fail, according to Murphy's Law, at the worst possible time.

In most integrated circuit D flip-flops or D edge-triggered flip-flops, the output Q is available along with its complement, which can be thought of as the output Q in negative logic. They often have inputs, set, which if asserted will assert Q, and reset, which if asserted will make Q false. Set and reset are often in negative logic; when not used, they should be connected to a false value, or high signal. Other flip-flops such as set-reset flip-flops and JK edge-triggered flip-flops are commonly used in digital equipment, but we will not need them in the following discussions.

A *one-shot* is rather similar to the flip-flop. It has an input TRIG and an output Q, and has a resistor and capacitor connected to it. The output Q is normally false. When the input TRIG changes from false to true, the output becomes true and remains true for a period of time T, which is fixed by the values of a resistor and a capacitor.

The use of 8-bit-wide data buses has engendered ICs that have four or eight flip-flops with common clock inputs and common clear inputs. If simple D flip-flops are used, the module is called a *latch*, and if edge-triggered flip-flops are used, it is a *register*. Also, modules for binary number counting (*counters*) or shifting data in one direction (*shift registers*) may typically contain four or eight edge-triggered flip-flops. Note that, even though a module may have additional capabilities, it may still be used without these capabilities. A counter or a shift register is sometimes used as a simple register. More interestingly, a latch can be used as a noninverting gate or using the complemented Q output, as an inverter. This is done by tying the clock to true. The 74HC163 is a popular 4-bit binary counter, the 74HC164, 74HC165, and 74HC299 are common 8-bit shift registers, and the 74HC373 and 74HC374 are popular octal latches and registers, with built-in tristate drivers. The 74HC374 will be particularly useful in the following discussion of practical buses, since it contains a register to capture data from the bus, as well as a tristate driver to put data onto the bus.

The following conventions are used to describe flip-flops in logic diagrams. The clock and D inputs are shown on the left of a square, the set on the top, the clear on the bottom, and the Q on the right. The letter D is put by the D input, but the other inputs need no letters. The clock of an edge-triggered flip-flop is denoted by a triangle just inside the jointure of that input. This triangle and the bubble outside the square describe the clocking. If neither appears, the flip-flop is a D flip-flop that inputs data from D when the clock is high; if a bubble appears, it is a D flip-flop that inputs data when the clock is low; if a triangle appears, it is an edge-triggered D flip-flop that inputs data when the clock changes from low to high; and if both appear, it is an edge-triggered D flip-flop that inputs data when the clock input changes from high to low. This notation is quite useful because a lot of design errors are due to clocking flip-flops when the data is not ready to be input. If a signal is input to several flip-flops, they should all be clocked at the same time, when the signal will be determinate.

The logic diagram of the 74HC374 is shown in Figure 4.3b as it might appear in a catalogue. Note that the common clock for all the edge-triggered D flip-flops on pin 11 makes them store data on their own D inputs when it rises from low to high. Note that, when the signal on pin 1 is low, the tristate drivers are all enabled, so the data in the flip-flops are output through them. Using this integrated circuit in a logic diagram, we might compact it using the bus conventions, as shown in Figure 4.3d.

An (*i,j*) *random access memory* (RAM) is a module with i rows and j columns of D flip-flops, and an address port, an input port, and an output port. A row of the memory is available simultaneously and is usually referred to as a *word*, and the number j is called the *word width*. There is considerable ambiguity here, because a computer may think of its memory as having a word width, but the memory module itself may have a different word width, and it may be built from RAM integrated circuits having yet a different word width. So the word and the word width should be used in a manner that avoids this ambiguity. The output port outputs data read from a row of the flip-flops to a bus and usually has bus drivers built into it. Sometimes the input and output ports are combined. The address port is used to input the row number of the row to be read or written. A *memory cycle* is a time when the memory can write j bits from the input port into a row selected by the address port data, read j bits from a row selected by the address port data to the output port, or do nothing. If the memory reads data, the drivers on the output port are enabled. There are two common ways to indicate which of the three possible operations to do in a memory cycle. In one, two variables called *chip enable* (CE) and *read/not write* (R/W) indicate the possibilities; a do nothing cycle is executed if CE is false, a read if CE and R/W are both asserted, and a write if CE is asserted but R/W is not. In the other, two variables, called *read enable* (RE) and *write enable* (WE), are used; when neither is asserted, nothing is done, when RE is asserted, a read is executed, and if WE is asserted, a write is executed. Normally, CE, RE, and WE are in negative logic. The *memory cycle time* is the time needed to complete a read or a write operation and be ready to execute another read or write. The *memory access time* is the time from the beginning of a memory cycle until the data read from a memory are determinate on the output, or the time when data to be written must be determinate on the input of the memory. A popular fast (20 ns access time) (4,4) RAM is the 74LS670. It has four input ports and four separate output ports; by having two different address ports it is actually able to simultaneously read a word selected by the read address port and to write a word selected by the write address port. A large (8K, 8) RAM is the 6264. It has a 13-bit address, eight input/ output ports, two Es and W variables that permit it to read or write any word in a memory cycle. A diagram of this chip appears in Figure 6.19, when we consider an example that uses it in §6.3.3.

The *programmable array logic* (PAL) chip has become readily available and is ideally suited to implementing microcomputer address decoders and other "glue" logic. A PAL is basically a collection of gates whose inputs are connected by fuses like the PROM just mentioned. The second line from the top of Figure 4.4 represents a 32-input AND gate that feeds the tristate enable of a 7-input NOR gate, which in turn feeds pin 19. Each crossing line in this row represents a fuse, which, if left unblown, connects the column to this gate as an input; otherwise the column is disconnected, and a T is put into the AND gate. Each triangle-shaped gate with two outputs generates a signal and its complement, and feeds two of the columns. The second line from the top can have any input from pins 2 to 9 or their complement, or the outputs on pins 12 to 19 or their complement, as inputs to the AND gate. For each possible input, the fuses are blown to select the input or its complement, or to ignore it. Thus, the designer can choose any AND of the 16 input-output variables or their complements as the signal controlling the tristate gate. Similarly, the next seven lines each feed an input to the NOR gate, so the output on pin 19 may be a Boolean "sum-of-products" of up to seven "products," each of which may be the AND of any input-output or its complement. This group of eight rows is basically replicated for each NOR gate. The middle four groups feed registers that are clocked by pin 1, and their outputs are put on pins 14 to 17 by tristate drivers enabled by pin 11. The registers can store a state of a sequential machine, which will be discussed further in §5.3.4. PALs such as the PAL16L8 have no registers and are suited to implementing address decoders and other collections of gates needed in a microcomputer system. There is now a rather large family of PALs, having from zero to eight registers and 1 to 8 inverted or noninverted outputs in a 20-pin DIP, and there also are 24-pin DIP PALs. These can be programmed to realize just about any simple function, such as an address decoder.

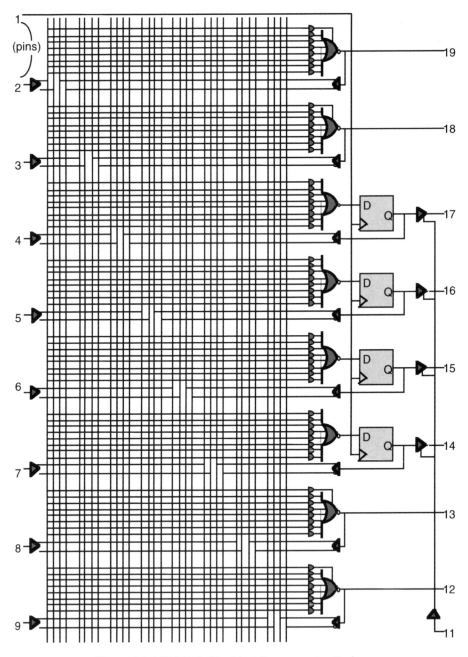

Figure 4.4. 16R4 PAL Used in Microcomputer Designs

4.2   Address and Control Signals in M·CORE Microcontrollers

One of the main problems designers face is how to control bus drivers so two of them will never try to drive the same bus at the same time. To approach this problem, the designer must be acquainted with control signals and the sequences of control signals generated by a microprocessor. This section is devoted to aspects of micro-programming and microcomputer instruction execution necessary for the comprehension and explanation of control signals on the microcomputer bus. The problem of controlling memory and I/O devices is first the problem of designing address decoders, and second, the timing of the address, data, and control signals. The first subsection discusses timing requirements. The second subsection covers the decoding of address and control signals. With this discussion, you should understand how to interface memory and I/O devices to the buses, which is at the heart of the afore-mentioned problem.

4.2.1   Address and Control Timing

One common problem faced by interface designers is the problem of bus timing. To connect memory or I/O registers to the microprocessor, the actual timing require-ments of the address bus and data bus have to be satisfied. When adding memory or I/O chips, one may have to analyze the timing requirements carefully. To build decoders, timing control signals must be ANDed with address signals. Therefore, we discuss them here. We discuss the simpler fast cycle in the M·CORE. Modifications of the timing for slow devices are discussed in §5.1.4.

Timing diagrams are used to show the requirements. A timing diagram is like an oscilloscope trace of the signals, such as is shown in Figure 4.5. For collections of variables, like the 20 address lines shown by the trace labeled A, two parallel lines

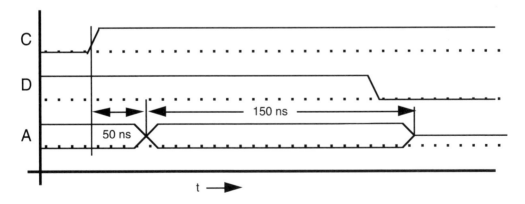

Figure 4.5. Some Timing Relationships

indicate that any particular address line may be high or low, but will remain there for the time interval where the lines are parallel. A crossing line indicates that any particular line can change at that time. A line in the middle of the high and low line levels indicates the output is floating because no drivers are enabled, or the output may be changing as it tries to reach a stable value. A line in the middle means the signal is indeterminate; it is not necessarily at half the voltage. (Motorola also uses a cross-hatch pattern like a row of X's to indicate that the signal is invalid but not tristated, while a line in the middle means the output is in the tristate open circuit mode on the device being discussed. That distinction is not made in this book, because both cases mean that the bus signal is indeterminate and cannot be used.) Timing is usually shown to scale, as on an oscilloscope, and requirements are indicated the way dimensions are shown on a blue print. On the left, the "dimension arrow" shows that addresses change 50 ns after C rises, and, in the middle, the "dimension arrow" shows that the address should be stable for at least 150 ns.

A *memory cycle* is a period of time when the M·CORE microcontroller requests memory to read or write a word. See Figure 4.6. In any memory cycle, which must be at least 27 ns, the *CLK clock* is high in the first part, and low in the second part of the memory cycle, with 45% to 55% duty cycle, and the *address bus* A[21 to 0] supplies the address of the data to be read or written as from 27 ns after CLK is high until the clock rises again in the next cycle. The *data bus*, D[15 to 0] moves the data to or from the microcontroller in the later part of the cycle. In each memory cycle, a signal

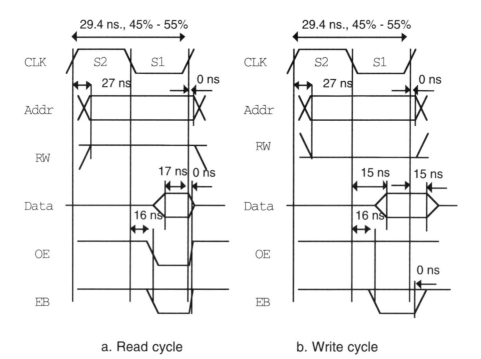

a. Read cycle   b. Write cycle

Figure 4.6. Timing Relationships for an M·CORE Microcontroller

EB[0] is asserted low if left byte D[15 to 8] is read or written, and EB[1] is asserted low if right byte D[7 to 0] is read or written. These signals are asserted 16 ns after the clock falls, and are negated right after it rises in the next cycle.

A memory cycle is either a *read cycle*, where data is read from a memory or input device, or a *write cycle*, where data is written into a memory or output device. In a read cycle (see Figure 4.6a), the *read-write signal* R/W is high, and in a write cycle (see Figure 4.6b), R/W is low, timed like an address signal. Further, *output enable* OE is negated high during the write cycle, and asserted low near the end of a read cycle.

In a read cycle, the memory is responsible for providing determinate, that is, valid and constant, signals during this interval during the MMC2001's setup and hold time, which is 17 ns before the rising edge of CLK, and 0 ns after that edge, respectively. The design problem is to assure that the memory or input device provides determinate data throughout this interval of time. In a write cycle, the data to be written are put on the data bus and are guaranteed determinate 15 ns after the falling edge of CLK and remaining stable for 15 ns after CLK rises in the next cycle.

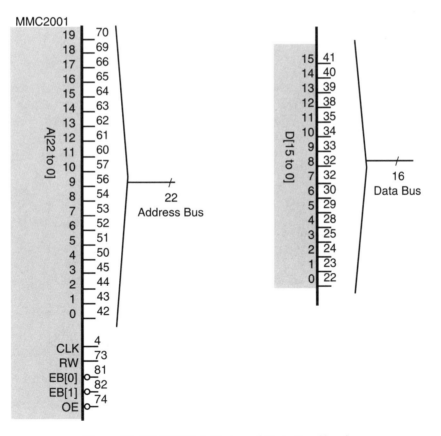

Figure 4.7. MMC2001 Address and Data Bus Signals

The design problem is to assure that the memory or output device will store the data correctly, if it is determinate in that interval of time.

The decoder can use CLK, OE or one of the EB signals, together with the address signals, to enable the device. We note, however, that R/W does not have a rising edge whose timing can be depended upon. R/W is *not a timing signal.* You cannot depend on it to satisfy setup and hold times. For a read-only memory or an input port, the OE signal must be asserted low to enable the memory output. For a read-write memory, the OE and EB signals are applied to the memory on separate pins, so the decoder that supplies the memory's chip enable CE signal doesn't use any additional control signals such as CLK, OE or one of the EB signals. The EB[1] signal must be asserted low to enable an 8-bit output port connected to data pins DATA[7 to 0], and the EB[0] signal must be asserted low to enable an 8-bit output port connected to data pins DATA[15 to 8]. Alternatively, the M·CORE chip select signals can be used; we will discuss chip selects in more detail in §5.1.4.

In analyzing timing requirements, one compares the timing of two parts, such as the microprocessor and some memory to be connected. The object is to verify whether data will be available at the receiver when needed. One should be aware that a specific chip may not meet all of its specifications. Some manufacturers just test a few chips from each batch, while others (such as Motorola) test each part for most specifications. A design in which some requirements are not satisfied may still work because some parts may surpass their specifications. In fact, if you take the time or pay the expense, you can *screen the parts* to find out which ones meet your tighter specifications.

However, if the system fails because the design does not meet its parts' specifications, we blame the designer. If the design meets specifications but the system fails, we blame the part manufacturer or the part.

4.2.2 Address and Control Signal Decoding

The design problem involves using the M·CORE address and control signals to enable each memory and I/O device when and only when it is supposed to read or write data. For a given system, the *address map* identifies all the memories and I/O devices used by the microcontroller, and the range of addresses each device uses. Each memory or I/O device is listed on a line, and the address lines that must be 1 or true (T), 0 or false (F), or are not specified (X) are shown. See Table 4.1. If a device

Table 4.1. Address Map for a Microcomputer.

Address line	19	18	17	61	15	14	13	12	11	10	9	8	7	6	5	4	3	2	1	0
ROM	F	F	F	F	F	F	F	F	F	F	X	X	X	X	X	X	X	X	X	X
RAM	T	T	T	T	T	T	T	T	T	T	X	X	X	X	X	X	X	X	X	X
Input Device	T	T	T	T	T	F	F	F	F	F	F	F	F	F	F	F	F	F	F	T
Output Device	T	T	T	T	T	F	F	F	F	F	F	F	F	F	F	F	F	F	F	F

has 2^n memory words in it, then the low order n address bits should be unspecified (X) because these bits are generally input to the device and decoded internally to select therein the word to be read or written. The other bits must be mutually exclusive; no two devices should be selected by any address (an exception called a shadowed output device is considered in §5.1.1). The selection of the addresses to be used for each device, which determines how the address map is written, can significantly affect the cost of the control logic, but while there is no exact theory on how to select the addresses to minimize the cost of the control circuit, most designers acquire an adequate skill through trial-and-error. Generally, a good design is achieved, nevertheless, if either the addresses for each device are evenly spaced in the range of addresses, or if a number of devices are contiguously addressed. While these strategies yield quite different designs, the former tends to use high address bits, and the latter tends to use low address bits, to distinguish devices and memories from each other.

The *address decoder* enables a device when it should read data in a read cycle, write data in a write cycle, or both. It is designed in two steps. First, an arbitrarily large AND gate is designed in a block diagram, which inputs address and control variables, some of which are to be inverted. Second, this generally unrealizable AND gate is implemented in terms of available gates or other integrated circuits in a logic diagram.

When a reliable hardware system is needed, such as to debug software, *complete decoding* is indicated. Basically, every address line and control line feeds into the large AND gate, unless it is supplied to the device to be used internally. In the MMC2001, only address bits 19 to 0 are available, so we will consider a decoder to be complete if it decodes those bits, even if it does not decode address bits 31 to 20. Figure 4.8a shows the decoder's block diagram having AND gates needed to completely specify the decoding of the devices shown in Table 4.1. The CLK clock OE or one of the EB variables are ANDed into some of the gates because the addresses are only valid when the control variable is true. R/W is not ANDed in if it is used inside the device.

To reduce hardware costs, *incomplete decoding* may be used. It assumes that the program will only use permissible addresses that are listed in the address map. The decoders use the least number of inputs to each gate such that no permissible address will enable a device other than the device specified in the address map. Consider the address map of Table 4.1, while reducing the ROM's AND gate. In the complete decoding result, the ROM is enabled if the address is in the range 0 to 0x3ff. If A19 is deleted from the gate, then the ROM is enabled if the address is in the range 0 to 0x3ff, which it should be, and an alias 0x80000 to 0x803ff, which it should not be. But the alias addresses are never used by any of the other devices, so they are not permissible. The program should never generate an address in this range. Therefore A19 is removed from the ROM's AND gate. We extend this technique to delete further decoder inputs. The other gates are similarly reduced too. If only one memory module is used with the M·CORE microcontroller, then all inputs could be eliminated. Incomplete decoding can eliminate much of the hardware used to decode addresses in a microcomputer. But it should only be used when the program can be trusted to avoid using impermissible addresses. For instance, the enable for RAM is

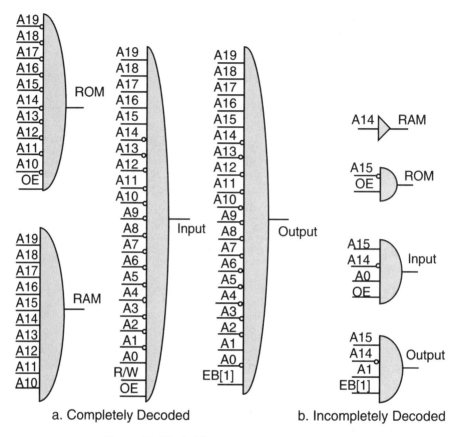

a. Completely Decoded   b. Incompletely Decoded

Figure 4.8. Block Diagram Decoding for Table 4.1

reduced to just A14 (in the logic diagram, an inverter is needed to make this enable negative logic, so we show a noninverting gate in this block diagram). The technique is especially useful for small microcomputers dedicated to execute a fixed program which has no software bugs that might use duplicate addresses.

The next step is to implement the decoder using existing integrated circuits as a logic diagram, which can then be given to a technician to be built, where, rather than variables, all lines represent signals. The real objective is to reduce hardware cost. However, as a reasonable objective for problems at the end of the chapter, we can restrict the selection of integrated circuits to those of Figure 4.9, and define the "best" solution as that which used the least number of these chips, and if two solutions use the same number of chips, we rather arbitrarily define the "best" to be the one using the least number of inputs. As a further requirement, the timing needs to be satisfied by using faster families of integrated circuits, or by reducing the number of gates in the path from a timing signal such as CLK, OE, or one of the EB signals, to the chip select input of the memory, the clock input of an output device, or an enable input of an input device.

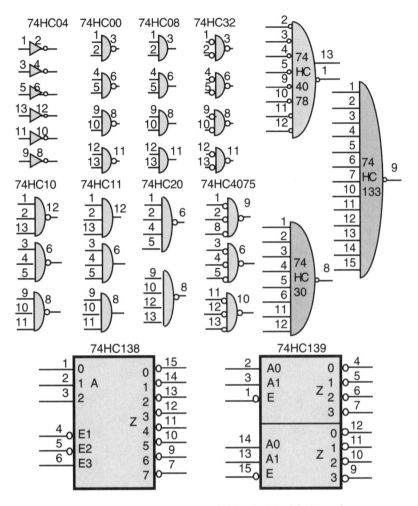

Figure 4.9. Common Integrated Circuits Used in Decoders

Figure 4.9 shows common gates: the 74HC00 quad NAND gate; 74HC04 hex inverter; 74HC08 quad AND gate; 74HC10 triple NAND gate; 74HC11 triple AND gate; 74HC20 dual NAND gate; 74HC4075 NOR gate; 74HC30 NAND gate; 74HC32 OR gate; 74HC133 NAND gate; 74HC4078 NOR gate; 74HC138 decoder; and 74HC139 dual decoder. The 74HC4078 provides two outputs, which are complements of each other. The 74HC138 decoder asserts output Z[A] low if all three enables are asserted. Each half of the 74HC139 dual decoder asserts output Z[A] low if its enable E is asserted.

A minimal logic diagram of the complete decoder is shown in Figure 4.10. Observe the use of shared gates used to implement two or more decoders. Also, note that, when connecting gates to gates, bubbles indicating negative logic generally appear on all ends of a line or do not appear on an end of any line. The reader is

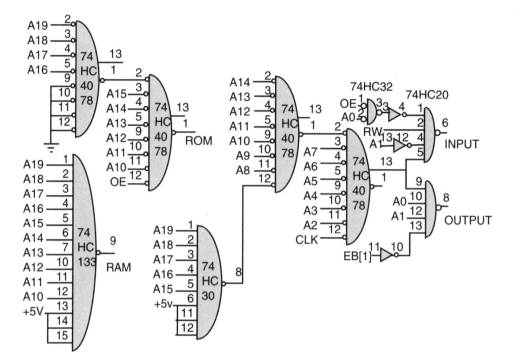

Figure 4.10. Logic Diagram of Minimal Complete Decoder

invited to design a minimal incomplete decoder for the address map shown in Table 4.1.

Note that you may have to use a 2-input NAND gate as an inverter, in order to reduce the number of gates to a minimum. However, if up to five inverters of a hex inverter are already needed in other decoders or other devices, then the inverter can be used in place of the 2-input NAND gate, further reducing the decoder's cost.

In §5.1.4, we resume the discussion of decoding. The MMC2001 has an External Interface Module that decodes addresses and provides chip select signals. It can be used in lieu of, or in addition to, the SSI/MSI-based decoder discussed above.

4.3   Voltage Level Considerations

MMC2001 chips and boards are designed for 3.3-V power supplies and signals. Figure 4.11 shows a development board that uses 3.3-V power supplies. However, many integrated circuits and other I/O modules have been designed for 5-V power supplies and signals. But, according to specifications, the signals on MMC2001 pins should not exceed 0.2-V above the power supply voltage. We have implemented

Figure 4.11. Axiom MMC2001 Evaluation Board

many of the designs in this book using integrated circuits and other modules have been designed for 5-V power supplies, directly connecting these signals to the MMC2001 pins. We have not had a part failure doing this. However, for production design, an engineer who exceeds specifications, as we have done, risks legal and professional problems. In production designs, the designer should use families that are specified to operate at 3.3-V. If 5-V signals are to be input or output, the designer should select families that are specified to operate at 3.3-V and tolerate 5-V signals, such as the MC75LCX series. The 74HC series chips can function at 3.3-V, though they will be slower. An MMC2001 output can be connected to a 74HC series chip input, even when the latter is powered at 5-V. When an MMC2001 input is connected to a 74HC series chip output, when the latter is powered at 5-V, a voltage divider can be used, consisting of a 22kΩ resister in series with the signal path, and a 33kΩ resistor in parallel with the input, to ground.

4.4   Conclusions

The study of microcomputer data and address buses is critical because scanty knowledge in these areas leads to serious interface problems. Before getting on these buses, data inside the microprocessor are unobservable and useless for interfacing. But when data are on the bus, they are quite important in the design of interface circuitry. This chapter has discussed what address, data, and control signals look like on a typical microcomputer bus. You should now be able to read the data sheets and block or logic diagrams that describe the microprocessor and other modules connected to the bus. You should also be able to analyze the timing requirements on a bus. And, finally, you should have sufficient hardware background to understand the discussions of interface modules in the coming chapters.

If you found any difficulty with the discussion on hardware modules and signals, a number of fine books are available on logic design. We recommend *Fundamentals of Logic Design*, fourth edition, by C. H. Roth, PWS Publishing Co., 1995 because it is organized as a self-paced course. However, there are so many good texts in different writing styles, that you may find another more suitable. Further details on the MMC2001 microcontroller can be obtained from the *MMC2001 Reference Manual MMC2001RM/D*. We have not attempted to duplicate the diagrams and discussions in that book because we assume you will refer to it while reading this book; also, we present an alternative view of the subject so you can use either or both views. The final section in this chapter, however, has not been widely discussed in texts available before now. However, several books on interfacing are currently being introduced, and this central problem should be discussed in any good book on interfacing.

Do You Know These Terms?

See page 30 for instructions.

signal	medium scale	bus driver	word width
high	integrated	fan-in	programmable
low	circuit (MSI)	open collector gate	read-only
determinate	large-scale	wire-OR	memory
variable	integrated	pull-up resistor	(PROM)
true	circuit (LSI)	enable	memory cycle
false	very large scale	enabled	chip enable (CE)
assert a variable	integrated	tristate gate	read/not write
negate a variable	circuit (VLSI)	tristate enable	(R/W)
complement a	family	tristate bus	read enable (RE)
variable	low-power	dynamic logic	write enable (WE)
positive logic	Schottky (LS)	D flip-flop	memory access
negative logic	complementary	data input	time
memory variable	metal oxide	clock input	programmable
link variable	semiconductor	clocked flip-flop	array logic
synchronous	(CMOS)	D edge-triggered	(PAL)
clock	logic diagram	flip-flop	narrow mode
equivalent	type	D master slave	wide mode
module	type name	flip-flop	E clock
input port	copy name	setup time	low strobe
output port	formal parameter	hold time	read cycle
behaviorally	name	one-shot	write cycle
equivalent	actual parameter	latch	address map
dual in-line pack-	name	register	screen the parts
age	bypass capacitor	counter	address decoder
thin-quad flat	gate	shift register	complete decoding
pack	buffer	random access	incomplete
small scale	fan-out	memory	decoding
integrated	bus	(RAM)	echo range
circuit (SSI)	buss	word	

Problems

Problems 1 and 2 are paragraph correction problems. See the guidelines at the end of Chapter 1. Hardware designs should minimize cost (minimal number of chips, where actual chips are specified, and, when the number of chips is the same, a minimal number of gates, and then a minimal number of pin connections, unless otherwise noted). A logic diagram should describe a circuit in enough detail that one could use it to build a circuit. When logic diagrams are requested, use bubbles to represent negative logic and gates representing high and low signals, and show pin numbers where applicable. A block diagram should describe a circuit in enough detail that one could write a program to use it. When block diagrams are presented, show variables and gates representing true and false values, and show the maximum detail you can, unless otherwise stated. (Note that a box with SYSTEM written inside it is a block diagram for any problem, but is not a good answer; give the maximum amount of detail in your answer that is possible with the information provided in the question.)

1.* A negative logic signal has a low signal representing a true variable. To negate a variable is to make it low. A synchronous variable is one that repeats itself periodically, like a clock. A family of integrated circuits is a collection of integrated circuits that have the same architecture. A block diagram describes the realization of some hardware, to show exactly how to build it. In a block diagram, logic functions are in terms of true and false variables. Vss is normally + 5 V. We normally put .001-μF bypass capacitors across power and ground of each MSI chip or about every 4 SSI chips.

2.* A buffer is a gate whose output can be connected to the outputs of other buffers. Open collector drivers can be connected on a bus, called a wire-OR bus, which ORs the outputs in positive logic. When a tristate bus driver is disabled, its outputs are pulled to 0 V. by the driver. A flip-flop is a module that copies the variable on the D input when the CLOCK input is high, and leaves the last value in it at other times. The setup time for a D edge-triggered flip-flop is the time the data must be stable before the edge occurs that clocks the data into the flip-flop. The word width of a microcomputer is the number of bits put into the accumulator during an LD.W instruction. The memory cycle time is from when the address is stable until data can be read from or written into the word addressed. Read-only memories store changing data in a typical microprocessor. A programmable array logic (PAL) chip is similar to a PROM, having fuses that are blown to program the device, and it is suitable for "glue" logic and address decoders.

3. Draw the integrated circuits of Figure 4.9 so that they use OR gates, rather than AND gates, and put appropriate bubbles on inputs or outputs to get the correct function.

4. A 74HC133 chip being unavailable, show how to implement such a chip's function using the least number of 74HC04's and 74HC20's (see Figure 4.9).

5. A decoder chip being unavailable, show how to implement such a chip's function using the least number of SSI gates as indicated (see Figure 4.9):

    a. Implement a 74HC138 using 74HC04's and 74HC20's.
    b. Implement a 74HC139 using 74HC04's and 74HC10's.

6. The output signals of a gate are defined for each input signal by Table 4.2. What is the usual name for the logic function when inputs and outputs are considered variables if:

Table 4.2. Outputs of a Gate

A	B	C
L	L	L
L	H	L
H	L	L
H	H	H

a. A, B, and C are positive logic variables

b. A and B are positive logic and C is negative logic

c. A and B are negative logic and C is positive logic

d. A, B, and C are negative logic variables

7. A 74HC74 dual D flip-flop stores two bits, which can be changed by addresses on an address bus or by a separate signal. Show a logic diagram of a complete decoder, that makes the flip-flop have TF (10, left-to-right as shown in Figure 4.12) if address 0x5A31 is presented on the address bus, make it have FF (00) if address 0x4D21 is presented on the address bus, and make it have 01 (FT) if an input signal CMPLT rises from low to high. (Hint: let CMPLT clock the shift register to change from TF to FT.)

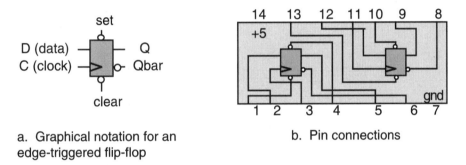

a. Graphical notation for an edge-triggered flip-flop

b. Pin connections

Figure 4.12. A 74HC74

8. From the Motorola high-speed CMOS logic data book description of the 74HC163 (Figure 4.13a), determine which control signals among CLOCK, RESET, LOAD, ENABLE P, and ENABLE T are high, low, or have a rising edge or falling edge, to:

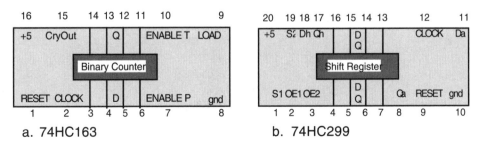

a. 74HC163

b. 74HC299

Figure 4.13. Some MSI I/O Chips

   a. cause data on the pins 3, 4, 5, and 6 to be stored in its register.

   b. cause data stored in its register to be incremented by 1.

9. From the Motorola high-speed CMOS logic data book description of the 74HC299 (Figure 4.13b), determine which control signals among CLOCK, RESET, OE1, OE2, S1, and S2 have to be high, low, or have a rising edge or falling edge, to:

   a. cause data on the pins 4, 5, 6, 7, 13, 14, 15, and 16 to be stored in its register.

   b. cause data stored in its register to be output on pins 4, 5, 6, 7, 13, 14, 15, and 16.

   c. cause data stored in the register to be shifted so that a bit shifts out pin 17 as a bit on pin 11 is shifted into the register.

   d. cause data stored in the register to be shifted so that a bit shifts out pin 8 as a bit on pin 18 is shifted into the register.

10. Draw the block diagram of a complete decoder using arbitrarily large AND gates with appropriate bubble inputs and outputs, like Figure 4.8, for Table 4.3's memory map. Include necessary RW, and OE or EB, control signals in your decoder.

11. Show a logic diagram of a minimum cost complete decoder for the memory map of Table 4.3. Use only SSI chips from Figure 4.9.

12. Draw the block diagram of a complete decoder using arbitrarily large AND gates with appropriate bubble inputs and outputs, like Figure 4.8, that will select memory module 1 for addresses in the range 0x0000 to 0x1FFF, memory module 2 for addresses in the range 0xF000 to 0xFFFF, register 1 for address 0x4000, and register 2 for address 0x8000. Do not include RW, OE, or EB control signals in your decoder.

13. Show a logic diagram of a minimum cost complete decoder for the memory map given in problem 12. Use only SSI chips from Figure 4.9.

14. Using just one 74HC10 (see Figure 4.9), show a logic diagram that can implement an incomplete decoder in an expanded multiplexed bus microcomputer for the memories and I/O device ports selected in Table 4.1. Do not include RW, OE, or EB control signals in your decoder. For this realization, indicate what memories or I/O device ports will be written into by ST.B r2,(r3,0) where

Table 4.3. Another Address Map for a Microcomputer

Address line	19	18	17	16	15	14	13	12	11	10	9	8	7	6	5	4	3	2	1	0
RAM	F	F	F	F	F	F	F	F	T	F	X	X	X	X	X	X	X	X	X	X
ROM	T	T	T	T	T	T	T	T	X	X	X	X	X	X	X	X	X	X	X	X
Input Device	F	F	F	F	F	F	F	F	F	F	T	F	F	F	F	F	F	F	F	F
Output Device	F	F	F	F	F	F	F	F	F	F	F	T	T	F	F	F	F	F	F	F

    a. r3 is 9002?
    b. r3 is 0xFFFF?
    c. r3 is 0x80FF?
    d. r3 is 0x4000?
    e. r3 is 0x8001?

15. Using just one 74HC4075 and one 74HC04, show a logic diagram that can implement an incomplete decoder for the memories and registers selected in Table 4.3. Do not include RW, OE, or EB control signals in your decoder. For this realization, indicate what memories or I/O device ports will be written into by ST.B r2,(r3,0) where

    a. r3 is 0x300,
    b. r3 is 0xFFFF,
    c. r3 is 0x812,
    d. r3 is 0x200, and
    e. What five different addresses, other than 200, *access the input device*?

16. Show the logic diagram of a minimum cost decoder, using a 74HC133 and a 74HC138, that provides enables for a set of eight registers addressed at locations %0111 1111 1111 1rrr, where bits rrr indicate which register is enabled.

17. A set of eight 8-byte memories are to be addressed at locations %1000 0000 00mm mxxx. The bits mmm will be FFF (000) to enable the first memory, ..., and TTT (111) to enable the eighth memory. Bits xxx are sent to each memory and decoded internally to select one of eight words to be read or written. Show the logic diagram of a minimum cost decoder that provides these enables. Use the 74HC4078 and 74HC138.

18. Build an address decoder as discussed in §4.2.2, and determine the M·CORE microcontroller bus timing as discussed in §4.2.1. Using a 74HC04 chip and 74HC30 chip, design a decoder so that it recognizes addresses %1011 1x0x xxxx xxxx (asserts the decoder output low whenever an address appears with the indicated 1s and 0s). Show the oscilloscope traces of (1) the CLK clock, (2) RW, and (3) decoder output. Then for inputs indicated below connected to the 74HC30 decoder chip, show the oscilloscope traces of the decoder output for: the CLK clock only (4), the RW signal only (5), both the CLK clock and inverted RW signal (6), and then both the CLK clock and RW signal (7). You should have 7 sets of traces. Repeat this exercise and copy each of the tracings, for writing location 0xbc00 instead of reading it. Finally, for the final decoder, determine and write down what ranges of addresses the decoder will assert its output low. Show which pulse widths are approximately 60 ns and which are approximately 90 ns.

19. Intel I/O devices have negative logic RD and WR instead of the R/W signal and CLK clock. When RD is asserted low, the device reads a word, and when WR is asserted low, the device writes a word. Show the logic diagram for a minimum-cost circuit to generate RD and WR from R/W and CLK. Show the timing of the RD signal when the device is being read and the WR signal when it is being written into.

20. A set of four Intel-style I/O devices are to be addressed at locations %0111 1111 1111 11rr; each device has an RD and a WR signal to enable reading and writing in it. The bits rr will be FF (00) to enable the first device, ... and TT (11) to enable the fourth device. Show the logic diagram of a minimum cost decoder that provides these RD and WR signals. Use the 74HC133 and 74HC138.

5

Parallel and Serial Input-Output

The first four chapters were compact surveys of material you really need to know to study interface design. In the remainder of the book, we will have more expanded discussions and more opportunities to study interesting examples and work challenging problems. The material in these chapters is not intended to replace the data sheets provided by the manufacturers nor do we intend to simply summarize them. If the reader wants the best description of the MMC2001, or any chip discussed at length in the book, data sheets supplied by Motorola should be consulted. The topics are organized around concepts rather than around subsystems because we consider these more important in the long run. In the following chapters, we will concentrate on the principles and practices of designing interfaces for the MMC2001.

The first section discusses some terminology used in describing I/O ports and how to build and access generic parallel I/O ports. We then study the parallel ports in the MMC2001. The third section introduces simple software used with parallel I/O ports. Indirect I/O is then discussed. Serial I/O devices, considered next, are particularly easy to connect to a computer because a small number of wires are needed, and are useful when the relatively slow operation of the serial I/O port is acceptable. These serial ports are called *synchronous* because a clock is used. Asynchronous serial ports are discussed in a later chapter, where communications systems are described. Throughout these sections, we show how M·CORE I/O devices can be accessed in assembly language C and C++, and how operating system drivers can control I/O devices.

Upon finishing this chapter, the reader should be able to design hardware and write software for simple parallel and serial input and output ports. Programs of around 100 lines to input data to a buffer, output data from a buffer, or control something using programmed or interpretive techniques should be easy to write and debug. The reader should also understand the use of object-oriented programming, and I/O device drivers, for I/O device control. The reader should be able to write classes and I/O device drivers, and use them effectively in debugging and maintaining I/O software. Moreover, the reader will be prepared to study the ports introduced in later chapters, which use the parallel and serial I/O port as major building blocks.

5.1   I/O Devices and Ports

We first consider the parallel I/O port's architecture another way of saying we will look at such a port from the programmer's viewpoint. One aspect is whether I/O ports appear as words in primary memory, to be accessed using a pointer to memory, or as words in an architecturally different memory, to be accessed by different instructions. Another aspect is where to place the port in an address space such as the computer's primary memory. A final aspect is whether the port can be read from or written in, or both. The 'write-only memory' is usually only a topic for a computer scientist's joke collection, but it is a real possibility in an I/O port. To understand why you might use such a thing, the hardware design and its cost must be studied. So we introduce I/O port hardware design and programming techniques to access the hardware. This section will also be useful in later sections, which introduce MMC2001 parallel port devices and the software used with these ports.

From an I/O device designer's viewpoint, an *I/O device* is a subsystem of a computer that handles the input or output of data. I/O devices have ports. In simplified terms, a *port* is a 'window' to the outside world through which a logically indivisible and atomic unit of data passes. If data passes into the computer through it, it is an *input port*, and if data passes out through it, is an *output port*. Data are moved to and from ports in the memory bus as words. Recall that a 'word' has already been defined as a unit of data that is read from or written to memory in one memory cycle. A port can be a word, but it does not have to be a word, because the unit of data read or written in a memory cycle need not be an indivisible unit of data that passes into or out of the computer, as we see later in this section.

All three of the above terms can be hierarchical. An I/O device can be composed of I/O devices, since subsystems can be composed of smaller subsystems. A port can be composed of ports because a unit of data passed indivisibly at one time can be subdivided and an indivisible subunit can be passed at a different time. Even a word accessed in a memory cycle can be the same as two words accessed in two memory cycles elsewhere.

There are two major ways in a microcomputer to access I/O, relative to the address space in primary memory. Using the first, *isolated I/O*, the ports are read from by means of *input instructions*, such as IN 5. This kind of instruction would input a word from I/O port 5 into a data register. Similarly, *output instructions* like OUT 3 would output a word from a data register to the third I/O port.

Using the second, *memory-mapped I/O*, the ports are read by means of LD.B, LD.H, or LD.W instructions, and output ports are written using ST.B, ST.H or ST,W instructions. Memory-mapped I/O uses the data and address buses just as memory uses them. The microprocessor thinks it is reading or writing data in memory, but the I/O ports supply the data read or capture the data written at specific memory locations.

We do have to worry about accidentally writing over the output ports when we use memory-mapped I/O. Memory-mapped I/O can be protected, however, by a *lock*. The lock is an output port which is itself not locked, so the program can change it. The lock's output is ANDed with address and other control signals to get the enable or clock signals for all other I/O ports. If the lock is F, no I/O ports can be read or written. Before reading an I/O port, the program has to store T in the lock, then store F in the lock after all I/O operations are complete. Larger micro-controllers, such as those of the M·CORE family, have supervisor and user modes. Some I/O ports, including the MMC2001 parallel ports, are only accessible in supervisor mode, so they are locked to programs running in user mode.

One of the most common faulty assumptions in port architecture is that I/O ports are 8 bits wide. For instance, in an M·CORE system, the byte-wide LD.B or ST.B instructions are often used in I/O programs. There are 8-bit I/O ports on a large number of chips that are designed for 8-bit microcomputers. But 8 bits is not a fundamental width. In this book, where we emphasize fundamentals, we avoid that assumption. Of course, if the port is 8 bits wide, the LD.B instruction can be used, and used in C by accessing a variable of type *char*. There are also 16-bit ports which can be read by LD.H instructions, or as a *short* variable in C or C++. A port can be 1 bit wide. Several ports are 2 or 3 bits wide. Many ports read or write ASCII data. ASCII data is 7 bits wide, not 8 bits wide. If you read a 10-bit analog-to-digital converter's output, you should read a 10-bit port. Whatever your device needs, consider using a port of the right width.

5.1.1 Generic Port Architecture

Generic parallel I/O device hardware, such as a tristate driver or a register, is enabled or clocked by a device's address decoder that decodes the address on the address bus. In Figure 5.1, we show as complete a decoder as possible, assuming that the 32-bit chip select port CS0CR at location 0x10004000 is 0x0000f861, as it is after reset. This chip select port will be discussed in §5.1.4. Given the memory map, inputs can be removed to implement an incomplete decoder, as discussed in §4.2.3. The MPU reads from the port or writes into the port when a specific memory address is sent out, and must not access it when any other address used by the program is sent out. (An exception, shadowed output, is discussed later in this section.) The following discussion is oriented to 16-bit ports aligned to 16-bit word boundaries. 8-bit ports, and other ports that are shorter than 16-bit ports, can be implemented by removing some of the tristate drivers or flip-flops.

Figure 5.1 illustrates the logic diagram of a 16-bit input device with an input port at 0x2d020000. The generic *input port* samples a signal from the outside world when the microcomputer executes an LD.B, LD.W, or LD.H instruction in memory-mapped I/O, and reads the sample into the MPU. Because most microcomputers use tristate bus drivers, the port must drive the sample of data onto the data bus exactly when the microprocessor executes a read command with this port's address. Since this port address has many zeros, the use of negative-logic-input NAND gates (positive logic OR gates such as the 74HC4078) often reduces the decoder's cost. The CS[0] signal is asserted low when the high address byte is 0x2d. It and output enable OE are ANDed with the address, in negative logic, to assert the tristate driver's enable only in the latter part of the cycle, and to prevent this device from responding when location 0x2d020000 is written into.

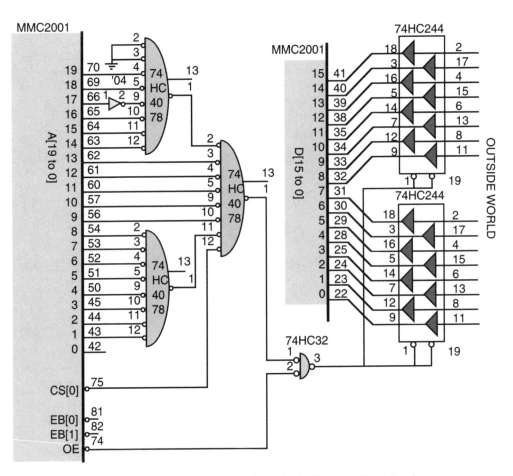

Figure 5.1. Logic Diagram for a Completely Decoded Input Device

The output port usually has to hold output data, and provide it as signals to the outside world, for an indefinite time — until the program changes it. The generic *basic output port* is therefore a latch or register which is capable of clocking data from the data bus whenever the microcomputer writes to a location in memory-mapped I/O. The register's D bus connects to the MMC2001's D bus, and its clock connects to an address decoder so that it is clocked when the microprocessor executes an ST.B, ST.W, or ST.H instruction at the address selected for this port. Figure 5.2 illustrates the logic diagram of a 16-bit output device with its port at 0x2d0f67fe, assuming CS0 is set up as in Figure 5.1. Since the port address has many ones, the use of NAND gates such as the 74HC30, often reduces cost. The CS[0] and EB[0] signal are ANDed into the decoder that asserts the high byte register's clock signal at the end of a memory cycle when it is written into, and the CS[0] and EB[1] signal are ANDed into the decoder that asserts the low byte register's clock signal at the end of a memory cycle when the low byte is written into.

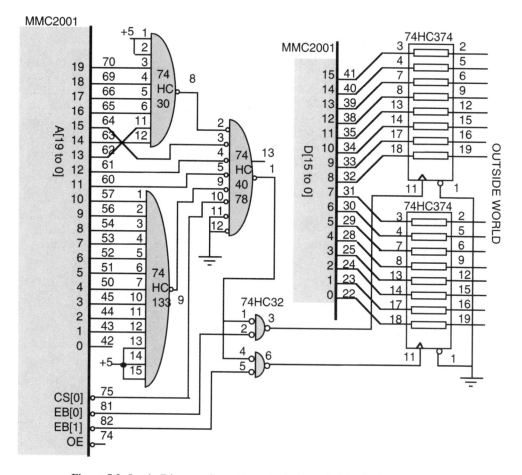

Figure 5.2. Logic Diagram for a Completely Decoded Basic Output Device

An output port can be combined with an input port at the same address, that inputs the data stored in the output port, to implement a more flexible but more costly generic *readable output port*. Figure 5.3 shows the block diagram of a readable output port at location 0x2d020000. Note that the bubbles are not shown in this block diagram gates represent functions of variables, and there is some ambiguity in the connections between the outside world to the registers, and registers to the data bus. But this block diagram conveys the information that the programmer needs to understand how the device operates. The decoder shown therein should be implemented with a minimal number of available gates, but this design is left as a problem at the end of this chapter. There, the reader can develop a detailed logic diagram, suitable to give to a technician, to build the device.

These generic ports can be accessed in assembly language. To read the data from an input port (Figure 5.1) or readable output port (Figure 5.3), use the following instructions:

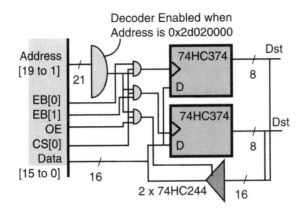

Figure 5.3. Block Diagram for a Readable Output Device

```
LRW r8, 0x2d020000
LD.H    r9, (r8,0)
```

When the LD.H instruction recalls a 16-bit word from location 0x2d020000, the decoder responds, asserting both 74HC244's negative-logic tristate enables. At that time, the 74HC244 drives the data bus with its input data, and the MPU puts the data on the bus into the least significant 16 bits of GPR r9, clearing the 16 most significant bits of r9.

The following instruction writes into a 16-bit basic output port (Figure 5.2):

```
LRW   r8, 0x2d0f67fe
ST.H      r9, (r8,0)
```

When the ST.H instruction memorizes a 16-bit word at location 0x2d0f67fe, the decoder responds, causing a rising edge on both the 74HC374 clock at the end of the memory cycle. The MPU has put the low-order 16 bits of GPR r9 onto the data bus, and the clock's rising edge causes this data to be stored in the 74HC374s. This data is available to the outside world until the MPU writes new data into it. A readable output port (Figure 5.3) can be written into, with address 0x2d020000 in place of 0x2d0f67fe.

These 16-bit ports can also be read and written as 8-bit ports. The instruction

```
LRW   r8, 0x2d020000>
LD.B      r9, (r8,0)
```

reads the 16-bit data from the input device, putting its high-order byte into GPR r9, and

```
LRW   r8, 0x2d020000
LD.B      r9, (r8,1)
```

reads the 16-bit data from the input device, putting its low-order byte into GPR r9. Also,

```
LRW   r8,0x2d0f67fe
ST.B      r9, (r8,0)
```

writes GPR r9 low-order byte into the upper 74HC374 in Figure 5.2 (by asserting EB[0] only), and

```
LRW   r8, 0x2d0f67ff
ST.B      r9, (r8,0)
```

writes GPR r9 low-order byte into the lower 74HC374 (by asserting EB[1] only).

These ports can be accessed in C or C++. The declaration or cast for a 16-bit port is usually *short* (to test the sign bit) or *unsigned short* (to prevent sign extension) and is further declared to be *volatile*, indicating that the data can change due to external activities, to prevent the compiler's optimizer from removing access to it. A constant, cast to a *volatile unsigned short* pointer, can be used to read a 16-bit port. Or a *volatile unsigned short* global variable in can be forced to have an address 0x2d020000 by means of an 'at' sign @, as *in volatile unsigned short in@ 0x2d020000*. Finally, a *volatile unsigned short* pointer *inPtr* can be loaded with 0x2d020000. To read the data from an input port (Figure 5.1) or readable output port (Figure 5.3) into a *short* variable *i*, use one of the following statements:

$$i = *(volatile\ unsigned\ short\ *)0x2d020000;$$
$$i = in;$$
$$i = *inPtr;$$

The statements *i = *(volatile unsigned short *)0x2d020000;* or *i = in;* (where *in* is 0x2d020000) generate LRW r8, 0x2d020000, then LD.H r9, (r8,0), and *i = *inptr;* generates an LD.H r9, (r8,0) if r8 is preloaded with 0x2d020000.

A constant, cast to a *volatile unsigned short* pointer, can write in a 16-bit port. Or a *volatile unsigned short* global variable *out* can be forced to have an address 0x2d0f67fe, as in *volatile unsigned short out@0x2d0f67fe*. Finally, a *volatile unsigned short* pointer *outPtr* can be loaded with 0x2d0f67fe. To write *short* variable *i* into a basic output port (Figure 5.1b) or readable output port (Figure 5.2), one can use any of:

$$*(volatile\ unsigned\ short\ *)0x2d0f67fe = i;$$
$$out = i;$$
$$*outPtr = i;$$

The statements *(volatile unsigned short *)0x2d0f67fe = i;* or *out = i;* (where symbolic name *out* is 0x2d0f67fe) generate *LRW 0x2d0f67fe,* then ST.H r9, (r8,0), and *outPtr = i;* generates ST.H r9, (r8, 0) if r8 is preloaded with 0x2d0f67fe.

The cast for an 8-bit port is usually *volatile char* (if we want to test the sign bit) or *volatile unsigned char* (if we want to prevent sign extension). A constant, cast to a *volatile unsigned char* pointer, can be used to read an 8-bit port. To read the data from the higher 74HC244 input port (Figure 5.1) into an *unsigned char* variable *i*, use:

*i = * (volatile unsigned char *) 0x2d020000;*

This statement generates LRW 0x2d020000 followed by LD.B r9, (r8, 0), To read the data from the lower 74HC244 input port into an *unsigned char>* variable *i*, use:

*i = * (volatile unsigned char *) 0x2d020001;*

Of course, a *volatile (unsigned) char* declaration, or a *volatile (unsigned) char* * pointer, can be used to read this port, in place of a cast. A constant, cast to a *volatile unsigned char* pointer, can be used to write into an 8-bit port. To write the data to the higher 74HC374 output port (Figure 5.2) into an *unsigned char* variable *i*, use:

*i = * (volatile unsigned char *) 0x2d0f67fe;*

This statement generates LRW 0x2d0f67fe followed by LD.B r9, (r8, 0), To write the data to the lower 74HC374 output port into an *unsigned char variable i*, use:

*i = * (volatile unsigned char *) 0x2d0f67fe;*

Of course, a declaration or pointer can be used in place of a cast.

A constant, cast to a *volatile unsigned short* pointer *portPtr*, can be used to read from, modify and then write into a 16-bit readable output port. A *volatile unsigned short* global variable *port* can be forced to have address 0x2d020000 by means of an 'at' sign @, as in *volatile unsigned short in@0x2d020000*, or a *volatile unsigned short* pointer *outPtr* can be loaded with 0x2d020000; then to increment a readable output shown in Figure 5.2, one can use the following:

*(volatile unsigned short *) 0x2d020000++;*

port++;

*(*portPtr)++;*

The statements *(* (volatile unsigned short *) 0x2d020000)++;* and *port++;* (where symbolic name *port* is 0x2d020000) generate the assembly language

```
LRW     r8, 0x2d020000    ; put port's addr. into GPR r8
LD.H    r9, (r8, 0)       ; get the data from the port to r9
ADDI    r9, 1             ; increment the value
ST.H    r9, (r8, 0)       ; write the data back in the port
```

Similarly the statements *port--;; port |= i; port &= i;* generally also generate load-modify-store instruction sequences to access a readable output port.

Obviously since an 8-bit port which is the upper or lower half of a 16-bit port can be read, and can be written, then such an 8-bit port can be accessed in a read-modify-write statement such as in:

*((volatile unsigned char *) 0x2d020000)++;*

Notice that an instruction sequence generated by *port++;* fails to work on a basic output port like Figure 5.2, which is not a readable output port. The instruction would read garbage on the data bus, increment it, and then write the incremented value of garbage into the output port. One must spend more money to build a readable output port if one wishes to use read-modify-write instruction sequences.

Alternatively, an output port can be at the same address as a word in RAM; writing at the address writes data in both the I/O port and the RAM and reading data reads the word in RAM. This technique is called *shadowed output*. This effect can be achieved, moreover, through software. If a basic output port is to be updated after being read, like a readable output port, a global variable can keep a duplicate of the data in the port so it can be read whenever the program needs to get the data last put into the port. In C or C++ for instance using a pointer *outPtr* to a port, declare also a global *volatile unsigned short portValue;* then write **outPtr = portValue = i;* whenever we output to the port. Then *i = portvalue;* reads what is in the port. Also, **outPtr = portValue++,* increments it, **outPtr = portValue--;* decrements it, **outPtr = (portValue |= i);* sets bits, and **outPtr = (portValue &= i);* clears bits. If we consistently copy output data written into the port into *portValue*, we can read *portValue* to read the port.

A port may be only part of a word or of two adjacent words. (See Figure 5.4.) A 'worst-case' 3-bit port, low bit of the byte at location 0x2d020000 and two high bits of the byte at 0x2d020001 (i.e. the middle 3 bits of the 16-bit word at 0x2d020000) occupies two consecutive 8-bit words in the memory map (Figure 5.4a) and parts of each word (Figure 5.4b). Generally, if an n-bit port ($n < 16$) is read as a 16-bit word, the other 'garbage' bits should be stripped off, and if it is not right-justified, logical shift instructions should align the port data. If such a port is written into, other ports that share word(s) written in order to write in this port must be read and then rewritten to be maintained. Consider an example: inputting from and outputting to a misaligned port.

Assuming we use the declaration *short *ptr = (short*)0x2d020000* and *0x2d020000* is the address of a 16-bit word containing a 3-bit input port just discussed, the port is read into a variable d with $d = (*ptr >> 6)$ & 7. The declaration, *(short*)*, causes reading or writing 16 bits at the address pointed to by *ptr*. The operator $>>$ is used to move data from the port to the least significant bits of d, and the operator & removes data that is not in the port from the words read from memory; this generates the following code:

```
30001094  8607  LD.W   R6, (R7, 0)   ; get *ptr
30001096  3A66  ASRI   R6, $06       ; shift port data to right-justify it
30001098  6075  MOVI   R5, $07;      ; put mask into r5
3000109A  1656  AND    R6, R5        ; clear bits not obtained from the port
```

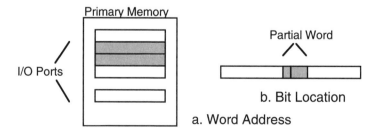

Figure 5.4. An Unusual I/O Port

Omit the shift operator if the port is aligned so that the least significant bit of the port is the least significant bit of a word. The AND operator is omitted if the port consists of whole words. For instance, assuming pointer *ptr=(short*)0x2d020000* is the address of a 16-bit port, *d=*ptr;* will read a 16-bit port at *0x2d020000* into *d*.

Similarly, consider the case where those three bits are a readable output port in words that also contain other readable output ports which must be unchanged. Assume pointer *ptr = (short*)0x2d020000* is the address of a pair of words containing a 3-bit output port just discussed. Then **ptr = *ptr & 0xFE3F | ((d << 6) & 0x1C0);* outputs *d* aligned into it; this generates the following segment

```
3000000E   3C66   LSLI    R6, $06      ; line up d with the bit loc. of the port
30000010   6705   MOVI    R5, $70      ; generate right-adjusted 0x1C0
30000012   3825   ROTLI   R5, $02      ; shift it over
30000014   1656   AND     R6, R5       ; clear other bits
30000016   0166   ZEXTH   R6           ; extend it to 32-bit word
30000018   C407   LD.H    R4, (R7,0)   ; get port data using *ptr
3000001A   7503   LRW     R5, [*+12]   ; abs = $30000028
3000001C   1645   AND     R5, R4       ; clear out bits in port
3000001E   0165   ZEXTH   R5           ; extend it to 32 bits
30000020   1E56   OR      R6,R5        ; combine the patterns
30000022   9607   ST.W    R6, (R7,0)   ; write it back in the port at *ptr
30000024   00CF   JMP     R15          ; main's return to caller
30000026   0000   BKPT                 ; padding to align word used by LRW
30000028   0000   BKPT                 ; hi 16-bit constant used by LRW
3000002A   FE3F   BSR     *-896        ; low 16-bit constant used by LRW
```

The expression *d << 6* moves the least significant bits of the data in *d* into position to put in the port, *& 0x1C0* removes parts of *d* that are not to be output, **ptr & 0xFE3F* gets data in the words not in the port that must be unchanged, and the OR instruction merges the data in *d* to be written into the port and data in the words not in the port together, to put this data into the bytes at 0x2d020000 and 0x2d020001. If the port is aligned so that the least significant bit of the port is the least significant bit of a word, the shift operator is omitted, and if the words being written into have no ports other than the port being written, the **ptr & 0xFE3F |* may be omitted. For instance, assuming pointer *ptr = (short*)0x2d020000* is the address of a word and all other ports within this word are readable, **ptr = d;* will write 16-bit data in *d* into a 16-bit output port at location *ptr*. If the output port is not readable, but the basic output port of Figure 5.2, then a copy of the data that is output to that and other ports in the words it writes into must be kept in memory, such as in *short* variable *portValue;* the statement **ptr = portValue = portValue & 0xFE3F | ((d << 14) & 0x1C0);* is not that much more complex than the statement for a readable output port.

The I/O port can also be accessed as an element of a vector. If an input port is at 0x2d020001 and *ptr* is 0x2d020000, then the expression *i = *(ptr + 1);* reads the port's data into *i*. Note that *ptr[1]* is the same as **(ptr + 1)*, so this statement can be written *i = ptr[1];* which is often considered easier to understand.

The I/O port can also be accessed as an element of a *struct*. A struct can be declared:

```
typedef struct{ volatile unsigned char A, B, short C; } F;
```

If we declare a pointer to this `struct` as follows: `F *fptr = (F *)0x2d020000;` then if an input port is at 0x2d020001, `d = fptr->B;` reads the port's data. Similarly, *fptr-> B = d;* writes data into the port at 0x2d020001. Further, if we declare `F` *f@0x2d020000* then *d = f.B;* reads the port's data. Similarly, `f.B = d;` writes data into the port. This use of structs is further considered in §5.3.5.5.

If the *struct* has bit fields, these statements generate code like that shown above for the misaligned port discussed under Figure 5.4, but are easier to understand; e.g.:

```
typedef struct{ volatile unsigned char A:2, B:3, C:3, D:2; } F;
```

In some compilers, bitfields can be coded eight bits to a byte, such that the left struct element corresponds to the leftmost bit, or they can be coded so that the left struct element corresponds to the rightmost bit. But in most compilers, if a bitfield were to overlap boundary between bytes, as in Figure 5.4, the potentially overlapping bitfield is put left justified on the next byte boundary instead. Check your assembly language code.

Some logic function can be implemented in hardware upon writing to a port. The data can set bits in a *set port*. (See Figure 5.5.) A pattern of ones and zeros is written by the processor via the data bus; wherever it writes a 1, the port bit is set, wherever it writes a 0, nothing is done. The data bit is ANDed with the decoded address enable, which is asserted only when a matching address appears and the EB[0] or EB[1] signal is asserted at the end of a memory cycle when we are memorizing. Note that if the data bit and enable are true (1), a bit in the flip-flop at the bottom of this figure will be set. If a data bit is false (0), nothing is done.

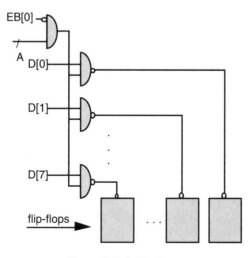

Figure 5.5. A Set Port

Data can clear bits in a *clear port*. In such a port, writing a true (1) clears the port bit, writing a false (0) does nothing. For instance, if a clear port at 0x2d020000 is to have bits 1 and 4 cleared, and *ptr* is *0x2d020000;* the statement **ptr = 0x12;* clears the two bits. The hardware is the same as in Figure 5.5, except that the clear input to the flip-flop is asserted low if the decoder enable is asserted and a data bit is true (1). Alternatively, a clear bit can be implemented in which writing a false (0) clears the bit, and writing a true (1) does nothing. This hardware is the same as in Figure 5.5, except that the clear input to the flip-flop is asserted low if the decoder enable is asserted and a data bit is 0.

These ports are set or cleared by just writing a constant or variable into them. If a set port at the address 0x2d020000 is to have bits 0 and 3 set, and *ptr* is declared as previously, the statement **ptr = 9;* sets the bits. If an 8-bit clear port at 0x2d020000 is to have bits 1 and 2 cleared, and writing zeros clears a bit, we can use **ptr = 0xf9;* or the statement **ptr = ~6;* to clear the bits. Set ports and clear ports can be readable; if so the data in the port are read without logic operations being done to them. A clear port is frequently used in devices which themselves set the port bit and the programmer only clears the bit. An interrupt request, discussed in the next chapter, is set by the device and can be cleared by the processor writing to a clear port. C and some assembly languages have means to OR or to AND data to memory, so these port functions are redundant. Nevertheless, the MMC2001 interrupt flags are clear ports; writing a 1 in a bit position clears an interrupt flag bit there.

Finally, we introduce two techniques that use the address bus without the data bus to output data. These are the address trigger used in many Motorola I/O devices, and the address output port that was used in the 68000-based Atari 520ST.

In an *address trigger*, an instruction's recalling or memorizing at an address causes the address decoder to provides a pulse that can be used to trigger a one-shot, clear or set a flip-flop, or be output from the computer. Figure 5.6a shows an example of an address trigger which produces a short pulse when location 0x2d020000 is read or written. An address trigger is, in effect, an output port in a word that requires no data bits in it. These instructions could simultaneously load or store data in another port even while the address triggers a one-shot in a manner similar to the shadow output port, or they might load garbage into the data register or store the data register into a nonexistent storage word (which does nothing). Some variations of an address trigger are the *read address trigger*, which produces a pulse only when a fetch or recall operation generates the address, a *write address trigger*, which generates a pulse only when the address is recognized in a memorize operation, and an *address trigger sequence*, which generates a pulse when two or more predefined addresses appear in a sequence. Instead of CLKOUT, a read address trigger should use OE, and a write address trigger should use EB[0] or EB[1].

An *address register output* was used in the 68000-based Atari 520ST, whose extension ports support only read-only memories so games can be plugged into the computer. The extension port is suitable for use only for an input port. But output can be done. The low-order 8 bits of the address are not decoded, as if a memory chip used these bits internally. The remaining high-order bits of the address are decoded. To use the port for 8-bit output, data are put on low-order 8 bits of the

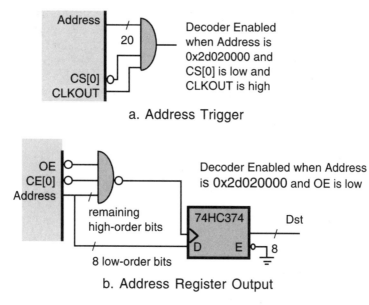

a. Address Trigger

b. Address Register Output

Figure 5.6. Address Output Techniques

address while the device address is put on the high-order bits of the address bus. This is easily handled in C; let a be a dummy variable, d be `volatile unsigned char` data, and ptr point to the port; the low-order 8 bits of ptr are 0s, `a=*(ptr+d)` puts the data in d into the port.

5.1.2 MMC2001 Parallel Port Architecture

The MMC2001 has parallel ports, shown in Figure 5.7. However, they are also actually designed for specific functions — keypad port *KPDR*, which is a 16-bit combination of two 8-bit keypad column *KCD* and keypad row *KRD*, is designed for a keypad, 16-bit edge port *EPDR*, in which only the low-order 8 bits are used, is designed for interrupt signal inputs. Ports *EPDR* and *EPDDR*, although they are 8-bit ports, must be accessed as 16-bit ports. 8-bit *U0DDR* and *U1DDR*, in which only the low-order 4 bits are used, are designed for serial UART communications. Thus they are not always available for parallel I/O. *KPDR* (which is *KCD* and *KRD*) is available as parallel ports if 16-bit *KPCR* at 0x10003000 is clear, *U0DR* are available as parallel ports if 16-bit *U0PCR* at 0x1000908a is clear, and *U1DR* are available as parallel ports if 16-bit *U1PCR* at 0x1000a08a is clear. These ports, *KPCR*, *U0PCR*, and *U1PCR* are clear after the MMC2001 is reset, until alternative uses of the pins require setting them.

Ports can be named in assembly language using EQU directives, or in C or C++ declaration statements, as shown below. A block diagram of these ports is shown in Figure 5.7. *KDDR* is *KCDD* catenated to *KRDD*, and *KPDR is KCD* and *KRD*.

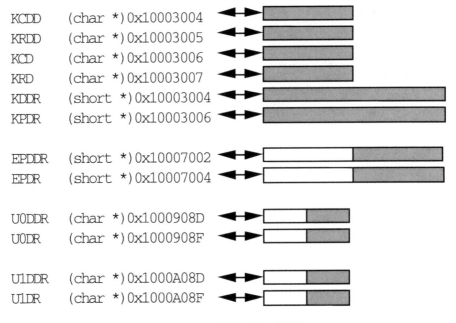

```
KCDD    (char *)0x10003004
KRDD    (char *)0x10003005
KCD     (char *)0x10003006
KRD     (char *)0x10003007
KDDR    (short *)0x10003004
KPDR    (short *)0x10003006

EPDDR   (short *)0x10007002
EPDR    (short *)0x10007004

U0DDR   (char *)0x1000908D
U0DR    (char *)0x1000908F

U1DDR   (char *)0x1000A08D
U1DR    (char *)0x1000A08F
```

Figure 5.7. MMC2001 Parallel Ports

```
volatile unsigned char KCDD@0x10003004, KRDD@0x10003005,
    KCD@0x10003006, KRD@0x10003007,
    U0DDR @0x1000908D, U0DR @0x1000908F,
    U1DDR @0x1000A08D, U1DR @0x1000A08F;
volatile unsigned short KDDR@0x10003004, KPDR@0x10003006,
    EPDDR@0x10007002, EPDR@0x10007004;
```

5.1.3   Programming of Keypad Parallel Ports

This section illustrates how to access the keypad port in assembly language and C. Throughout this chapter we use this port in our examples. So we introduce techniques to program it here. But these techniques can be applied to other MMC2001 parallel ports.

All these ports have a *direction port*. For *KPDR*, for each bit position, if the keyport data direction register (*KDDR*) bit is F (0), as it is after reset, the corresponding port *KPDR* bit is an input, otherwise if the *KDDR* direction bit is T (1) the corresponding port *KPDR* bit is a readable output port bit. The other ports and their direction port exhibit the same relationship. The keypad column *KCDR* port is directed by *KCDDR*, and keypad row *KRDR* port is directed by *KRDDR*. The edge port *EPDR* is directed by *EPDDR*, *U0DR* is directed by *U0DDR*, and *U1DR* is directed by *U1DDR*.

A direction port is an example of a *control port*, which is an output port that controls the device, but does not send data outside it. Writing device's control ports is called an *initialization ritual*. This *configures* the device for a specific use.

We illustrate the use of the 8-bit *KPDR* in assembly language first, and in C or C++, after that. To make *KPDR* an output port, we can write in assembly language:

```
BMASKI    r8,8         ; generate all ones
LRW       r9,KCDD      ; address of column direction port
ST.B      r8,(r9,0)    ; put them in direction bits for output
```

Then, any time after that, to output r8 to *KCD* we can write

```
LRW       r9,KCD       ; address of column data port
ST.B      r8,(r9,0)    ; output r8
```

To make the keypad data port an input port, we can write:

```
MOVI      r8,0         ; generate all zeros
LRW       r9,KCDD      ; address of column direction port
ST.B      r8,(r9,0)    ; put them in direction bits for input
```

Then, any time after that, to input the keypad data port into accumulator B we can write:

```
LRW       r9,KCD       ; address of column data\ port
LD.B      r8,(r9,0)    ; input to r8
```

It is possible to make some bits, for instance the rightmost three bits, readable output bits and the remaining bits input bits, as follows:

```
MOVI      r8,7         ; generate five 0s, followed by three 1s
LRW       r9,KCDD      ; address of column direction port
ST.B      r8,(r9,0)    ; to direction bits for input and output
```

The instruction ST.B r8,(r9,0) writes the rightmost three bits of r8 into the readable output port bits. The instruction r8,(r9,0) reads the left five bits as input port bits and the right three bits as readable output bits into r8. A minor feature also occurs on writing the 8-bit word: the bits written where the direction is input are saved in a register in the device, and appear on the pins if later the pins are made readable output port bits by setting the direction bits.

The equivalent operations in C or C++ are shown in what follows. To make the 8-bit keypad column data port an output port, we can write:

$$KCDD = 0xff;$$

Note that *KCDD* is declared an `unsigned char` variable. Then, any time after that, to output an `unsigned char` variable *i* to *KCD*, put:

$$KCD = i;$$

Note that *KCD* is declared an `unsigned char` variable. To make *KCD* an input port, we can write:

$$KCDD = 0;$$

Then, any time after that, to input *KCD* into an `unsigned` char variable *i* we can write:

$$i = KCD;$$

Generally, the direction port is written into in assembly language or C before the port is used the first time, and need not be written into again. However, one can change the direction port from time to time, as shown in the IC tester example in a later section.

KCD and *KRD* together, and their direction ports *KCDD* and *KRDD* together, can be treated as a 16-bit port because they occupy consecutive locations. Therefore they can be read from or written into using LD.H and ST.H instructions. Further, they can be accessed in C and C++ as shown below. To make the entire keypad data port *KPDR* an output port, we can write:

$$KDDR = 0xffff;$$

Note that *KDDR* is declared a *short* variable. Then, any time after that, we can output a *short* variable *i*, high byte to the keypad data port and low byte to *KRD* with the statement:

$$KPDR = i;$$

Note that *KPDR* is declared a *short* variable. To make the keypad data port *KPRD* an input port, we write:

$$KDDR = 0;$$

Then, any time after that, to input the entire keypad data port to a *short* variable *i*, use:

$$i = KPDR;$$

5.1.4   External Interface Module

The MMC2001's external interface module (EIM) provides four chip selects CS0, CS1, CS2, and CS3, for external devices. CS0, CS1, and CS2, are negative logic chip selects. Only CS3 is in positive logic. These pins are controlled by *CS0CR*, *CS1CR*, *CS2CR, and CS3CR*. A configuration port, *EIMCR*, similarly controls internal memories. CS0 is asserted when the address bus ADDR[31 to 24] is 0x2d; CS1 when it is 0x2f; CS2 when it is 0x2e; CS3 when it is 0x2c. These can be further decoded externally.

CS0CR and *EIMCR* are shown in Figures 5.8a and 5.8b, and *CS1CR, CS2CR,* and *CS3CR* are essentially like *CS0CR*, at consecutive locations. We use the declarations:

```
volatile unsigned long CS0CR@0x10004000, CS1CR@0x10004004,
    CS2CR@0x10004008, CS3CR@0x1000400C, EIMCR@0x10004018;
enum{ WSC=12, WWS=0x800, EDC=0x400, CSA=0x200, OEA=0x100, WEN=0x80,
        EBC=0x40, DSZ=4, SP=8, WP=4, PA=2, CSEN=1};
```

Chip select control registers determine the shape of the CSi, OE, and EB[1,0] pins. In the *i*th chip select register, *CSiCR*, least significant bit, *CEN*, enables the chip select (otherwise the pin can be used as a one-bit output port). The next bit, *PA*, is the output bit if *SEN* is zero. The next bit, *WP*, write-protects the selected device, and the

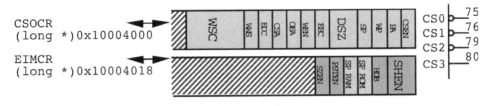

Figure 5.8. MMC2001 EIM Control Ports

next bit, *SP*, supervisor-protects it. For instance, if both *WP* and *SP* are set, the chip select is only asserted during an ST instruction executed in supervisor mode. The next two bits, *DSZ*, indicates the port size: a value 00 indicates an 8-bit port attached to data bus bits 15 to 8; a value 01 indicates an 8-bit port attached to data bus bits 7 to 0; and a value 10 indicates a 16-bit port. Negating the next bit, *EBC*, indicates pins EB[1,0] are asserted during reads (they are always asserted during writes). Asserting the next bit, *WEN*, indicates pins EB[1,0] are asserted half a clock cycle earlier during writes. Asserting the next bit, *OEA*, indicates the OE pin is asserted half a clock cycle later during reads. Asserting the next bit, *CSA*, indicates the CS*i* pin is asserted a clock cycle later, and that an extra cycle is inserted during consecutive external accesses. Asserting the next bit, *EDC*, indicates that an extra cycle is inserted during consecutive external read cycles that use different chip selects. Asserting the next bit, *WWS*, indicates that an extra cycle is inserted during write cycles. Finally the wait state control, *WSC*, indicates, for this chip select, the number of wait state clock cycles that a memory cycle normally takes. Zero wait state results in a memory access that is one clock cycle long.

In the EIM configuration register, *EIMCR*, asserting the fourth least significant bit, *SPROM*, supervisor-protects the internal ROM, asserting the next least significant bit, *SPRAM*, supervisor-protects the internal RAM. Other bits are used to provide logic analyzer access to information, and to provide alternate uses for some UART pins.

These ports are generally initialized just after the microcontroller comes out of reset, and are not further modified as I/O devices are intialized or the application is executed. An example of initializing chip select 0 for a 40 ns 16-bit static RAM, which requires two wait states when used with a 32-MHz clock, is:

CS0CR = (2 << WSC) | (2 << DSZ) | CSEN;

5.1.5   Parallel Port Classes

Although a high-level language greatly simplifies I/O interfacing, objects further simplify this task through encapsulation, information hiding, polymorphism, inheritance, and operator overloading. This section shows how they can be used.

Objects can be used as part of the core of a program, as well as for I/O interfacing. The original idea of an object was to tie together a data structure with all the operations that can operate on it. For instance, we define a data structure such as a queue, and we tie to it the operations like push, pull, and so on. Further, and independently, the I/O software they use can be organized in terms of objects.

The I/O device can be encapsulated into an object, as discussed in §2.3.3. In effect, we will handle all access to parallel I/O devices using object function members. Since all these alternative techniques can now be implemented in the same way, replacing a device using one technique with a device using another technique is easy. We can separate the design into a part having to deal with algorithms and interpreting a data structure, and a part having to deal with the I/O, and test each part separately, or mix and match different alternatives of both parts of the design at will. We can make I/O operations *device-independent*, meaning that their calls are basically identically written at compile time in the main program, regardless of the implementation technique. By using different procedure arguments that *declare* or *bless* an I/O *object*, a different I/O device that uses the same techniques or one that uses different techniques can be substituted. When this can be done at run time, blessing pointers to objects, this is called *I/O redirection*.

We can define classes of objects for I/O devices in many ways, having increased power and sophistication. We can get as sophisticated as a device driver, discussed in the next section. However, this sophistication also increases overhead. When a class has a function member with the same name as a function member of its base class, the class's function member over-rides the base class's function member, as discussed in §2.3.3. However, the compiler generally loads both function members into memory, because it may be unable to determine if the base class's function members will ever get used. We must be cautious about what we put into classes and base classes or we will fill memory with a lot of unnecessary function members. (By the way, a similar kind of overhead appears when we use device drivers.) Having tried a number of different approaches, we have concluded that the following rules should be used for I/O objects.

1. There will be a hierarchical library of classes designed for interchangeable I/O devices.
2. All 8-bit, 16-bit and 32-bit I/O devices are objects of a templated root class *Port* or a derived class. A data member *port*, points to an 8-bit, 16-bit or 32-bit port. The constructor initializes the I/O device as well as the object data members; its first parameter specifies the port address. There should be function members `get` for element and vector input and `put` for element and vector output, and possibly a destructor. Overloaded operators will be used for IoStreams and set and clear port functions, and assignment and cast overloaded operators can be used instead of element `get` and `put`. A general purpose function member `option` will be used for all other operations.

The templated root class `Port` is written below:

```
#define MAXCHARS 10
#define VIRTUAL virtual
enum { clear1Port = 4, clear0Port = 8, setPort = 0x10, offset2 = 0x20,
       offset1 = 0x40, basicOut = 0x80};
```

```
template <class T> class Port{
protected : unsigned char attr, curPos; T *port, value;

public : Port (unsigned long a = 0, unsigned char attr curPos = 0; T mask = 0) {
    this->attr = attr; port = (T *)a; |
    if (attr & (offset2 | offset1)) option (1, mask);
}
VIRTUAL int option (int mode = 0, int data = 0) { char i;
    if (mode == 0) { i = attr & 3; attr &= ~3; return i; }
    if (mode == 1) { /* direction port */
      if (attr & offset2) port [-2] = data;
      if (attr & offset1) port [-1] = data;
    }
    return 0;
}
VIRTUAL T get (void)
    { if (attr & basicOut) return value; else return *port; }

VIRTUAL void put (T data) { *port = value = data; }

VIRTUAL void get (T *v, short n) { while (n--) *v++ = get (); }

VIRTUAL void put (T *v, short n) { while (n--) put (*v++); }

T operator = (T data) { put (data); return data; } // overloaded assignment

operator T () { return get (); } // overloaded cast

void operator |= (T data) // overloaded OR =
      { if (attr & setPort) put (data); else put (data |= get ()); }

void operator &= (T data) { // overloaded AND =
    if (attr & clear0Port) put (data);
    else if (attr & clear1Port) put (~data);
    else put (data &= get ());
}
Port &operator << (char c) { put (c); return *this; }

Port &operator << (char *s) { while (*s) put (*s++); return *this; }
Port &operator >> (char &c) { c = get (); put (c); return *this; }

Port &operator >> (char *b) {
    do put (value = get ()); while (value <= ' '); // skip c.r, space, tabs
    do { /* input alphanumerics, permit backspace */
      if (value == 0x08) {if (curPos) { backspace (); curPos--; }} // bspc.
      else { b[curPos++] = value;
      if (curPos >= MAXCHARS)
            {attr |= 1; return *this; }} // error reported, in lsb of attr
    put (value = get ());
    } while ((value > ' ') || (value == 8));
    b[curPos++] = '\0'; curPos = 0; return *this; /* null terminate */
}
```

```
VIRTUAL void backspace(void) { put('<'); } // app. can override bksp.
};
```

The constructor performs the initialization ritual for a parallel port. Its parameters, which are an address, attributes, and the direction port value, initialize data members *port* and `attr`. *Port's* constructor then calls function member `option` with `mask`, to set the device's direction port, because the port's direction is set up before using the port.

The `option()` function member provides access to control and status ports and variables. If `option`'s first operand is zero, it returns the error indicator and then clears it. If its first parameter is 1, its second parameter is loaded into the direction register. If `attr` bit 6 is set, it is put into the next lower vector element (if *T* is `short`, at the address `port` minus two byte locations), and if `attr` bit 5 is set, `mask` is put into the second next lower vector element (if *T* is `char`, at the address `port` minus two locations). If neither `attr` bits 6 or 5 are set, it does nothing. Bits 6 or 5 of `attr` should be cleared for a generic input or output port, and one of them should be set for a port that has a direction port, depending on the relative location of the data port and the direction port. To facilitate generation of these constants, we utilize the `enum` statement in the above code.

Consistent with C++ conventions, all input function members are called `get`, and output function members are called `put`. The functions that have scalar operands simply use the pointer `port`, initialized by the constructor, to read data from or write to the port. These basic input and output function members have been slightly enhanced to correctly handle a basic output port, to implement device independence. If the most significant bit of data member `attr` is set, `get` reads the data member `value` which is consistently loaded whenever `put` writes data to the port, otherwise the `get` function member reads the port. The most significant bit of `attr` should be set for a basic output port, and cleared for readable output, shadowed output, and input ports.

Before we introduce more complex function members and overloaded operators, we show some examples of the simple function members introduced so far. An object for Figure 5.1's 16-bit parallel input port is initialized and used as follows:

```
void main(){short i; Port<short> port(0x2d020000); // declare the port
    i = port.get();       // read data from port
}
```

The last line of the procedure represents the rest of the program. In a real application, the rest of the program might be several pages of code with interspersed I/O operations. An object for Figure 5.2's 16-bit parallel basic output port is initialized and used as follows:

```
void main(){ short i;
    Port<short> port(0x2d0f67fe, BasicOut); // declare the port
    port.put(5); i = port.get(); // write into port, read last data written
}
```

An object for MMC2001's 16-bit parallel *KDDR* is initialized for input as follows:

```
void main () { short i;
    Port<short> port (0x10003006, offset1, 0); // declare the port
    i = port.get ();    // write into port, read last data written
}
```

An object for MMC2001's 8-bit parallel *KCD* is used for a readable output port by:

```
void main () { char i;    // local variable
    Port<char> port (0x10003006, offset2, 0xff); // declare the port
    port.put (5); i = port.get (); // write into port, read last data written
}
```

Alternatively, a pointer *ptr* can be blessed for class *Port*, and then used as shown below. Note that the low-level *get* and *put* procedures must be declared *virtual* if we wish to be able to redirect the object at run time. It is then possible for a program to have several output paths, for instance to an LCD, to a serial port, or a parallel port, so that the end user can select which path the program's data is sent to.

```
void func (char mode) { unsigned char i; Port<char> *ptr;
    if (mode) ptr = new Port<char> (0x2d020000);
    else ptr = new Port<char> (0x2000);
    ptr->put (5); i = ptr->get ();
}
```

Functions to input or output whole vectors are useful because direct memory access (DMA) and floppy disk sector input and output can directly input or output whole vectors. It is better to pass them a vector rather than passing one byte at a time from a vector, then combining the bytes into a vector, to input or output the vector as a whole. In the event that your I/O operation is ever redirected to a DMA or disk, user-callable vector functions will allow efficient input or output of whole vectors. So even though your current design might not transfer whole vectors, using vector input and output functions and overloaded operators may make it possible at a later date, if the I/O should be redirected to a disk or DMA device, to take advantage of such devices. Besides, using these vector functions factors the code used to step through the vectors into one place, in *Port* function members rather than having this code throughout the program. The vector *get* and *put* functions input and output whole vectors, without the application program managing each byte transfer.

The cast operator, and assignment '=' operator, can be overloaded to be directly used with I/O ports. Using these overloaded operators, input or output operations throughout the remainder of the program appear as I/O operations using simple data types in §5.1.1. Do not overlook the advantage of using these overloaded operators over using *put* and *get* function members; a program using simple input and output software as discussed in §5.1.1 can be converted to classes without modifying the main program, except for the inclusion of the class declaration and the constructor. Conversely, a program written using *Port* classes can be changed back to one using simple data types without rewriting the larger part of it, after the initialization.

Overloaded ORing |=, and ANDing &=, are useful for set and clear ports, by setting one of bits 2 to 4 in *attr* in the constructor. Then, using overridden | = and &=, if *port* is a readable output port the expression *port | = i*; will OR the data *i* into the the data stored in the device's register, but if *attr* bit 4 is set because the port is a set port, the software appears to output the operand, and the hardware sets the bit. Similarly, clear ports directly handle &=. For instance, an I/O subsystem (a device) might be written and tested on an MMC2001 using a clear port, and *attr* bits 3 or 2 will be set to let the hardware clear the register data. Then if it is implemented on an MMC2001 using external hardware such as a 74HC74 which readily implements an ordinary output port, *attr* bits 3 and 2 will be clear and the software will perform the ANDing operation. A program written to clear a port's bits will work exactly the same way in both the MMC2001 environments.

Vector input and output functions are often called 'raw I/O' because data are not interpreted as control characters. *IoStreams* use the overloaded operators >> for 'cooked I/O' input and << output of ASCII data, honoring control characters like carriage return. A string of << operators provides a convenient output mechanism for terminals and keyboards similar to C's *printf*. *Port* supports a minimal subset of conventional C++ IoStream functions. An example of using 'formatted' output is:

```
void main(){ short x = 0x1234; Port<char> c(0x2d020000);
    c << "x is " << itoa(x) << "or in hexadecimal: "<<htoa(x)<<'\r';
}
```

The segment << '\r' outputs a single character using the *Port* class's overloaded operator *Port &operator* << *(char c)*. The segment '*or in hexadecimal:*' uses the *Port* class's overloaded operator *Port &operator* << *(char *s)*

Using library or user-defined procedures that return a character pointer to a null-terminated ASCII character string, such as *itoa(x)* and *htoa(x)*, numbers can be converted to character strings, and the << operator can output the strings, as in <<*itoa(x)* or <<*htoa(x)*. Since a function returning a character string *itoa()* appears to the right of the << operator, the *Port* class's overloaded operator *Port &operator* << *(char *s)* also is used to output the string returned by *itoa()*. A library can supply conversion routines such as *itoa* or *htoa*, or the programmer can write conversion routines for special applications.

For 'cooked input,' IoStreams use the overloaded operators >>. The single character input *Port &operator* >> *(char &c)* inputs one character and outputs it. The character string input *Port &operator* >> *(char *b)* also echos data as it inputs a character string if the data member is nonzero, but further handles the special character backspace. This function also ignores, but echos, any 'nonalphanumeric' characters below the character '0', such as carriage returns, tab, space, and comma characters. The overloaded operator then assembles all other 'alphanumeric' characters, until a nonalphanumeric character is met again.

Such collected characters obtained by using >> can be passed to library or user-defined functions like *atoi* or *atoh*, which convert the ASCII character string to a signed integer or a hexadecimal number corresponding to it. For instance,

```
void main(){
    char s1[MAXCHARS], s2[MAXCHARS], i; short x, y;
    Port<char> c(0x2d020000);
    c >> s1 >> i >> s2; x = atoi(s1); y = atoh(s2);
}
```

will skip nonalphanumeric characters, enter alphanumeric characters into *s1*, put the next (nonalphanumeric) character into *i*, skip further nonalphanumeric characters, then enter alphanumeric characters into *s2*, until it encounters a nonalphanumeric character. *s1* converts to binary number x as a signed decimal number, and *s2* converts to binary number y as a hexadecimal number. If ...123 ,456(c.r.) is typed, then x becomes 123, *i* becomes ASCII ',', and y becomes 0x456.

The advantage of object-oriented programming for I/O should be somewhat apparent from the preceding examples. However, object-oriented programming has further very useful features in designing a state-of-the-art microcomputer's I/O devices, as proposed by Grady Booch in his tutorial *Object-Oriented Computing*. Encapsulation is extended to include not only instance variables and methods, but also the I/O device, digital, analog, and mechanical systems used for this I/O. An *object* is these parts considered as a single unit. For instance, suppose you are designing an automobile controller. An object (call it *PLUGS*) might be the spark plugs, their control hardware, and procedures. Having defined *PLUGS,* you call function members (for instance, call *PLUGS.SetRate(10);*), rather like connecting wires between the hardware parts of these objects. The system takes shape in a clear intuitive way as the function members are defined. In top-down design, you can specify the arguments and the semantics of the methods that will be executed before you write them. In bottom-up design, the object PLUGS can be tested by a driver as a unit before it is connected to other objects.

An object can be replaced by another object, if the function calls are written the same way (polymorphism). If you replace your spark plug firing system with another, the whole old *PLUGS* object can be removed and a whole new *PLUGS1* object inserted. You can maintain a library of classes to construct new products by building on large pretested modules. Having several objects with different costs and performances, you can insert a customer-specified one in each unit. Factoring can be used to save design effort.

Factoring can be used in a different way to simplify programming rather complex M·CORE I/O systems. In order to use the M·CORE's ISPI module for external I/O devices, some basic routines, available in a library of classes, will be needed to initialize it, to repetitively exchange data with the device, or to exchange data with it only on the program's or the device's command. Then as larger systems are implemented, such as *PLUGS,* that use the ISPI, new classes can be defined as derived classes of these existing classes, to avoid rewriting the methods inherited from the classes in the library.

Putting these two notions together might produce an incompatible notion of factoring, but actually appears to work synergetically. The hierarchy of classes at the root end can implement the factoring of routines needed to control an M·CORE's I/O system that prevents duplication of code. In this book, we will build up this infrastructure. For instance we will build an object for an I/O device that includes all the methods needed to initialize and use it. The leaf-ward part of the hierarchy can be used to add special functions to the basic I/O system to meet a specific application's requirements. For instance, an object for a robot controller might be coupled to the M·CORE system by means of an RS232 serial link as discussed in Chapter 9. The object *ROBOT* can be a member of a newly defined class *ROBOTDevice* that has additional methods, or it can use the methods of its base class(es) to correctly and efficiently function calls sent to *ROBOT* or received from *ROBOT*. The control of *ROBOT* will be high-level, because all the lower-level operations are invisible to the writer of the function calls (information hiding), which substantially reduces the design cost and improves system reliability.

The various functions and overloaded operators extend all the I/O class's capabilities to any derived class, so they are included in the base class *Port*. However, all these functions and overloaded operators generate a significant amount of code, which Hiware's current linker does not remove. Therefore in the file *Device.c*, we have put conditional compilation preprocessor commands around most of these functions and overloaded operators, to remove them when they are not needed. The user can *#define* a constant *USAGE* so as to compile only the functions and overloaded operators needed in an application. In future HIWARE linkers, which should be able to load only the functions actually used, these conditional compilation preprocessor commands won't be needed.

5.1.6   Debugging Tools

Object-oriented programs for I/O devices, which separate I/O procedures from the rest of the program, can be debugged using techniques described in this section. An object driver can exercise the object, object stubs can replace the I/O device object, and function member checking can make the function inform the designer of improper actions (at run time if the device is redirected, so the error isn't discovered until run time).

An *object driver* can execute the object function members, to test the passing of parameters between the object and the rest of the program, and to test the passing of data between the object function members and the hardware. A simple driver for an output device shown below simply inverts the output pattern each time it executes an output operation. It is shown with an object *dataPort* declared to be a member of class *Port*.

```
Port<char> dataPort(0x2d020000); char pattern;
void main(){ do dataPort.put(pattern ^= 0xff); while(1); }
```

The expression *pattern ^= 0xff;* inverts the value of *pattern* and passes that value to the output function member *put* each time it executes. The output port should have a square wave on each bit, each bit having the same period. A slightly better driver for an output device simply increments the output pattern each time it executes an output operation. It is shown with an object pointer *ptr* blessed to make the object *ptr* point to a member of the class *Port*.

```
void main()
{Port *ptr = new Port(0x2d020000); do ptr->put(pattern ++); while(1);}
```

The output port should have a square wave on each bit, but each bit should have a period that is twice as long as its next less significant bit. Using an oscilloscope and this output driver, a technician can expose hardware errors due to shorting outputs together or miswiring outputs. A simple input device's driver shown below uses an output port to provide data which are read through the input port and checked with the data that was sent. It is shown with an object pointer *ptr* blessed as a member of the class *Port*.

```
void main() { char error = 0;
    Port *ptr = new Port(0x2d020000); Port outPort(0x2d0f67fe);
    do {outPort.put(++pattern;if(ptr->get()!=pattern)error=1;
    while(1);
}
```

Clearly, an easily written object driver can effortlessly check the hardware. However, if the output is connected to hardware that acts differently to some patterns, more complicated object drivers are needed. Also, if an input port is connected to something, so that during testing it can not be easily connected to an output port, a specialized object driver compatible with the I/O device's environment must be written.

An object can be blessed as a member of the *Stub* class to verify the program that uses an I/O device. A *Stub* class is defined below.

```
template <class T> class Stub : public Port<T> {
    public: Stub(T * a) : Port<T>((short)a) { } // constructor
    virtual T get(void) { return *port++; }// input
    virtual void put(T data) { *port++ = data; }// output
    virtual T operator = (T data){put(data);return data;} // assignment
};
```

Note that the input function member merely gets consecutive items from an input vector, presumably a vector initialized with constant values that are a useful input test pattern. The output function member merely puts consecutive items into consecutive elements of a vector, presumably a variable vector. This vector can be examined after the program stops. Examples of defining an object as a member of this *Stub* class in lieu of *Port* are shown in two *main* functions shown below.

```
unsigned char outStream[5], inStream[] = { 1,2,3,4,5 };

void main() { unsigned char i; // declaration of objects
    Stub<unsigned char> in(inStream), out(outStream);
    out = 5; i = in;
}
void main() {short i, j; Port<unsigned char> *inPtr, *outPtr;
    inPtr = new Stub<unsigned char>(inStream);
    outPtr = new Stub<unsigned char>(outStream);
    *outPtr = 5; i =*inPtr;
}
```

The stub function members can also have function calls to output data in them to verify that they are executed. Constructor parameters can be output, or *put* function member parameters can be output to verify that the right data are being sent to them. However, calls to output data can slow down execution, which can interfere with debugging real time programs. The use of input and output vectors as shown in the example above interferes less with real-time programming; that is why we recommend using these stubs.

Another tool in debugging object-oriented I/O programs is to use *function member checking*. Here, the class is expanded to include illegal calls to function members, which set bits of *Port*'s data member *attr* when they are executed. Alternatively *printf* can be used to indicate an error. These errors can detect when hardware is asked to do something it can't do, such as loading arbitrary data into a set port, and when the function member's parameters are illogical, such as when a pointer to a *char* is passed in place of a *char* that is expected by the function member. The example below illustrates the *SetPort* class with some function member checking.

```
template <class T> class SetPort : public Port { char errors;

    public: SetPort(long l) : Port<T>(l) { } // constructor
    virtual T operator |= (T data) {put(data); return data |= get();}

    virtual void put(T data) { attr |= 1; }return data;} // illegal

    virtual T operator &= (T data) {attr |= 1; return data;} // illegal

    virtual T operator |= (T *data) {attr |= 1; return 0;}// illegal
};
```

One can list every data type and operation that is illegal in this class, to let the compiler tell the programmer when an improper operation is requested. Further, new classes can be defined with checks for improper requests. We used *Port* for input ports as well as readable output ports. An input port could use a derived class of *Port* with a function member put that sets *attr* to indicate that the operation was not completed.

Normally, the use of function member checking does not warrant the effort needed to define additional classes or function members; a programmer can simply not use the overloaded operator &= with *SetPort* or the function member *put* for input ports. We do not use function member checking in most examples in this book, so they are easier to understand. However, classes in an **M·CORE** library should insert or remove function member checking using conditional compiling, so it can be used to catch run-time errors.

While this discussion of debugging is specific to C++ object-oriented I/O interfacing, it can be adapted in part to conventional C programming by using #*define* statements that expand into different macros depending on how the #*define* statements conditionally compile the program. For instance if *debug* is #*defined*, then a message is printed on the screen, but if *debug* is not #*defined*, the program reads a port's data.

```
#ifdef debug
#define inputPort (printf ("read data") & 0);
#else
#define inputPort (*(unsigned char *) 0x2d020000)
#endif
```

The *Port* class does not provide function member checking, but that can be added as shown in a problem at the end of the chapter. Another interesting class at the end of the chapter is for one-bit input or output. The constructor gets an *id* whose low-order 3 bits designate a bit number, and high bits designate a port (0 is the keypad data port, etc.).

Object-oriented programming simplifies the use of the MMC2001 parallel ports. I/O device designers can work as independently from applications programmers because the application program need not be rewritten if an I/O device, such as a direction port, is substituted with another, such as a basic output port, except where the I/O object is blessed or declared. Object-oriented programming makes possible a library of programs that use a port with a block of hardware; a program and block of hardware can be taken from the library and easily modified to adapt it to another port. One merely has to change the constructor's arguments. Finally, the class function members are easy to test in a function member driver, and to be substituted for by a stub. Function member checking, if implemented, also catches illegal use of an inappropriate device for an application.

5.1.7   A Device Driver for a Parallel Direction Port

An alternative to an I/O class is the device driver, which also exhibits I/O independence and run-time redirection. The first subsection, following §3.2.6, shows how a parallel port device driver can be dynamically created, called, and used in a task. The second subsection shows how to write this simple driver that is called in the first subsection.

5.1.7.1   Calling Device Drivers

The device driver has six service calls (Figure 3.5). We now examine the most important of these, *_pio_create()* and *_pio_cmd()*. The example below initializes the driver called "*CMP0*" in main, and calls *put(char *, short)*, and *get(char *, short)* procedures, to output and input bytes to *KCD* as a directed parallel port. Figure 5.9 shows the service call arguments and the structures they point to.

This device driver is initialized by *_pio_create()* in *main*. This service call uses a *unit control data* (UCD) structure, which is *ucd0* in this example. *ucd0*'s first argument is its external name of the device driver and its second argument is a *struct* of function entry points of the driver, which is called *functions*. *ucd0*'s fifth argument points to *driver-dependent parameters* (DDP). *ucd0*'s other arguments

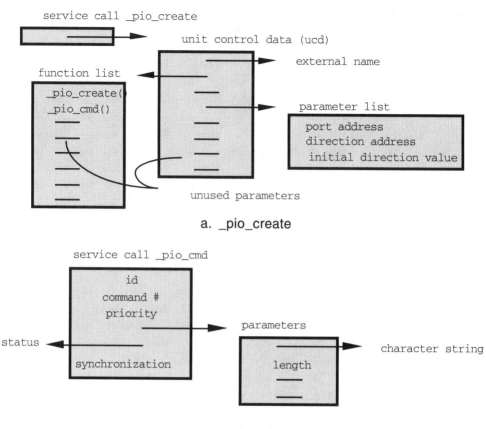

Figure 5.9. Driver Arguments and Associated Structures

are not needed in this simple example. The first two entries of *functions* are the entry point of the function which is called when the driver is initialized in the constructor, and the entry point of the function which is called whenever the *_pio_cmd()* service is called. *_pio_create()* sets the port's direction as we discuss near the end of this subsection.

```
#include "os.h"        /* Ariel general definitions */
#include "sysdef.h"    /* Ariel structure definitions */
#include "drvdat.h"    /* Ariel driver definitions */

#define _PIO_DIRECTION ((u_int32) 0x0000000a) /* comnd to set dir of port */

extern int32 initFun(void); extern void svcFun(ucd_t *,int,cdev_t *);

uxid_t id; /* unit id */ rv_t status; /* error status */
```

```
const drv_t function0 = { /* Driver procedure entry point addresses */
    initFun, /* Initialization function called when _pio_create is run*/
    svcFun, /* entry point of func called when _pio_cmd is executed */
    0, 0, 0, 0, 0 /* other function entry points which are not used here */
};


typedef struct param0_s { char *port, *dir, initdirection; } param0_t;
param0_t param0 = {(char *)0x10003006, (char *)0x10003004, 0xff};

ucd_t ucd0 = {                    /* Unit control data (UCD) for this device */
    _EXTNAME('C','M','P','O'),/* external name */
    &function0,                   /* struct of function names */
    0,&param0,                    /* address of driver-dependent paramters for device */
    0,0,0,0                       /* other fields of UCD are unused */
};


cdev_t charDev = {0,1,0,0};/* arg. for svc calls, initialized to fix direction reg */

void put(char *data, short n){
    charDev.aub = data; charDev.lub = n; /* reinitialize changed parameters */
    _pio_cmd(
        id,                /* unit id */
        _PIO_WRITE,        /* specific command (to write) */
        0L,                /* priority - use task's priority */
        &charDev,          /* parameters */
        &status,           /* error status reporting */
        _CONT_NOCOOR       /* synchronization (no coordination) */
    );
}


void get(char *data, short n){ /* get is essentially the same as put */
    charDev.aub = data; charDev.lub = n;
        _pio_cmd(id, _PIO_READ, 0L, &charDev, &status, _CONT_NOCOOR);
}


main() { char s[10];
    id = _pio_create(&ucd0); /* service call */
    if(!ISSVCERR(id)) {        /* checking errors from _pio_create */
        put("Hi There",9); get(s,10);
    }
}
```

We first examine the *put* procedure in some detail. We call the service
_pio_cmd with first parameter *id*, which was returned by *_pio_create*. The
second parameter is the specific command to be executed – *_PIO_WRITE* in this
case. The priority of this request is given next; *0L* indicates that we use task's priority
as the I/O priority. The next parameter *&charDev* points to the device's parameters,

which we discuss below, the next parameter &status provides for reporting errors, and the last parameter _CONT_NOCOOR indicates the synchronization requested for the service call. The structure charDev contains the address and length of the data to be output. The last two fields of this structure are for parameters that we do not use in this simple example.

The void get(T *data, short n) procedure calls the service _pio_cmd in essentially the same way as the void put(T *data, short n) service call described above. Whereas the above service calls for output and input are standard, _pio_cmd can also be used for special operations that are unique to the device. Another _pio_cmd with the specific command _PIO_DIRECTION and a vector of one element, whose value is the desired direction, can set the direction register. This operation is also done in _pio_create, but can be called by the user program whenever the direction changes.

5.1.7.2   Writing Device Drivers

The simple device driver called in §5.1.7.1 has just two functions, initFun which is used to initialize the hardware, and svcFun which is used to access the device. For this simple device, we implement the initialization function as shown below.

```
int32 initFun( ucd_t *ucp ) { /* arg. pointer to unit control data */
    param0_t *ddptr = (param0_t *)ucp->ddp;/* -> device-specific parameters */
    *(ddptr->dir) = ddptr->initdirection; /* move init. direct. from DDP */
    return SUCCESS;
}
```

This function sets the initial direction from DDP, using the address provided in DDP.

The function svcFun is used to input and output data, and set the direction.

```
void svcFun(
    ucb_t *ucb, /* -> unit control block */
    int fun, /* specific func. requested */
    cdev_t *dparams /* data parameters passed to funct. */) {
    ucd_t *ucd = &ucb->ucd; /* find initFun's argument */
    param0_t *pptr = (param0_t *)(ucd->ddp); /* get device-specific args */
    switch(fun){
       case _PIO_READ:
          while(dparams->lub--) *dparams->aub++ = *pptr->port; break;
       case _PIO_WRITE:
          while(dparams->lub--) *pptr->port = *dparams->aub++; break;
       case _PIO_DIRECTION: *pptr->dir = *(dparams->aub); break;
          default: _drv_pio_done(ucb, EOS_BAD_PARAM, 0); return;
    }
    _drv_pio_done(ucb, SUCCESS, 0); /* special call to end service */
}
```

Its first argument ucb is a pointer to the *unit control block* (UCB) which holds variables needed for the device driver routines (not to be confused with the unit control data). Its second parameter fun is used in a *switch* statement to select the desired service. The UCB has in it a copy of the UCD, from which we can get the DDP, including the port and direction addresses. For an input service call, we fill the vector with consecutive samples of bytes read from the port, and for ouput, we feed consecutive elements of the vector to the port. For the service call to set the direction, we put only the argument's first vector element in the direction port.

This section illustrates a simple Ariel device driver. As opposed to the class technique this technique requires the programmer to follow a lot of rules. The structures used in this procedure, which is part of Ariel's kernel, are very specific and cannot be altered by the device driver programmer. However, being similar to the mailbox mechanism, a device driver can handle I/O from multiple tasks without confusion. The device driver technique may be overkill for a simple parallel port, but it will become more suited to more complex I/O interfaces including console devices and disks.

5.2   Input/Output Software

Software for input and output devices can be very simple or quite complex. In this section, we look at some of the simpler software. We show C programs to make the discussion concrete. The software to use a single input and a single output device to simulate (replace) a wire will be considered first, because it provides an opportunity to microscopically examine what happens in input and output instructions. It illustrates a simplest example of input and output, and introduces simple timing concepts. We next discuss input to and output from a buffer, by analogy to a movie. Programmed control of external mechanical and electrical systems is discussed next. We will discuss the control of a traffic light and introduce the idea of a delay loop used for timing. Then, in a more involved example, we will discuss a table-driven traffic light controller and a linked-list interpreter, which implement a sequential machine. Finally we discuss an IC tester.

5.2.1   A Wire

The program $main()$ following this discussion will move data from a 16-bit input port at $0x2d020000$ (Figure 5.1) to a 16-bit output port at $0x2d0f67fe$ (Figure 5.2) repetitively. Although this program it is not very useful, it illustrates reading an input and writing an output port in C. Observe how the port addresses are set up. The cast $(short\,*)$ is needed because the C compiler checks data types, and objects to assigning integers to addresses. Alternatively, the addresses can be specified when the pointers are declared, as we will show in the next section. This program is worth running, to see how data are sampled and how they are output, using a square-wave generator to create a pattern of input data and an oscilloscope to examine the output data. You may also wish to read the assembly language code that is produced by the

C compiler and count the number of memory cycles in the loop, and also the number of cycles from when data are read to when they are output (the latency). Its timing is hard to predict for all C compilers, and the best way to really determine it is to run it and measure it.

```
void main() { register unsigned short *src; unsigned short *dst;
    src=(unsigned short *)0x2d020000;dst=(unsigned short
    *)0x2d0f67ff;
    do *dst = *src; while(1);
    }
```

The assembly language generated for this C procedure is shown below:

```
30000004  7E04 LRW     R14,[*+16]     ; abs = $30000014 addr of input port
30000006  7D04 LR      R13,[*+16]     ; abs = $30000018 addr of output port
30000008  C70E LD.H    R7,(R14,0)     ; read input port
3000000A  D70D ST.H    R7,(R13,0)     ; write to output port
3000000C  F7FD BR      *-4            ; abs = $30000008 loop back
...
30000014  2D20 BMASKI  R0,$12         ; hi 16 bits of constant for 1st LRW
30000016  0000 BKPT                   ; lo 16 bits of constant for 1st LRW
30000018  2D3F BMASKI  R15,$13        ; hi 16 bits of constant for 2nd LRW
3000001A  67FF MOVI    R15,$7F        ; lo 16 bits of constant for 2nd LRW
```

The MPU actually reads the input port in the instruction LD.H, and then writes into the output port in ST.H about 31.25 ns later. The loop can execute in 93.75 ns. However, loop timing is dependent on the compiler and the programmer's style. Timing is best determined by measuring chip enable pulses on an oscilloscope.

An object-oriented 'wire simulator' appears below. Although we can use *get* and *put* function members for input and output, we use overloaded cast and assignment operators to show their ability to make the rest of the program, after the initialization ritual, essentially the same as the program just shown, using pointers.

```
void main() { Port<char>*dst; Port<char> *src; char i;
    dst = new Port<char>(0x2d0f67fe);
    src = new Port<char>(0x2d020000);
    do *dst = i = *src; while(1);
}
```

Overloaded cast and assignment operators make this program look almost like the first wire example, but *src calls the overloaded cast operator which calls the get function member, and *dst calls the overloaded assignment operator which calls put. The local variable i is used herein to force the use of overloaded cast and assignment operators; otherwise *dst = *src; would call the copy constructor instead, to make the object dst a copy of the object src. Objects have some overhead, especially in using virtual function members, but this overhead is usually not a problem. But to simplify debugging, or change the entire program to work with parallel ports, just change *main's* first three lines to bless the object as a member of the Stub or another class.

5.2.2 A Movie

We may wish to input data to a buffer. The declaration `buffer[0x40]` creates a vector of length 0x40 bytes to receive data from an input port at 0x2d020000. Observe that the address of the input port is initialized in the declaration of the pointer.

```
void main() { unsigned short *src, buffer[0x40], i;
    src = (unsigned short *) 0x2d020000;
    for(i = 0; i < 0x40; i++) buffer[i] = *src;
}
```

The assembly language generated by this C *for* loop is shown below:

```
3000109A   600D   MOVI    R13,$00      ; clear i
3000109C   F009   BR      *+20         ; abs = $300010B0
3000109E   12D6   MOV     R6,R13       ; copy i
300010A0   016D   ZEXTH   R13          ; fill to 32 bits
300010A2   3C1D   LSLI    R13,$01      ; double
300010A4   1207   MOV     R7,R0        ; copy stack pointer
300010A6   1C7D   ADDU    R13,R7       ; get address of buffer[i]
300010A8   C70E   LD.H    R7,(R14,0)   ; read input port
300010AA   D70D   ST.H    R7,(R13,0)   ; store in buffer[i]
300010AC   2006   ADDI    R6,$01       ; increment i
300010AE   126D   MOV     R13,R6       ; copy i
300010B0   12D6   MOV     R6,R13       ; copy i
300010B2   016D   ZEXTH   R13          ; fill i to 32 bits
300010B4   6407   MOVI    R7,$40       ; put constant into r7 for compare
300010B6   0C7D   CMPHS   R13,R7       ; if i less than 0x40
300010B8   126D   MOV     R13,R6       ; restore i
300010BA   EFF1   BF      *-28;        abs = $3000109E loop
```

Finally we may wish to output data from a buffer. Observe that the address *pnt* of the buffer is initialized in the for loop statement, and is incremented in the *for* loop statement rather than the third expression of the *for* statement, which is missing. Note the ease of indexing a vector or using a pointer in a *for* loop statement. This operation, emptying data from a buffer to an output port or filling a buffer with data read from an input port, is one of the most common of all I/O programming techniques. It can used pointers to read from or write in the buffer, or indexes to read from or write into the buffer. The programmer should try both approaches, because some architectures and compilers give more efficient results with one or the other approach.

```
unsigned char buffer[0x40];
void main() { unsigned char pnt;
    volatile char *dst = (char *)0x2d0f67ff;
    for(pnt = buffer; pnt < buffer + 0x100;) *dst = *(pnt++);
}
```

The assembly language generated by this C *for* loop is shown below:

```
30001096  7E06  LRW    R14,[*+24]    ; abs = $300010B0
30001098  F003  BR     *+8           ; abs = $300010A0
3000109A  A70E  LD.B   R7,(R14,0)    ; read buffer
3000109C  B70D  ST.B   R7,(R13,0)    ; output to port
3000109E  200E  ADDI   R14,$01       ; increment buffer pointer
300010A0  7705  LRW    R7,[*+20]     ; abs = $300010B4 get addr. of buff end
300010A2  0C7E  CMPHS  R14,R7        ; if lower then end
300010A4  EFFA  BF     *-10          ; abs = $3000109A then loop
. . .
300010B0  3000  BCLRI  R0,$00        ; hi constant for 1st LRU instruction
300010B2  0100  XTRB3  R1, R0        ; lo constant for 1st LRU instruction
300010B4  3000  BCLRI  R0,$00        ; hi constant for 2nd LRU instruction
300010B6  0200  MOVT   R0,R0         ; lo constant for 2nd LRU instruction
```

The direct implementation of the movie examples, using the simple *Port* class defined in §5.2.3, is shown below. Note the use of the *Port* class vector input and output function members, which does all the work of moving the data.

```
char b[0x10);
void main(){ Port<char> src(0x2d020000); src.get(b, 0x10); }
void main(){ Port<char> dst(0x2d0f67fe); dst.put(b, 0x10); }
```

These movie examples illustrate input and output from a buffer. This very important genera of I/O software occur very often in application software.

5.2.3   A Traffic Light Controller

Microcomputers are often used for *logic-timer* control. In this application, some mechanical or electrical equipment is controlled through simple logic involving inputs and memory variables, and by means of delay loops. (Numeric control, as opposed to logic-timer control, uses A/D and D/A converters.) A traffic light controller is a simple example, in which light patterns are flashed on for a few seconds before the next light pattern is flashed on. Using LEDs instead of traffic lights, this controller can be used in a simple and illuminating laboratory experiment. Moreover, techniques used in this example extend to a broad class of controllers based on logic, timing, and little else.

In the following example, a *light pattern* is a collection of output variables that turns certain lights on and others off. (See Figure 5.10a.) Each bit of the output port LIGHTS turns on a pair of LEDs (see Figure 5.10b) if the bit is T. For example, if the north and south lights are paralleled, and the east and west lights are similarly paralleled, six variables are needed; if they are the rightmost 6 bits of a word, then TFFFFF would turn on the red light, FTFFFF would turn on the yellow light, and FFTFFF would turn on the green light in the north and south lanes. FFFTFF, FFFFTF, and FFFFFT would similarly control the east and west lane lights. Then TFFFFT would turn on the red north and south and green east and west lights. We

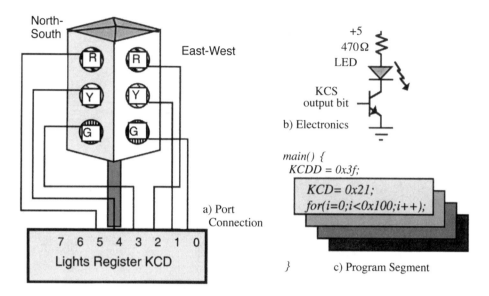

Figure 5.10. Traffic Light

will assume that the output port is connected so its right 6 bits control the lights as just described. The right 6 bits of KCDD are set to make these bits of the keypad data port outputs. The left 2 bits of the 8-bit output port need not be connected at all. Also, for further reference, TIME will be a binary number whose value is the number of seconds that a light pattern is to remain on. For example, the pair LIGHT = TFFFFT and TIME = 16 will put the red north and south and green east and west lights on for 16 s. Finally, a *sequence* of light patterns and associated times describes how the traffic light is controlled. In this example, this is a *cycle*, a sequence that repeats itself forever.

In this technique, as the program in Figure 5.10c is executed, it supplies multiple instances of immediate operands to the output port (as in *KCD = 0x21;*) and immediate operands to control the duration of the light pattern. A loop, such as *for(i=0;i<0x100;i++);*, is called a *delay loop*. It is used to match the time of the external action with the time needed to complete the instruction. Delay loops are extensively used in I/O interface programs. The usual loop statement after the *for(;;)* and before the ending semicolon (Figure 2.3c) is missing because the control part of the *for* statement provides the required delay. The constant *0x100* that must be put in the statement to get a specific loop delay is hard to predict analytically and varies from one compiler to another, but can be empirically determined.

Figure 5.10c's output and delay loop is coded into assembly language as follows:

```
0000000A  6217    MOVI    R7,$21
0000000C  7609    LRW     R6,[*+36]  ; abs = $00000030 addr. KCD
0000000E  B706    ST.B    R7,(R6,0)  ; output the pattern
```

```
00000010  6007   MOVI    R7,$00        ; clear loop counter
00000012  127E   MOV     R14,R7        ; copy loop counter
00000014  F001   BR      *+4           ; abs = $00000018 (skip increment)
00000016  200E   ADDI    R14,$01       ; increment loop counter
00000018  12E6   MOV     R6,R14        ; save loop counter
0000001A  017E   SEXTH   R14           ; sign extend loop counter
0000001C  3287   BGENI   R7,$08        ; generate 2**8
0000001E  0D7E   CMPLT   R14,R7        ; compare count to limit
00000020  126E   MOV     R14,R6        ; restore loop counter
00000022  E7F9   BT      *-12          ; abs = $00000016
...
00000030  0000   #KCD
```

A better way than programming a control sequence using immediate operands is described in the following paragraphs. This function member makes it easier to write and modify the control sequences and to store them in a small microcomputer memory. These advantages are so great that the technique introduced in this section is usually recommended for most applications.

An *interpreter* is a program that reads values from a data structure such as a vector, a bit or character string, list, or a linked-list structure to control something, like drill presses or traffic lights, or to execute interpretive high-level languages like BASIC, LISP, or JAVA. You might like to scan §2.2 to review data structures before looking at interpreters. Table and linked-list interpreters are particularly useful in interface applications. The table interpreter is described first, then the linked-list interpreter is introduced by modifying the table interpreter.

A traffic light cycle might be described by Table 5.1. It can be stored in a table or array data structure. Recall from §2.2.1 that arrays can be stored in row major order or column major order. C accesses arrays in row major order. The array has two columns — one to store the light pattern and the other to store the time the pattern is output — with one row for each pair. Consecutive rows are read from the array to the output port and to the delay loop.

```
void main() { unsigned char i, j, tbl[4][2]; int k;
KCDD = 0x3f; /* output right 6 bits of port KCD */
    tbl[0][0] = 0x21; tbl[0][1] = 16;
    tbl[1][0] = 0x22; tbl [1][1] = 4;
    tbl[2][0] = 0x0c; tbl [2][1] = 20;
    tbl[3][0] =0x14; tbl [3][1] = 4;
    do
        for(i=0;i<4;i++){
            KCD = tbl[i][0];
            for(j = 0; j < tbl[i][1]; j++) for(k = 0;k < 0xffff;k++) ;

        }
    while(1);
}
```

Table 5.1. Traffic Light Sequence

LIGHT	TIME
TFFFFT	16
TFFFTF	4
FFTTFF	20
FTFTFF	4

Structures can implement tables more efficiently than arrays can, so their columns can have different data types and sizes. You can use a pointer to point to the structure. Pointers are discussed next. We first introduce a very simple link mechanism using the index in an array, and then the use of pointer variables and structures to implement links.

The data structures and their (interpreter) operations can be encapsulated like arrays and their operations are often encapsulated, using objects. We will illustrate the traffic light controller object in the following example.

```
Port<char> P(0x2d020000,0x3f);
const char table[8] ={
      0x21, 16,
      0x22, 4,
      0xc,  20,
      0x14, 4
};
class traffic_table {
    char *tbl; short rows; char errors;
    public : char error; traffic_table(const char *t, short size);
    void Execute(void), install(char, short);
}   T(table, 4);

void main(){ if(!(T.error || P.option())) T.Execute(); }


traffic_table::traffic_table(const char *t, short size){
    tbl = (char *)allocate(size << 1); rows = size; errors = 0;
    for(short i = 0; i < (rows << 1); i++) install(t[i], i);
}

void traffic_table::install(char data, short location){ // verify acceptable
    parameters
    if(((location & 1) == 0) && !((((data & 7) == 1) ||
       ((data & 7) == 2)||((data & 7) == 4)) && (((data & 0x38) == 8)
      || ((data & 0x38) == 0x10) || ((data & 0x38) == 0x20))))
        errors = 1;
    if(((location&1)==1) && ((data<4) || (data>24))) errors = 1;
    tbl[location] = data;
}
```

```
void traffic_table::Execute() { char i, j; long k;
   while(P.option() == 0) {
      for(i = 0; i < rows; i++) {
         P.put(j = tbl[i << 1]);
         for(j = 0; j < tbl[(i << 1) + 1]; j++)
            for(k = 0; k < 0x100; k++) ;
      }
   }
}
```

Observe that before *main* is executed, constructors of global objects T of class
`traffic_table` and P of class `Port<char>` are executed. All `main` does is call
`Execute` if no errors occur in the constructors. All `Execute` does is follow the
procedure of the earlier traffic light example. But T's constructor allocates room for
a copy of the traffic light table from an external constant vector `table` (observe that
both are vectors, but the constant global vector `table` is written spaced out to look
like a two-dimensional array for ease of checking against the table provided to the
user (table 5.1). This use of an external vector and an internal vector illustrates
protection provided by object-oriented programming. The external global vector is
an 'initial' value, copied into space provided by the `allocate` procedure, into a
'working copy' whose pointer, and therefore whose contents, are protected. Input
data is verified by `install` to ensure that exactly one light is on in a north-south
lane and in an east-west lane, and the delay time is reasonable. `install` could
conceivably be used after initialization is complete and the interpreter is running, to
change a light pattern. The internal 'working' copy of the traffic light pattern is
changeable but protected against illegal patterns, while the global 'initial' vector is
really used just to set up this internal copy without having to use a large number of
parameters to the constructor, or numerous calls to a build function member. Such
duplication of data structures is common in operating system device drivers, where a
constant data structure is used to initialize a working copy of the data structure,
which permits modification but is protected against improper modification.

5.2.4   A Sequential Machine

Linked-list interpreters strongly resemble sequential machines. We have learned that
most engineers have little difficulty thinking about sequential machines, and that
they can easily learn about linked-list interpreters by the way sequential machines are
modeled by a linked-list interpreter. (Conversely, programmers find it easier to learn
about sequential machines through their familiarity with linked-list structures and
interpreters from this example.) Linked-list interpreters or sequential machines are
powerful techniques used in sophisticated control systems, such as robot control.
You should enjoy studying them, as you dream about building your own robot.

A *Mealy sequential machine* is a common model for (small) digital systems.
While the model, described soon, is intuitive, if you want more information, consult
almost any book on logic design, such as *Fundamentals of Logic Design*, by C. H.

Roth, West Publishing Co., Chapter 14. The machine is conceptually simple and easy to implement in a microcomputer using a linked-list interpreter. Briefly, a Mealy sequential machine is a set S of internal states, a set I of input states, and a set O of output states. At any moment, the machine is in a *present internal* state and has an *input state* sent to it. As a function of this pair, it provides an *output state* and a *next internal* state. In the next time step, the next internal state is the present internal state.

The Mealy sequential machine can be shown in graph or table form. (See Figures 5.11a and 5.11b for these forms for the example below.) In this example, the machine has internal states S = {A,B,C}, input states I = {a,b}, and output states O = {0,1}. The graph shows internal states as nodes, and, for each input state, an arc from a node goes to the next internal state. Over the arc, the pair, input state/output state, is written. In the table, each row describes an internal state and each column, an input state. In the table, the pair, next internal state/output state, is shown for each internal state and input state. In this example, if the machine were in state A and received input a, it would output 0 and go to state B; if it received input b, it would output 1 and go back to state A.

Consider a simple example of a sequential machine operation. If the machine starts in internal state A and the input a arrives, it goes to state B and outputs a 0. In fact, if it starts in state A and receives the sequence abbaba of input states, it will go from internal state A through the internal state sequence BCABCA, and generate output 000001.

```
enum{A, B, C};
char tbl[3][2][2] = {{{B,0},{A,1}},
                {{A,0},{C,0}},
                {{A,1},{A,0}}};
void main() { char i,j;
    KCDD = 1; KRDD = 0;
    while(1){
        j = KRD & 1;
        KCD = tbl[i][j][1]; i = tbl[i][j][0];
    }
}
```

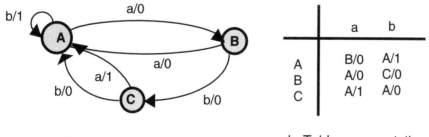

	a	b
A	B/0	A/1
B	A/0	C/0
C	A/1	A/0

a. Graph representation   b. Table representation

Figure 5.11. Mealy Sequential Machine

As shown above, the table representation can be stored in a microcomputer in a three-dimensional array. The interpreter for it would read an input, presumably from the least significant bit of KRD, and send the output to an output, the least significant bit of the keypad data port. The input state 'a' is the value 0x00, when read from the input port, and 'b' is 0x01. The internal state is associated with the leftmost array index being read. If the initial internal state is A, then the program implements this by initializing an index to index 0 associated with state A.

For low-end C compilers, two- or three-dimensional arrays are not implemented. However, the traffic light and sequential machine examples can be implemented with one-dimensional vectors. See problems at the end of the chapter. To assist checking such a vector, it can be laid out like a higher-dimensional array; an example of this layout will be given when an object-oriented traffic light controller is discussed.

The program using structures is a bit more complex but it is a more correct use of linked lists in C. In either program, an input port senses the input state and an output port provides the output state to some external system. We introduced the linked-list structure by comparing it to a row of the table. The structure is accessed (read from or written in) by a program, an interpreter. The key idea is that the next row to be interpreted is not the next lower row, but a row specified by reading one of the table's columns. For example, after interpreting the row for state A, if a 'b' is entered, the row for A is interpreted again because the address read from a column of the table is this same row's address. This view of a list is intuitively simple. More formally, a *linked-list structure* is a collection of *blocks* having the same *template*. A block is a list like the row of the table and the template is like the column heading. Each block is composed of *elements* that conform to the template. Elements can be 1 bit to 10s of bits wide. They may or may not correspond directly to memory words, but if they do, they are easier to use. In our example, the block (row) is composed of four elements: The first is an 8-bit element containing a next address, the second is an 16-bit output element, and the third and fourth elements are like the first and second. Addresses generally point to the block's first word, as in our example, and are loaded into the address register to access data in the block. Elements are accessed by using the offset in indexed addressing. Another block is selected by reloading the address register to point to that block's first word. Rather than describe blocks as rows of a table, we graphically show them, with arcs coming from address fields to the blocks they point to, as in Figure 5.12. Note the simple and direct relationship between Figure 5.12 and Figure 5.11a. This intuitive relationship can be used to describe any linked-list structure, and, without much effort, the graph can be translated into the equivalent table and stored in the microprocessor memory.

Linked-lists generally have elements that are of different sizes. Also, pointers which are addresses to memory may be needed because they are not multiplied and added to compute memory addresses, as array indexes are, and thus are faster. Such linked-lists should be stored as structures. Recall that in order to point to an element *e* of a structure *s*, we used *s.e* in earlier discussions. If a pointer is moved to different copies of a structure as the current internal state in the sequential machine, we can put it in the structure pointer variable `ptr,` and `(*ptr).e` is the element *e* of the structure pointed to by `ptr`. As a shorthand using operator `->; ptr->e` is

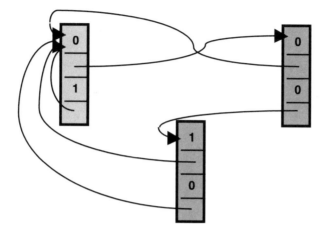

Figure 5.12. A Linked-List Structure

equivalent to *(*ptr).e.* Using the the previous program's *enum* statements, it can now be rewritten:

```
void main() {
    struct state{struct state *next; char out;} A[2],B[2],C[2],*ptr=A;
    A[0].next = B ; A[0].out = 0 ; A[1].next = A ; A[1].out = 1;
    B[0].next = A ; B[0].out = 0 ; B[1].next = C ; B[1].out = 0;
    C[0].next = A ; C[0].out = 1 ; C[1].next = A ; C[1].out = 0;
    KRDD = 1;
    while(1){ ptr += KCD & 1; KRD = ptr -> out; ptr = ptr -> next; }
}
```

Note that a data type, *struct state,* stores the next internal and the output states for a present internal and input state combination. There is a vector for each internal state; *A[2]* has, for each input state, a pointer to a *struct state,* and the entire vector represents the internal state A. We have to initialize the structures in the program, not in the declaration, because locations of structures must be declared before they are used as entries in a structure. The initial internal state is initialized at the end of the declaration of the structure to be state A. In the while loop, the input number is read from the input port and added to the pointer *ptr*. If a '1' is read, *ptr* is moved from *A[0]* to *A[1],* from *B[0]* to *B[1],* or from *C[0]* to *C[1].* Then the structure's element *out* is output, and the structure's element *next* is put in the pointer *ptr*.

The structure and pointer are very useful in I/O programming. An I/O device may have many ports of different sizes, which can be best described by structures. The address of the device can be put in a pointer *ptr,* and *ptr->port* will access the port element of it. Linked-list structures are especially useful for the storage of the control of sophisticated machines, robots, and so on. You can model some of the operations as a sequential machine first, then convert the sequential machine to a linked-list structure and write an interpreter for the table. You can also define the operations solely in terms of a linked-list and its interpretive rules. Some of our hardware colleagues seem to prefer the sequential machine approach, but our software friends insist that the linked-list structure is much more intuitive. You may use whichever you prefer. They really are equivalent.

Interpreters are useful for logic-timer control. A table best represents a straight sequence of operations, such as the control for a drill press that drills holes in a plate at points specified by rows in the table. A linked-list interpreter is more flexible and can be used for sequences that change depending on inputs. An interpreter can also use *switch* statements or *if then else* statements. Interpreters are useful in these ways for driving I/O ports. Their use, though, extends throughout computer applications, from data base management through operating systems, compilers, and artificial intelligence.

5.2.5   An IC Tester

In this subsection we consider a design problem to be able to test standard 14-pin ICs, about 30% of the ones we use, at the behavior level. We want to be able to put an IC into a socket, then run a test program that will determine whether the IC provides the correct sequence of outputs for any sequence of inputs; but we are not testing the delays, the input and output electrical characteristics, or the setup, hold, rise, or fall times of signals. Such a tester could be used to check bargain mail-order house ICs.

In principle, there are two design strategies: top-down and bottom-up. In top-down design, you try to understand the problem thoroughly before you even start to think about the solution. This is not easy, because most microcomputer design problems are said to be *nasty*; that means it is hard to state the problem without stating one of its solutions. In bottom-up design, one has a solution — a component or a system — for which one tries to find a matching 'problem.' This is like a former late-night TV show character, Carnack the Magnificent. Carnack reads the answer to a question written inside an envelope, then he opens the envelope and reads the question. This is bottom-up design. We do it all the time. The answer is microcomputers, now what was the question? A good design engineer *must* use top-down design!

We now approach the design of this IC tester in a top-down manner. We need 12 I/O bits to supply signals to all the pins and to examine the outputs for all the pins except power and ground. But the pins are not standard from chip to chip. Pin 1 may be an input in one chip and an output in another chip. An MMC2001's KPDR, would be more suitable than the simple parallel I/O device, because a line to these ports can be made an input or an output under control of software for different chips. Note that this is not always the case, and a simpler I/O device (a basic output device using 74HC374 or a input device using a 74HC244) may be indicated because it may be cheaper and use up less board space. Assuming these ports are available, we choose them. We examine the use of *KPDR*.

We will configure the devices so the *KPDR* bits will input or output data to the chip-under-test pins in the same order. A rugged (ZIF) socket will be used for 14-pin ICs, with power and ground connections permanently wired to pins 14 and 7, and other pins connected to the port bits as shown in Figure 5.13a, making it impossible to connect a port's output pin to $+5$ V or ground, which may damage it. The user will plug a 14-pin IC into the 14-pin socket to test it. Another rugged (ZIF) 16-pin socket can be used for testing 16-pin dual in-line packages.

The general scheme for programming will be as follows. A direction pattern will be set up once, just before the chip is inserted, and a sequence of patterns will be tested, one at a time, to check out the chip. A pattern of T and F values will be put into the direction ports, an F if the corresponding pin is an output from the test IC (and an input to the ports) and a T if the corresponding pin is a chip input (output from the ports). Then a test pattern will be put in the data port to set up inputs to the IC under test wherever they are needed. The test pattern bits corresponding to the IC's output pins are, for the moment, 'don't cares.' Data will be read from the I/O ports, and the bits corresponding to the test chip's output pins will be examined. They will be compared against the bits that should be there. The bits corresponding to the input pins on the test chip are supposed to be exactly the bits that were output previously. The other bits of the pattern, which were don't cares, will now be coded to be the expected values output from the IC under test. Summarizing, if an ICs pin is a chip input, its corresponding port bit's direction bit is T, and its data bit in the test pattern is the value to be put on the test IC's input pin; otherwise, if the IC's pin is a chip output, its corresponding direction bit is F, and its data bit is the value that should be on the pin if the chip is good. The test sequences are read from a vector by a vector-driven interpreter.

Constants for chip testing are #defined; these 16-bit values are finally put into ports A (high byte) and B (low byte). From Figure 5.13a we construct the definitions:

```
#define pin1 1       #define pin5 0x10     #define pin10 0x100
#define pin2 2       #define pin6 0x20     #define pin11 0x200
#define pin3 4       #define pin8 0x40     #define pin12 0x400
#define pin4 8       #define pin9 0x80     #define pin13 0x800
```

We illustrate the general scheme by showing concretely how a quad 2 input NAND gate, the 74HC00, containing four independent gates, can be tested. Figure 5.14a is the truth table for one of the four gates in the 74HC00. We use the above definitions to construct a value which is the number of testing iterations, a value to be put in the direction ports and values to be tested by the vector interpreter. From the 74HC00 chip pin connections shown in Figure 5.14b, we recognize that the

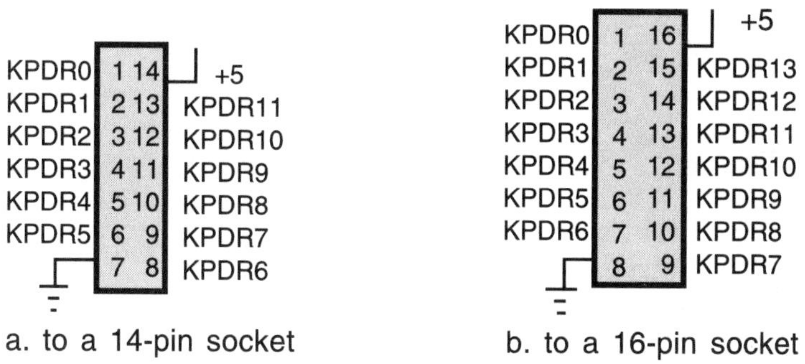

a. to a 14-pin socket   b. to a 16-pin socket

Figure 5.13. MMC2001 Port Connections for a Chip Tester

A	B	Z
L	L	H
L	H	H
H	L	H
H	H	L

a. Truth table

74HC00

#define A (pin1| pin4| pin10 | pin13)
#define B (pin2| pin5| pin9 | pin12)
#define Z (pin3| pin6| pin8 | pin11)

b. Pin connections

Figure 5.14 The 74HC00

truth table 'A' value should be put on pins 1, 4, 10, and 13; the truth table 'B' value must be put on pins 2, 5, 9, and 12; and the Z result will appear on pins 3, 6, 8 and 11. So we easily write the #*define* statements shown to the right of Figure 5.14.

These corresponding contents of the vector, which are actually evaluated to be $v[6]=\{4,\ 0x264,\ 0x264,\ 0x6f6,\ 0xb6d,\ 0xd9b\}$; are used in the procedure check(). (The advantages of using #*define* statements can be appreciated if you try to construct these constants manually.) The first element of v, the number of iterations, is used in the for loop; the second element is used to initialize the direction port. The next element, corresponding to the top row of the truth table, has 1's exactly where Z appears because the truth table so indicates, so we initialize it to the value Z; the next element of v, corresponding to the second row of the truth table, has 1s exactly where B and Z appears, so we initialize it to the value $B|Z$, and so on. The program sets up ports A and B to be inputs where $v[1]$, appears to be true (1), so their direction is initialized to $\sim v[1]$. Then the vector is read, element by element, the values of bits A and B are output, and the value returned is checked to see if it matches the element value. For a particular element, the vector value is output; wherever the direction bit is T (1) the element's bit is output and wherever the direction bit is F (0) the element's bit is ignored. The ports are read, and wherever the direction bit is F (0) the element's bit is compared to the bit from the port. $v[1]$ is a mask to check only the bits read back from the chip. The procedure check() returns 1 if the chip agrees with test inputs, and 0 if it fails to match the patterns in v.

The procedure above tests the 74HC00 chip. Other combinational logic chips can be tested in an almost identical manner, requiring only different vectors v. Chips with memory variables require more care in initialization and testing. Also, a thorough test of any chip can require a lot of patterns. If a combinational chip has a total of n input pins, then 2^n patterns must actually be tested.

```
unsigned short v[6]={4,Z,Z,B|Z,A|Z,A|B};/* construct the test vector for the 7400 */
short check(){ unsigned short i, bits;
    KDDR = ~v[1]; /* initialization ritual for ports A and B */
    for(i=0; i<v[0]; i++){ /* for all rows of truth table */
        KPDR = bits = v[i + 2]; /* output bits to chip, save pattern for testing */
        if((bits&v[1]) != (KPDR&v[1])) return 0; /* if no match, exit with 0 */
    }
    return 1; /* if all match, return 1 */
}
```

A very powerful message brought home by this example is the ability of high-level languages to abstract and simplify a design. By #*define* statements that are in turn defined in terms of other #*define* statements, we are able to utilize the ports of the MMC2001 in a manner that is easy to understand and debug. It is easy to develop vectors for other chips like the 7408 that has a different truth table, or the 7404 that has a different pin configuration and a different truth table. It is easy to modify this program to use, for instance, ports *EPDR*, *U0DR*, or *U1DR*, which have fewer bits. This modification is not unlike porting the program to a machine with a different I/O architecture. High-level languages make it easier to port from one to another machine.

We conclude with an object-oriented example of the IC tester. We use the same #*define* statements as in the earlier IC tester example; it incorporates similar concepts to those used in the previous traffic light controller. Streamed output to an object *cout* indicates the test result. You should therefore study this last example on your own.

```
Port <short> P(0x10003004, 0);
unsigned short p00[6]={4, Z, Z, B|Z, A|Z, A|B};
class ICTest { unsigned short *pattern;
    ICTest(unsigned short *pattern) {
        this->pattern = (unsigned short *)allocate(2 + *pattern);
        for(short i=0;i<(2+*pattern);i++)this->pattern[i]=pattern[i];
    }
    public: short check(void);
} Test00(p00);
short ICTest::check(void){ register unsigned bits, i;
    P.direction(pattern[1]); /* initialize ports A and B */
    for(i = 0; i < *pattern; i++){
        P = pattern[i + 2];
        if((P&pattern[1])!=(pattern[i + 2] & pattern[1])) return 0;
    }
    return 1; /* if all match, return 1 */
}
void main(){
    if(Test00.check()) cout << "good";
    else cout << "bad"; cout << '\r';
}
```

5.3   Input/Output Indirection

In this section, parallel ports are to simulate a memory bus signals, and shift registers are used as intermediaries to input and output data. These use indirect I/O. In this section, we examine I/O indirection and examine some issues a designer should consider regarding I/O indirection. We will cover indirect I/O in the first subsection, followed by serial I/O, and will conclude with object-oriented programming and a discussion of design issues.

5.3.1   Indirect Input/Output

Up to now, ports have been attached to the address and data buses. We shall call this *direct I/O*. Alternatively, in *indirect I/O,* one or more parallel ports can be used to connect to another device's address, data, and control pins that are normally connected to the memory address, data and control buses. Explicit bit setting and clearing instructions, often called 'bit-banging,' can raise and lower the control signals for the I/O chip. This is vaguely like indirect addressing. Incidentally, a coprocessor actually is an I/O device that is read or written in microcode, which is one level below normal I/O, analogous to immediate addressing.

We want to keep track of the time of day when the MMC2001 is powered down, so we choose to incorporate the MC6818A or MC146818A time-of-day clock chip. Figure 5.15a shows the memory organization of the MC6818A. The current time is in locations 0 to 9, except for locations 1, 3, and 5, which hold an alarm time to generate an interrupt. Control ports at locations 0xA to 0xD allow different options. Locations 0xE to 0x3F are just some CMOS low-power RAM. After an initialization, the time may be loaded into locations 0 to 9, and then 0x8 is put into control port C to start the timekeeping. Locations 0 to 9 can be read after that to get the current time.

The MC6818A can be indirectly controlled through MMC2001's keypad data port and *KRD* as shown in Figure 5.15d. M6818A control signal timing, and address and data sequencing, are taken from Motorola data sheets. Figure 5.15b shows the write cycle and in Figure 5.15c shows the read cycle. Control signals (address strobe *as*, data strobe *ds*, read-write *rw*, and chip select *cs*) are set high or low in the MMC2001 *KRD,* to write a word. Except for *ds,* they are initially high. We first raise *as* high, put the address into the keypad data port, make *cs* and *rw* low, drop *as* low, make *ds* high, put data to the keypad data port, drop *ds* low, and raise *cs*, *rw,* and *as* high. Reading is essentially the same, except that *rw* remains high and data are read from the keypad data port. Control signals are defined in the *enum* statement by having a 1 in the bit position through which they connect to *KRD*.

The C procedure *outa* accesses the chip. Observe that outa rather tediously, but methodically, manipulates the MC6818A's control signals. A call *outa(d,6)* in indirect I/O writes d to location 6 in the MC6818A, which stores the day of the week. High-level language programs are easy to write. It is generally possible to write the procedure *outa* in assembly language, while the main program is in C, to regain some speed but keep most of the advantages of high-level languages. While the program shows how MC6818A's memory can be written into, similar routines can read it.

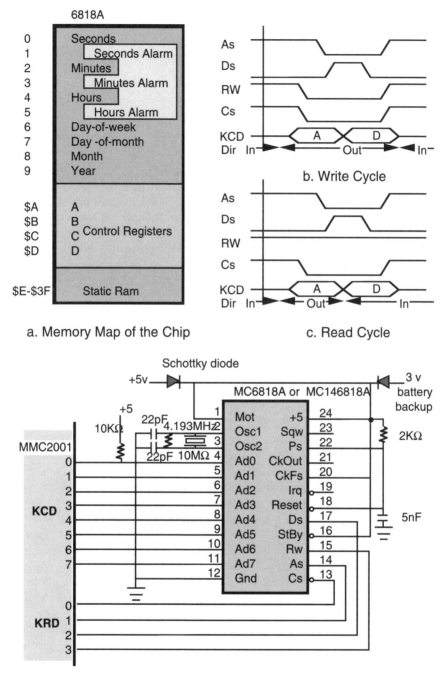

a. Memory Map of the Chip

b. Write Cycle

c. Read Cycle

d. Connecting an MC6818A Time-of-Day Chip Using Indirect I/O

Figure 5.15. MC6818A Time-of-Day Chip

```
enum { cs=1, as=2, ds=4, rw=8};

void main() {short yr,mo,dm,dw,hr,mn,se;
    KCDD = 0; KRDD = 0xf; KRD = as+rw+cs;
    outa(0x80,0xb);outa(0xf,0xa);outa(yr,9);outa(mo,8);/*set up time */
    outa(dm,7);outa(dw,6);outa(hr,4);outa(mn,2);outa(se,0);
        outa(8,0xb);
}
void outa(char d, char a) {
    KCDD = 0xff; KCD = a; KRD = as+cs; KRD = as; /* output the address a */
    KRD = 0; KRD = ds; KCD = d; KRD = 0; KRD = as; /* output d data */
    KRD = as+rw; KRD = as+rw+cs; KCDD = 0; /* make port KCD an input */
}
```

The main point of this section is the concept of indirect I/O, which we now elaborate on further. Besides being a good way to connect complex I/O devices to a single-chip computer, indirect I/O is a very good way to experiment with an I/O chip. The main advantage is that the connections to the chip are on the 'other side' of an I/O port, rather than directly on the MPU's address and data buses. Therefore, if you short two wires together, the MPU still works sufficiently to run a program. You have not destroyed the integrity of the microcomputer. You can then pin down the problem by single-stepping the program and watching the signals on the ports with a logic probe. There is no need for a logic analyzer. Indirect I/O is also a good way to implement some completed designs because it generally does not use external SSI chips, but rather it uses software to control a device. Indirect I/O can be implemented in the MMC2001 because the address, data, and control buses are available as KPDR.

We used this technique to experiment with a floppy disk controller chip and a CRT controller chip set in Chapter 10. We got these experiments to work in perhaps a quarter of the time it would have taken us using direct I/O. That experience induced us to write a whole section on this technique here in Chapter 5. There is a limitation to this approach. Recall from Chapter 4 that some chips use 'dynamic' logic, which must be run at a minimum as well as a maximum clock speed. The use of indirect I/O may be too slow for the minimum clock speed required by dynamic logic chips. However, if the chip is not dynamic, this indirect I/O technique is very useful to interface to complex I/O chips.

5.3.2   LCD Display Interfacing

The Liquid Crystal Display (LCD) has become the display device of choice for microcontrollers. An LCD display features low power, full ASCII character displays of one to four lines, from 16 to 40 characters per line, and low cost. Many inexpensive LCD modules use the Hitachi HD44780 LCD controller chip. The LCD displays such as OPTREX's DMC series, which uses this Hitachi controller, can display a 16-column 1-row, a 16-column 2-row, a 20-column 1-row, a 20-column 2-row, a 20-column 4-row, or a 40-column 2-row ASCII message. Essentially all displays use a standard interface that can be connected to MMC2001's keypad data port, as shown in Figure 5.16.

We show procedures for the 16-column 1-row display. *main*'s initialization ritual selects cursor blinking and movement (see Table 5.2). Its second line duplicates a command to configure its input port to 4 bits. The *put* procedure outputs a command or a character, using a delay loop to wait for the command's execution, and the *putStr* procedure outputs up to 16 characters. The constant *d410* puts a 410-µs delay and *d10* puts a 10-µs delay after the command is given. For an inexpensive 16-by-1 display, the cursor must be repositioned after outputting 8 characters, with the statement *if(i==7) put(0xc0, 0, d10);* To control other size displays, this statement is deleted.

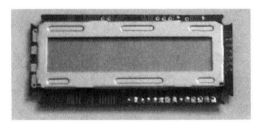

a. A 16–Character 1–Line LCD Display

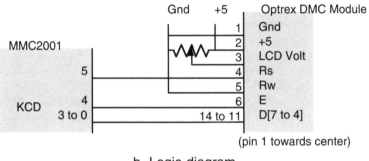

(pin 1 towards center)

b. Logic diagram

Figure 5.16. An LCD Display

Table 5.2. LCD Commands

Binary Code	Command		0	1
00000001	Clear Display*	x	don't care	don't care
0000001x	Home Cursor*	i	autodecrement	autoincrement
000001is	Set Entry Mode	d	display off	display on
00001dcb	Set Display Mode	c	cursor off	cursor on
0001srxx	Shift Cursor/Display	b	blinking off	blinking on
001wnfxx	Set Function	s	shift cursor	shift display
01aaaaaa	Set Character Gen Addr	r	shift left	shift right
1aaaaaaa	Set Display Address	w	4-bit port	8-bit port
a...aa is an address		n	1-line display	2-line display
*needs 410 µs delay (otherwise needs 10 µs)		f	5x7 font	5x10 font

```
enum{rs=0x20, e=0x10, d10=20, d410=800}; /* you might adjust d10, d410 */
void main(){ char i, j;
    KCDD= 0x3f; /* prepare keypad data port for output */
    put(0x28, 0, d10); put(0x28, 0, d10); /* use 4-bit interface */
    put(6, 0, d10); /* set entry mode to autoincrement *
    put(0xe, 0, d10); /* set display mode: display and cursor on */
    put(1, 0, d410); /* clear display */
    putStr("Hello world,Hello world!"); /* print a message */
}
void put(char c, char a, short d) /* output hi nibble and lo nbl, then delay */
    {put4(((c >> 4) & 0xf) | a); put4((c & 0xf) | a); while(d) d--;}
void put4(char c) { KCD = c + e; KCD = c; } /* display on falling edge of e */
putStr(char *s) { short i;
    put(0x80, 0, d410); /* clear display */
    for(i = 0; *s; s++, i++) { /* output until null */
      put(*s, rs, d10); /* output a character */
      if(i == 7) put(0xc0, 0, d10); /* after 8th character, reposition cursor */
    }
}
```

5.3.3   Synchronous Serial Input/Output

Except when it comes with a personal computer or is laid out inside a micro-controller chip, a parallel port and its address decoder takes a lot of wiring to do a simple job. Just wire up an experiment using them, and you will understand our point. In production designs, they use up valuable pins and board space. Alternatively, a serial signal can be time-multiplexed to send 8 bits of data in eight successive time periods over one wire, rather than sending them in one time period over eight wires. This technique is limited to applications in which the slower transfer of serial data is acceptable, but a great many applications do not require a fast parallel I/O technique. Serial I/O is similar to indirect I/O covered in §5.4.1, but uses yet another level of indirection, through a parallel I/O port and through a serial shift register, to the actual I/O device.

This subsection considers the serial I/O system that uses a clock signal in addition to the serial data signal; such systems are called synchronous. Asynchronous serial communication systems (Chapter 9) dispense with the clock signal. Relatively fast (4 Mbits/s) synchronous serial systems are useful for communication between a microcomputer and serial I/O chips or between two or more microcomputers on the same printed circuit board, while asynchronous serial systems are better suited to slower (9600 bits/s) longer-distance communications. We first examine some simple chips that are especially suited for synchronous serial I/O. We then consider the use of a parallel I/O port and software to communicate to these chips.

Although serial I/O can be implemented with any shift register, such as the 74HC164, 74HC165, 74HC166, and 74HC299, two chips - the 74HC595 parallel output shift register and the 74HC589 parallel input shift register - are of special value.

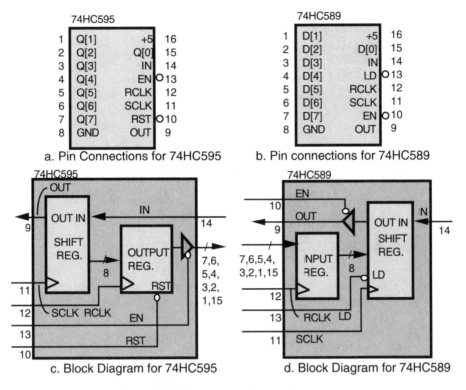

Figure 5.17. Simple Serial Input/Output Ports

The 74HC589 is a shift register with an input port and a tristate driver on the serial output of the shift register. (See Figures 5.17b and 5.17d.) Data on the parallel input pins are transferred to the input port on the rising edge of the register clock RCLK. Those data are transferred to the shift register if the load signal LD is low. When LD is high, data in the shift register are shifted left on the rising edge of the shift clock SCLK and a bit is shifted in from IN, as in the 74HC595, but the data shifted out are available on the OUT pin only if EN is asserted low; otherwise it is tristated open.

The 74HC595 is a shift register with an output port and tristate driver on the parallel outputs. (See Figures 5.17a and 5.17c.) We consider the shift register to shift left rather than right. A shift occurs on the rising edge of the shift clock SCLK. A bit is shifted in from IN and the bit shifted out is on OUT. On the rising edge of the register clock RCLK, the data in the shift register are transferred into the output port. If the output enable EN is asserted low, the data in the output port are available to the output pins; otherwise they are tristated open.

These chips can be connected in series or parallel configurations. (See Figure 5.18.) The 74HC589 can be connected in a series configuration to make a longer register, as we see in the 24-bit input port diagrammed in Figure 5.18a.

We will use the direction port KCD because we need various numbers of inputs and outputs. In Figure 5.18a, The outputs OUT of each chip are connected to the

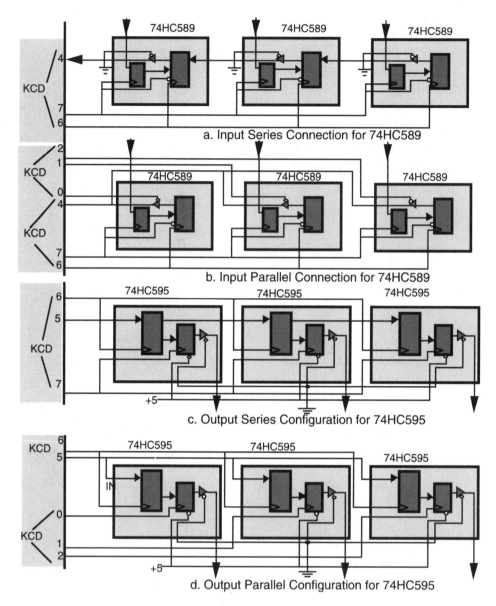

a. Input Series Connection for 74HC589

b. Input Parallel Connection for 74HC589

c. Output Series Configuration for 74HC595

d. Output Parallel Configuration for 74HC595

Figure 5.18. Configurations of Simple Serial Input/Output Registers

inputs IN of the next chip to form a 24-bit shift register. Each 589's RCLK AND LD pins are connected to KCD bit 7 to clock the input ports and to load the shift registers together, and each chip's SCLK pins are connected to KCD bit 6 to clock the shift registers together. The EN pins are connected to ground to enable the tristate drivers. In the software considered later, we pulse KCD bit 7 twice, to load the input registers at one time to get a consistent 'snapshot' of the data, then transfer this data into the shift register at one time by making LD low, and then, with LD high, send 24 pulses on SCLK to shift the data into the MMC2001.

The 74HC589 can be connected in a parallel configuration to make several separate input ports, as we see in the three 8-bit input ports of Figure 5.18b. Each chip's output OUT is connected to a common tristate bus line, and each chip's tristate enable EN of is connected to different keypad data port bits: 2 to 0, LD and RCLK connect to *KCD* bit 7, and bits 4 and 6, are connected as in Figure 5.18a. Any of the input ports may be selected by asserting its tristate enable low, the others being negated high. Then a sequence similar to that discussed in the previous paragraph inputs the chip's data, using eight pulses on *KCD* bit 6. While this configuration requires more output pins, software can choose any chip to read its data without first reading the other chip's data.

Figures 5.18c and 5.18d show the corresponding series and parallel configurations for the 74HC595. Reset can be connected to the MPU reset pin which resets the system when it is turned on or when the user chooses, but here it is merely connected to +5 to negate it. The output enable EN is connected to ground to assert it. The series configuration makes a longer shift register. The parallel configuration makes separate ports that can output data by shifting the same data into each port, but only pulsing the RCLK on one of them to transfer the data into the output register.

Series-parallel configurations, rather than simple series or simple parallel configurations, may be suited to some applications. The 74HC595 RCLK signals can come from the data source's logic, rather than from the microcomputer, to acquire data when the source is ready. The 74HC589 output enable EN can connect the output to a parallel data bus, so it can be disabled when other outputs on that bus are enabled. These configurations suggest some obvious ways to connect serial ports.

Serial I/O chips can use parallel I/O port bits to control the lines to the chips using indirect I/O. We discuss the general principles after we consider this example: sending 24 bits of data to a series configured output, as shown in Figure 5.18c, following the flow chart in Figure 5.19.

```
void serialOut(unsigned char *s) {unsigned char i, j;
  KCDD = 0xe0;
  for(j = 0; j < 3; j++){
    for(i = 7; i >= 0; i--){
      KCD = 0; /* make data and clock bits false (0) */
      if(0x80 & s[j]) KCD |= 0x20;/*if msb 1, make data true (1)*/
      KCD |= 0x40; KCD &= ~ 0x40; /* pulse shift clock */
      s[j] <<= 1;/* shift data */
    }
  }
  KCD |= 0x80 ; KCD &= ~ 0x80; /* pulse output register clock */
}
```

The outer loop of the procedure *serial_Out()* above reads a word from a buffer and an inner loop shifts 1 bit at a time into *KCD* bit 5, clocking *KCD* bit 6 after each bit is sent, and then pulsing *KCD* bit 7 to put the data into the output

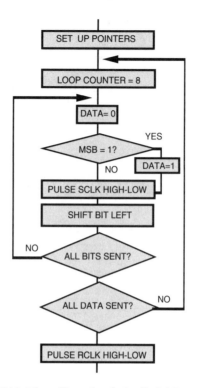

Figure 5.19. Flow Chart for Series Serial Data Output

buffer register. Procedures for the other configurations in Figure 5.18 are similar to this one (problem 24). The basic concept is that the individual signals needed to control the external chips can be manipulated by setting and clearing bits in parallel I/O ports. It is easy to write programs that will interface to serial I/O devices via a parallel I/O port.

serial_Out() can be trivially modified to input data from three 74HC589s, connected as in Figure 5.18b. See *serialIn()* on the next page. *KCD* bit 7, normally high, is pulsed to move data from the 74HC589s' input pins into their parallel holding register on the rising edge before the shift procedure is executed, and is again pulsed low to transfer the data to the shift register; shifting, when *KCD* bit 6 is high, puts that data serially into *KCD* bit 4. Further, combining hardware in Figures 5.18a and 5.18c, three bytes of data can be output to '595s as three bytes are input from '589s at the same time.

Parallel input or output connections can be similarly implemented. Parallel output (Figure 5.18d) can get data to only one output register, without taking the time to shift data through all of them, as the serial connection requires. *KCD* should be initialized to output T on *KCD* bits 2 to 4, and after shifting 8 bits, to pulse just one of them low. Parallel input (Figure 5.18b) similarly holds exactly one of *KCD* bits 2 to 4 low while the shifting takes place as in *serialIn* below. The procedure's parameter a is 0 to input a byte from the leftmost '589, 1 from the middle '589, and 2 from the right '589.

```
void serialIn(unsigned char a) { unsigned char i, value;
   KCDD = 0xe0; // output clock and enable signals, input data
   KCD &= ~ 0x80; KCD | = 0x80; // clock data into first register
   KCD &= ~ 0x80; KCD \ = 0x80; // load data into second register
   KCD &= ~( 1 << a ); // assert KCD's ath bit low to enable the ath '589
   for(i = 0; i < 7; i++){
      value <<= 1; /* shift data */
      KCD | = 0x40; KCD &= ~ 0x40; /* pulse the shift clock before reading */
      if(KCD & 0x10 ) value | = 1; /* get a bit from the port, insert it */
   }
   KCD | = 7; // negate KCD's ath bit high to disable the ath '589
}
```

The first line sets the direction to output control signals and input returned data. The next two lines pulse *KCD* bit 7 to clock data into '589's first register on its rising edge and then load the shift register when it is low. Then, when *KCD* bit 7 is high, one of the tristate drivers is enabled; then the data shifts into the microcontroller.

As an example of serial I/O, we consider a digital thermometer. The Dallas Semiconductor 1620 has a CONFIG register and a TEMPERATURE register (among others), and uses a serial three-wire interface (see Figure 5.20). Data are shifted in and out, least significant bit first, on D (pin 1). Each message consists of sending an 8-bit command, optionally followed by sending or receiving eight or nine data bits. RST (pin 3) must be high from before the command is sent until when the data have been completely sent or completely received, and there must be a 5 ms delay between issuing commands, while RST is low. Temperature is measusred about once per second.

The following program sends out a command to write 2 into the 8-bit 1620 CONFIG register to initialize it for temperature measurement, then it reads the 9-bit TEMPERATURE register, which is a two's complement number, in units of 1/2 °C.

```
enum { D = 0x20, CLK = 0x40, RST = 0x80, WrCnfg = 0xc, RdTemp = 0xaa,
   start = 0xee } ;
```

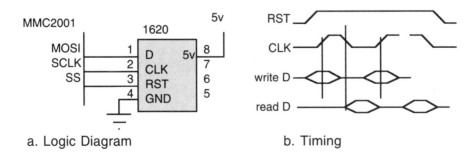

a. Logic Diagram   b. Timing

Figure 5.20. Dallas Semiconductor 1620 Digital Thermometer

```
void main() { short i;
   KCDD = 0xe0; KCD = 0;
   KCD |= RST; send(WrCnfg, 8); send(2, 8); KCD &= ~RST; // wait 5 ms
   wait(1); KCD |= RST; send(start, 8); KCD &= ~RST; wait(20); // wait 1 s
   KCD | = RST; send(RdTemp, 8); i = receive(9); KCD &= ~RST;
}

void send(short d, char n) { char i;
   for(i = 0; i < n; i++){
      KCD &= ~D; if(d & 1) KCD | = D;
      KCD | = CLK; KCD &= ~CLK; d >>= 1;
   }
}
short receive(char n) { short d; char i;
   KCDD &= ~D;
   for(i = d = 0; i < n; i++)
      { d >>= 1; if(D & KCD) d+=0x100; KCD |= CLK; KCD &= ~CLK; }
   KCDD | = D; if(n == 8) d >>= 1; return d;
}
void wait(short n){short i,j; for(j=0; j<n; j++) for(i=0; i<2105; i++
   );}
```

5.3.4   The MMC2001 ISPI Module

The MMC2001 was designed to exploit serial modules like the 74HC589 and
74HC595, using an *Interval (Mode) Serial Peripheral Interface* (ISPI) that takes care
of shifting serial data to and from the MMC2001. In this section we introduce the
ISPI module, performing essentially the same function as the previous section's
serialOut.

Figure 5.21 shows the ISPI data, control, and status ports. Writing to the data
port *SPDR* will cause the data to be shifted out sent msb first, and data shifted in can
be read from the same address *SPDR* (but a different port). The control port *SPCR*

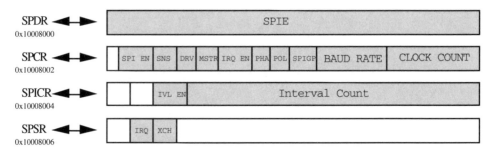

Figure 5.21. ISPI Data, Control, and Status Ports

has the following bits. *SPI_EN* must be set to use the ISPI, and *MSTR* is set to indicate the ISPI is a master. Although the ISPI module can also act as a slave to another computer, it is usually a master and we will consider it so until the end of this section. Thus master *MSTR* and *SPE* must be T. *SPI_EN* exclusive-ORed with *SNS* is available on a pin *SPI_EN*. This pin signal can effect a transfer of data in the external device which shifts data into or out of the MMC2001. The *DRV* bit, if set, enables open-drain rather than tristate outputs. Open drain is often used when a single wire is used for data input and output. *IRQ_EN* is set to enable interrupts. *PHA* and *POL* determines the shape of the clock. *SPIGP* is output as a simple one-bit output bit on a pin. The 3-bit baud rate port sets the shift clock rate (a value of zero gives a fast 4 MHz shift clock, if the CLK is 32 MHz), and the four-bit clock count established the number of bits that are transferred upon each shifting of the data register (a value of 7 shifts 8 bits). The control port *SPICR* has the bit *IVL_EN* which enables the interval mode, and a 13-bit interval count. The status port contains the *IRQ* bit that indicates that the ISPI requested an interrupt, and the *XCH* bit that indicates that an ISPI shift is in progress.

The procedure *serialXfr* simultaneously inputs an 8-bit word and outputs an 8-bit word. Writing data into *SPDR* begins the shifting out of that data. This is an example of a write address trigger. Even for input, a statement like *SPDR = 0;* is needed to start a shift, even if no data need be output. The *XCH* flag becomes clear when all bits are shifted; it can be tested to wait for shifting to be done. In the next chapter this will be called a gadfly loop. This *ISPIF* flag is an example of a *status port*, an input port that lets the programmer input information from the device, but this information is from the device itself, not from the outside world. The first statement of the *main* procedure initializes the *ISPI* so that the procedure *serialOut* can be executed. Then *main* shifts 3 bytes out to hardware shown in Figure 5.18d, and then pulses the SPI_GP (pin 128) to transfer the shifted data to the output register.

```
volatile unsigned short SPDR@0x10008000, SPCR@0x10008002,
    SPICR@0x10008004, SPSR@0x10008006;
enum{SPI_EN=0x4000,SNS=0x2000,DRV=0x1000,MSTR=0x800,IRQ_EN=0x400,
  PHA=0x200,POL=0x100,SPIGP=0x80,IVL_EN=0x2000,
  IRQ=0x4000,XCH=0x2000};

short serialXfr(unsigned short d = 0)
  { SPDR = d ; while(SPSR & XCH) ; d = SPDR; return d; }


main(){unsigned char j;
    SPCR = SPI_EN+MSTR + 7; for(j=0; j<3; j++) s[j] = serialXfr();
    SPCR | = SPIGP ; SPCR &= ~ SPIGP; /* pulse output port clock */
}
```

SPI_CLK (pin 126) is connected to the 74HC589 or 74HC595 SCLK pin to effect shifting of the data. SPI_MOSI (pin 124) connects to 74HC595 IN (pin 14). SPI_MISO (pin 121) is connected to 74HC589 OUT (pin 9). SPI_GP (pin 128) is connected to 74HC595 RCLK (pin 12), or to 74HC589 RCLK (pin 12) and LD (pin 13).

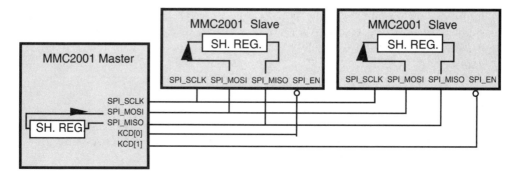

Figure 5.22. Multicomputer Communication System Using the ISPI

The ISPI interface can be used to communicate among several MMC2001s as shown in Figure 5.22. One MMC2001 is made a *master,* and all the others are *slaves.* In slaves, SPI_EN is an enable input. The SPI_MOSI, SPI_MISO, and SPI_SCLK pins are connected together. In contrast to the master, the slave clears its $SPCR$ bit $MSTR$. As one ISPI is initialized and inadvertently pulses its output port bits which input to the other ISPIs, other ISPIs may have to ignore the resulting spurious input byte. The master controls the slaves through a parallel output port like KCD bits 0 and 1, by asserting exactly one slave SS input low. Then the ISPI exchanges the data of the master ISPI shift register with the data in the selected slave ISPI shift register using a program like that just shown. To send data, the slave microcontroller writes its data into its $SPDR$ before the master writes its data into its $SPDR$, since the latter begins the shift operation. The slave's $SPSR$ bit XCH is tested until it becomes negated, indicating data have been exchanged. The slave reads the master's data from its $SPDR$.

This ISPI has an enhancement to accommodate serial devices such as D-to-A converters used for compact disk players. By setting the port $SPICR$ bit IVL_EN and writing a value N in the same port's 13-bit interval count, shifting of bits of data is repeated, and a delay of N shift clock cycles is inserted between each shifting of data.

The serial port is a valuable alternative to the parallel port. It requires substantially fewer pins and wires. The MMC2001's ISPI interface makes it easy to use these devices, but, with a modest amount of software, any parallel I/O port can be used to control them. However, a parallel port is required where speed is needed, because the serial port implemented with indirect I/O, is considerably slower than §5.1's parallel port.

5.3.5   Accessing Devices Using Vectors and Structs

The ISPI has a fair number of ports; such a collection of ports can be accessed using a vector or a $struct$, and #define statements can be used to clearly name its ports. Vector notation is useful in accessing neighboring ports having the same width. $struct$ notation is useful in accessing ports having different widths, and can be useful in accessing ports of the same width. Vector notation can also be used in simple compilers that do not support structs to handle ports that have different widths. We illustrate these concepts in following subsections by writing §5.3.4's *main* and *serialXfer*.

5.3.5.1   Vector Access to Ports

A vector can assign names to ISPI ports by a global declaration, as shown below:

```
volatile short spi[4]@0x10008000;
char serialXfr(unsigned char d)
    { spi[0] = d ; while(spi[3] & XCH) ; d = spi[0]; return d; }

main(){ char j, s[3];
    spi[1]=SPI_EN+MSTR+7; for(j=0; j<3; j++) s[j]=serialXfr();
    spi[1] | = SPIGP ; spi[1] &= ~ SPIGP; /* pulse output port clock */
}
```

5.3.5.2   Vector Pointer Access to Ports

Alternatively, a global or local variable pointer to the ISPI ports can be initialized in the first statement of *main*. Then treating this pointer with offsets as a vector with indexes, we write into the ISPI ports in the remaining lines.

```
volatile short *spi;

char serialXfr(unsigned char d)
    { *spi = d ; while(spi[3] & XCH) ; d = *spi; return d; }
main(){ char j, s[3]; spi = (short *)0x10008000;
    spi[1]= SPI_EN+MSTR+7; for(j=0; j<3; j++) s[j]=serialXfr();
    spi[1] |= SPIGP ; spi[1] &= ~ SPIGP; /* pulse output port clock */
}
```

5.3.5.3   Using #defines to Name Ports

The above example illustrates the use of vectors in handling devices with a lot of identically wide ports. However, #define statements can make access to these ports more self-documenting. (These appear in four columns to save space.)

```
#define SPDR *spi           #define SPCR spi[1]
#define SPICR spi[2]        #define SPSR spi[3]

volatile short *spi;
char serialXfr(unsigned char d)
    { SPDR = d ; while(SPSR & XCH) ; d = SPDR; return d; }
```

```
main() { char j, s[3];
    spi = (short *)0x10008000;
    SPCR = SPI_EN+MSTR+7; for(j = 0; j < 3; j++) s[j] = serialXfr();
    SPCR |= SPIGP ; SPCR &= ~ SPIGP; /* pulse output port clock */
}
```

Then we write into the ISPI ports in the remaining lines. Note that the procedure looks like the first *serialOut* in §5.3.4 except that the pointer *spi* must be initialized, but the assembly language code actually uses this pointer to access the ports.

5.3.5.4   Struct Access to Ports

The `struct ISPI` illustrates how `structs` access ports of different widths, and our procedure `serialOut` shows how these `struct` members can be used. In HIWARE's compiler, the 'advanced options' setting for 'code generation' must have 'bit field byte allocation' set to 'most significant bit in byte first,' for `struct` members to correspond to ports in the same order in the declaration as they are in the M·CORE data sheet.

```
struct SPI { unsigned short SPDR, :1, SPI_EN:1, SNS :1, DRV :1, MSTR :1,
    IRQ_EN :1, PHA :1, POL:1, SPIGP:1, BAUD_RATE:3, CLK_CNT:4;
    unsigned short :1,:1, IVL_EN:1, INTERVAL_COUNT:13, :1, IRQ:1,
    XCH:1};

volatile SPI spi@0x10008000;
char serialXfr(unsigned char d)
    { spi.SPDR = d ; while(spi.XCH) ; d = spi.SPDR; return d; }

main() { char j, s[3];
    spi.SPI_EN = 1; spi.MSTR = 1; spi.CLK_CNT = 7;
    for(j=0; j<3; j++) s[j]=serialXfr(); spi.SPIGP=1; spi.SPIGP=0;
}
```

5.3.5.5   Struct Pointer Access to Ports

Alternatively, if we globally declare an ISPI pointer, as in *SPI *spiPtr = (SPI *)0x10008000;* we can write the same procedure:

```
char serialXfr(unsigned char d)
    {spiPtr->SPDR = d ; while(spi.XCH) ; d = spiPtr->SPDR; return d; }

main() { char j, s[3];
    spiPtr->SPI_EN = 1; spiPtr->MSTR = 1; spiPtr->CLK_CNT = 7;
    for(j=0;j<3;j++)s[j]=serialXfr();spiPtr->SPIGP=1;spiPtr->SPIGP=0;
}
```

Either a pointer is initialized to the port address, or else a globally defined
struct is forced to be at a predefined location using the @ symbol in the global
declaration. *Structs* can give self-documenting names to ports, especially where
the ports have a variety of widths, or even identical width ports otherwise described
by vectors.

5.3.6   Indirect and Serial I/O Classes

Indirect, and serial I/O using the ISPI, are suited to object-oriented programming.
The classes *Indirect, SerialOut,* and *Lcd,* shown in this section, illustrate
encapsulation, inheritance, overriding functions and operator overloading.

Indirect I/O for the 6818 uses two parallel ports. This creates a minor dilemma:
how do we define a class that might inherit function and data members from two
classes? While C++ provides multiple inheritance to handle such cases, we couldn't
find a simple way to express our design using multiple inheritance. Instead, we define
one of these, the port through which address and data are time-multiplexed, as
Indirect's base class which is *Port*. The other port, carrying control information, is a
Port blessed in *main* as an argument to *Indirect's* constructor. This class's *put* and
get function members execute the same algorithms as C procedures *outa( )* described
in §5.4.1 and *ina( ),* a problem at the end of this chapter, which is very similar to
outa( ).

```
class Indirect:public Port<char>{public:char index;Port<char>
    *control;

  Indirect(long id, Port<char> *control) : Port<char>(id, 0x20)
    { this->control = control; *control = as + rw + cs;}

  void put(char value){
     option(1, 0xff); /* recall that option(1,.. sets the direction port */
     Port<char>::put(index);*control  =  as+cs;  *control=as;
     *control=0;
      *control = ds; Port<char>::put(value); *control = 0; *control = as;
      *control = as + cs; *control = as + rw + cs; option(1, 0);
  }
  char get(void){
     option(1,0xff); /* make this port an output */
     Port<char>::put(index); *control = as + rw; *control = rw;
     option(1,0);/* make this port an input */ *control = as; *control = 0;
     *control = rw + ds; value = Port<char>::get(); *control = rw;
     *control = rw + cs; *control = as + rw + cs; return value;
  }
  virtual char operator = (char data) { put(data); return data; }
  virtual Indirect &operator [] (char data){index=data;return *this;}
};
```

In the program above, the overloaded index operator *[]* illustrates another C++ feature that is very well suited to indirect I/O. It is called whenever the compiler sees an index *[]* to the right of an object, such as the object *clk[0]* that appears later in this example, whether the object and index are on the left or right of an assignment statement =. It executes the overloaded operator *[]* before it executes the overloaded cast or overloaded assignment operator. This overloaded operator simply stores what is inside the square brackets, *0* in our example, in an object data member *index*. Then the following overloaded cast calls *get*, or assignment operator calls *put*, which can use this saved value to supply an address, such as the address of a 6818 port or memory location.

main() declares object *clk* to be of class of type *Indirect*. Its constructor's first parameter, *0x10003004*, is passed to parent class *Port*'s constructor, so *Indirect* is essentially an object connected through the 6818 data and addresses buses. *Indirect's* constructor's second parameter is a pointer to an object of class *Port;* it is blessed as *main* declares the object *clk,* to be used to output the control signals to the 6818.

```
void main() {
    Indirect clk(0x10003004, new Port<char>(0x10003005,0x20,0xf));
    char i, yr = 1, mo = 2, dm = 3, dw = 4, hr = 5, mn = 6, se = 7;
    clk[0xb]=0x80; clk[0xa]=0xf; clk[9]=yr; clk[8]=mo; clk[7]=dm;
    clk[6]=dw; clk[4]=hr; clk[2]=mn; clk[0]=se; clk[0xb]=8; i=clk[0];
};
```

The index operator *[]* used with assignment and cast operators, accesses 6818 ports using vector notation. The statement *clk[8]* = *mo;* puts local variable *mo* into 6818's location 8. Actually, *[8]* calls the *Indirect* class's overloaded operator *[]* which just stores *8* in the data member *index*. Then because an object is on the left of an assignment, *Indirect* class's overloaded operator = is called, which calls *put*. *put* uses *index* to provide the address sent to the 6818A. Similarly, *i* = *clk[0];* uses *Indirect* class's overloaded *[]* operator to store 0 in member variable *index*. Then because an object is on the right of an assignment, *Port* class's overloaded *cast* operator is called; it calls *get* which uses *index* to provide the address sent to the 6818A. The code looks like last section's access to a device's multiple ports using vectors. *#defines* could also be used here to make the code more self-documenting.

Our next example shows again how the overloaded index operator *[]* can be used with objects. The templated class *SerialOut* outputs a *char*, a *short*, or a *long*, to one of three groups of '595s. Each group is connected in series (Figure 5.18c), but the whole groups are connected in parallel (Figure 5.18d). The class *SerialOut's put* function member sends a byte out at at time though the ISPI, for as many bits as there are in the template data type, and then through *KCD* it pulses the clock of the group that loads the shifted data into its output registers.

```
char B[8] = {0x7f, 0xbf, 0xdf, 0xef, 0xf7, 0xfb, 0xfd, 0xfe};
template <class T> class SerialOut : public Port<T> { Port *p;
    char serialXfr(char data){*port=data;while(port[3]&0x    0));return
        *port;}
```

```
public:SerialOut(short format; Port *p) : <short>Port (0x10008000)
  { this->p = p; port[1] = format; p->direction(0xff) }

virtual void put(T c) { unsigned char i,j;
 for(i=0, j = sizeof(T) - 1; i<sizeof(T);i++, j--)
 { serialXfr(c<<(8 * j)); c=c<<8; } *p = 0xff;
 }

 virtual T get(void) { unsigned char i; T c;
  for(i=0;i<sizeof(T);i++){serialXfr(c>>(sizeof(T)-1)*8);c<<=
  8;}
  *p = 0xff; /* negate register clock asserted in overloaded [] */ return c;
  }

 virtual T operator = (T data) { put(data); return data; }
 virtual Port<T> operator [] (char data){*p = B[data]; return *this;}
};

void main(){ char i ;
 SerialOut<char> sDevice(0x4807,
  new(char>Port(0x10008000,0x20,0xff));
 for(i=0;i<5;i++)s[i]=sDevice[i];  for(i=0;i<5;i++)  sDevice[i] =
  s[i];
}
```

In a second and different use of vectors, *SerialOut*'s constructor uses *Port's*
data member *port* to point to the ISPI ports. There, vector notation can be used to
access the ports. *SerialOut*'s constructor's *short* argument initializes the ISPI
control port. *put* always provides bytes most significant byte first. In a problem at
the end of the chapter, *put* will be modified so that it shifts data least significant bit
first.

    The LCD display can use C++'s *Port* class for formatted output; this class
is:

```
enum { rs = 0x20, e = 0x10, d10 = 20, d410 = 800};

class Lcd : public Port<char> {

 void put4(char c) { Port<char>::put(c + e); Port<char>::put(c) }

 virtual void backspace()
   {put(0x10, 0, d10); put(' ', rs, d10); put(0x10, 0, d10);}
 public: Lcd(short a; unsigned char attr) : Port(a, attr, 0x3f) {
  put(0x28, 0, d10); put(0x28, 0, d10); put(6, 0, d10);
  put(0xe, 0, d10); put(1, 0, d410); col = 0; lcr = 1;
  }
```

```
virtual void put(char data) { put(data, rs, d10); };
virtual void put(char c, char a, char d)
  {put4 (((c>>4)&0xf)+a); put4((c&0xf)+a); if(a)col++; while(d)d--
  ;}
} cout;
```

```
void main()
  {cout<<"x is" <<itoa(x)<< "or in hexadecimal:" <<htoa(x)<<'\r'; }
```

Port's overloaded operator $<<$ is used to output different data types as described in §5.3.6. But backspacing is taken care of by calling the virtual *back-space()* function, which for this class, by shifting the cursor left, writing a space, and shifting the cursor left again, clears the character which the backspace removed, then moves the cursor back one position. This class can be used to output data on the LCD display. *main* displays the ASCII character stream 'x is', then prints the value of x in decimal because *x* is of type *short,* then displays the message 'or in hexadecimal:' and displays the value of *x* in hexadecimal, using library or user-defined functions *itoa(x)* and *htoa(x).*

5.3.7   Indirect and Serial Device Drivers

We now implement §5.3.6's classes as Ariel device drivers. We first illustrate a simple driver for the 6818, using fixed ports *KCD* and *KRD* and means to read and write a vector of data at different addresses. Then we illustrate an LCD driver that can be shared by all tasks. Finally, we show an ISPI driver that illustrates access to its control ports.

The 6818 driver's constants, for the *outa()* and *ina()* procedures, and global variables for the Ariel device driver, are as follows:

```
enum{ cs = 1, as = 2, ds = 4, rw = 8};
```

```
uxid_t id; rv_t status; cdev_t charDev;
```

_pio_create()'s structures (as in Figure 5.9a) are as follows:

```
const drv_t function6818 = {initFun, svcFun, 0, 0, 0, 0, 0};
```

```
ucd_t ucd6818=
  {_EXTNAME('6', '8', '1', '8'), &function6818, 0, 0, 0, 0, {0, 0}};
```

main(), shown below, calls *_pio_create()*, and if no errors occur, it calls the *put()* procedure and the *get()* procedure. *_pio_create()* calls *initFun()* shown below to initialize the *KCD* and *KRD* direction ports for the 6818 (Figure 5.14d)

```
char d[10]={0x80,0xf,1,2,3,4,5,6,7}, a[9] = {0xb,0xa,0,2,4,6,7,8,9};
void main() {
   if(!ISSVCERR(id=_pio_create(&ucd6818))) {put(d,a,9); get(d,a,9);}
}

int32 initFun(ucd_t *ucp)
   {KCDD = 0; KRDD = 0xf; KRD = as + rw + cs; return SUCCESS;}
```

`get()` and `put()` illustrate the use of an auxiliary buffer pointer that is used here to hold 6818 addresses. In this example, global variable `d[0]`, which happens to be 0x80, is input or output from address `a[0]`, which happens to be 0xb, and so on. This example outputs the registers as in §5.3.1's `main()`, and then inputs them. `get()` and `put()` merely set up `charDev` as in Figure 5.9b, and call on `_pio_cmd()` as in §5.1.7.1.

```
void get(char *data, char *addresses, short n){
   charDev.aub = data; charDev.axb = addresses; charDev.lub = n;
   _pio_cmd(id, _PIO_READ, 0L, &charDev, &status, _CONT_NOCOOR);
}

void put(char *data, char *addresses, short n){
   charDev.aub = data; charDev.axb = addresses; charDev.lub = n;
   _pio_cmd(id, _PIO_WRITE, 0L, &charDev, &status, _CONT_NOCOOR );
}
```

`_pio_cmd()` calls `svcFun()`. Its case statement executes either the input or output routines by calling either `ina()` or `outa()` for each member of the strings passed in the `struct` argument `charDev` (Figure 5.9b).

```
void outa(char a, char d) {
   KCDD = 0xff; KCD = a; KRD = as+cs; KRD = as; KRD = 0; KRD = ds;
   KCD=d; KRD=0; KRD=as; KRD=as+rw; KCDD=0; KRD=as+rw+cs; KCDD=0;
}
char ina(char a){ char value;
   KCDD = 0xff; KCD=a; KRD = as+rw; KRD = rw; KCDD=0; KRD = as; KRD=0;
   KRD=rw+ds; value=KCD; KRD=rw; KRD=rw+cs; KRD=as+rw+cs; return value;
}

void svcFun( ucb_t *ucb, int fun, cdev_t *dparams ) {
   switch(fun){
     case _PIO_READ:
       while(dparams->lub--)*dparams->aub++=ina(*dparams->axb++);
       break;
     case _PIO_WRITE:
       while(dparams->lub--)outa(*dparams->aub++,*dparams->axb++);
       break;
     default: _drv_pio_done(ucb, EOS_BAD_PARAM, 0); return;
   }
   _drv_pio_done(ucb,SUCCESS,0);/* end-service call to enable other tasks' access */
}
```

This particular example does not offer much advantage for using a device driver as opposed to using an object-oriented interface. The user probably only accesses the 6818 during startup to initialize MMC2001 timers. However, it serves to show how indirect I/O is easily handled using a device driver. A device driver would have considerable advantages over an object-oriented interface if a plurality of tasks access an I/O device indirectly. The driver would ensure each task of undivided use of the device until the driver executes _drv_pio_done(), which permits other tasks to access the device.

The LCD driver shown below illustrates a write-only device that puts a string of characters, such as 'hello world!', onto the LCD. Accessing the LCD through a device driver permits multiple tasks to access the display. Text from one task is not garbled with text from another task because one task has exclusive use of the LCD until the first task transfers all of its data. This driver's constants and global variables are shown first.

```
enum{ rs = 0x20, e = 0x10, d10 = 20, d410 = 400 };

uxid_t id; rv_t status; cdev_t charDev;

const drv_t function_lcd = {initFun,svcFun,0,0,0,0,0,0};

ucd_t ucd_lcd =
    {_EXTNAME('L','C','D','0'),&function_lcd,0,0,0,0,{0,0}};
```

The main() procedure creates the driver and, if no errors occur, writes 'Hello world!'.

```
main(){if(!ISSVCERR(id=_pio_create(&ucd_lcd))){put("Hello
    world!",12);}}
```
_pio_create() initializes the LCD as in §5.3.2, using put8() shown later.
```
int32 initFun( ucd_t *ucp ){
    KCDD = 0x3f; put8(0x28, 0, d10); put8(0x28, 0, d10); put8(6, 0, d10);
    put8(0xe, 0, d10); put8(1, 0, d410); return SUCCESS;
}
```

The put() procedure shown below calls up the device driver procedure svcFun().

```
void put(char *data, short n){
    charDev.aub = data; charDev.lub = n;
    _pio_cmd( id, _PIO_WRITE, 0L, &charDev, &status, _CONT_NOCOOR );
}
```

svcFun() calls on put8() to output each character. svcFun() ends with a call to _drv_pio_done() to permit other tasks to use the LCD device.

```
void svcFun( ucb_t *ucb, int fun, cdev_t *dparams ) {
    if(fun!=_PIO_WRITE){_drv_pio_done(ucb, EOS_BAD_PARAM, 0); return;}
    put8(0x80,0,d410);while(-dparams->lub)put8(*dparams->aub++,rs,d10);
    _drv_pio_done(ucb, SUCCESS, 0); /* special call to end service */
}
```

```
void put4 (char c) { KCD = c + e; KCD = c; }

void put8 (char c, char a, char d)
    { put4 (((c >> 4) & 0xf) | a); put4 ((c & 0xf) | a); while (d) d-- ;
}
```

The ISPI device driver below permits tasks to access '589's or '595's using vectors to guarantee consistent access through these chips. Since the ISPI data are 16 bits, we define a structure *sdev* which has pointers to *shorts*. A feature of this driver, which shows how driver local variables are stored, is that one of the frame structure bits is saved in *shift_direction* (up to 46 bytes of the device driver's local data can be saved in its *UCD* structure). This variable is used in the *svcFun()* to reverse the data before it is shifted out, to handle chips in which data is sent lsb, rather than msb, first. This driver's definitions and global variables are shown below.

```
#define _PIO_EXCHANGE ((u_int32) 0x0000000b) /* exchange with device */
#define _PIO_BAUD ((u_int32) 0x0000000c) /* set baud rate */
#define _PIO_FRAME ((u_int32) 0x0000000d) /* set frame structure */
#define _PIO_INTERVAL ((u_int32) 0x0000000e) /* set interval size */
#define UCB_GPR ((ugpr_t *) &ucb->gpr) /* parameters overlay GPR */

enum{SPI_EN=0x4000,SNS=0x2000,DRV=0x1000,MSTR=0x800,IRQ_EN=0x400,
   PHA=0x200,POL=0x100,SPIGP=0x80,IVL_EN=0x2000, IRQ=0x4000,
     XCH=0x2000};

uxid_t id; rv_t status;

typedef struct { char shift_direction; } ugpr_t;/* driver local data */

typedef struct sdev {
   short *aub; u_int32 lub; short *axb; u_int32 spl; } sdev_t;

sdev_t shortDev;

const drv_t function_ispi={initFun,svcFun,0,0,0,0,0,0};

ucd_t ucd_ispi =
    {_EXTNAME('I','S','P','I'),&function_ispi,0,0,0,0,{0,0}};
```

The *main()* procedure initializes the ISPI and, if no errors occur, exchanges ten 16-bit elements with the external '589 and '595 chips by calling the *exchange()* procedure.

```
short d[10], a[10] = {0,1,2,3,4,5,6,7,8,9};

main(){if(!ISSVCERR(id=_pio_create(&ucd_ispi))){exchange(-
    d,a,10);}}
```

_pio_create() calls *initFun()* to initialize the ISPI, essentially as in §5.3.4.

```
int32 initFun( ucd_t *ucp )
   { SPCR = SPI_EN + MSTR + 0xf; KCDD = 0xff; return SUCCESS; }
```

To exchange data we call *exchange()* which calls *_pio_cmd()*

```
void exchange(short *data, short *address, short n){
  shortDev.aub = data; shortDev.axb = address; shortDev.lub = n;
  _pio_cmd(id, _PIO_EXCHANGE, 0L, &shortDev, &status, _CONT_NOCOOR);
}
```

Alternatively, *control()* sets the ISPI's baud rate, interval, and frame structures.

```
void control(short data, long command){ shortDev.aub =
  _pio_cmd( id, command, 0L, &shortDev, &status, _CONT_NOCOOR );
}
```

These procedures, *control()* and *exchange()*, call the driver's *svcFun()*.

```
void svcFun( ucb_t *ucb, int fun, sdev_t *dparams ) { short i;
  switch(fun){
    case _PIO_EXCHANGE:
      while(-dparams->lub) {
        KCD=*dparams->axb++;/*enable exactly one (series of) external chip(s)*/
        if(UCB_GPR->shift_direction) SPDR = reverse(*dparams->aub);
        else SPDR = *dparams->aub; while(SPSR & XCH) ;
        if(UCB_GPR->shift_direction) *dparams->aub++=
           reverse(SPDR);
        else *dparams->aub++ = SPDR; KCD = 0xff;/* all enables high (F) */
      }
      break;

    case _PIO_BAUD: SPCR=(SPCR&0xff8f)|((*dparams->aub<<4)& 7);
    break;
    case _PIO_FRAME: SPCR = (SPCR & 0x10) | (*dparams->aub&0xff8f);
      UCB_GPR->shift_direction = (*dparams->aub >> 4) & 7; break;
      case _PIO_INTERVAL:
        if(*dparams->aub)SPICR=*dparams->aub| 0x2000; else SPICR = 0;
        break;
        default: _drv_pio_done(ucb, EOS_BAD_PARAM, 0); return;
  }
  _drv_pio_done(ucb, SUCCESS, 0); /* special call to end service */
}
```

In the *_pio_cmd()* service call with command *_PIO_EXCHANGE*, the auxiliary string pointer *shortDev.axb* puts out values on *KCD* to enable one or more external chip while the corresponding elements of *shortDev.aub* are exchanged with external chips. We assume each element of *shortDev.aub* has exactly one bit asserted low. This service call calls driver procedure *svcFun()* to execute case *_PIO_EXCHANGE*. If local variable *UCB_GPR->shift_direction* is nonzero, the procedure *reverse()*, shown below, reverses the order of the bits being sent, and of the bits being received.

```
short reverse(short i){ asm {
      brev r2 /* reverse bits (note: puts in high-order 16 bits */
      lsri r2,16 /* put in low-order 16 bits */
      jmp r15
} return 0; }
```

In *_pio_cmd()* with command *_PIO_BAUD,* the 16-bit data pointed to by *_pio_cmd()* argument's *axb* is put into the ISPI baud rate port. In *_pio_cmd()* with command *_PIO_INTERVAL,* the 16-bit data pointed to by *_pio_cmd()* argument's *axb* is put into the ISPI *SPICR,* to control the interval. In *_pio_cmd()* with command *_PIO_FRAME,* the 16-bit data pointed to by *_pio_cmd()* argument's *axb* is put into the ISPI's *SPCR,* except that the bits in the position of the baud rate port are not inserted, but are put in the device driver's local variable *shift_direction* to determine whether data are reversed when they are exchanged.

In addition, the *control()* procedure can have commands *_PIO_RESERVE* and *_PIO_RELEASE.* The former command prevents other tasks from using the device until the latter command is executed. These two commands are executed within Ariel without calling *svcFun().* They permit a task calling a sequence of *control()* and *exchange()* procedures without another task switching the effect of these calls.

5.4   A Designer's Selection of I/O Ports and Software

The parallel I/O device is the most flexible and common I/O device. When designing a parallel I/O device, the first step is to decide on the architecture of the port.

First we select the address. The address is often selected with an eye to minimizing the cost of address decoders. Second, we select the data transfer mode of the port. One of the major design questions is whether the port should be directly or indirectly coupled to the microcontroller, or serially coupled through a 'three-wire interface'.

Indirect I/O is a mode where one I/O device is used to provide the address, data, and control signals for another I/O device. Software emulates the microprocessor controller and generates its signals to the I/O device. It is generally an order of magnitude slower than direct I/O. But it is very useful when a parallel I/O device is available anyhow, such as in single-chip microcomputers like the MMC2001, or in personal computers that have a parallel port — often used for a printer. It is not necessary to attach devices to the address, data, and control buses within the computer, which might destroy the integrity of the computer and render debuggers inoperable.

Serial I/O, whether we use indirect I/O or the ISPI, provides an obvious advantage over direct I/O, by requiring fewer pins or wires. If the computer needs to be isolated from the external world, serial I/O uses fewer opto-isolators. Many manufacturers of specialty I/O devices, such as D-to-A converters, select this least expensive data transfer mode, and 'three-wire interface' devices are becoming widespread.

The main factor affecting the design decision is the speed of the I/O device. Sometimes the speed is dictated by the external system's speed, as when data are sent to or from a fast communication network using light pipes, and sometimes it is dictated by the process technology used to build the chip, as when dynamic logic is used, and requires a maximum time between events. Generally, the slower the required speed, the simpler the system. Many I/O devices are overdesigned with respect to speed. You should carefully determine the minimum allowable speed for the device and then choose the technique that fits that requirement. Then look at the system and determine if it has the needed mechanisms — for instance a parallel port of sufficient width for indirect I/O. We suspect that a lot of cases where indirect I/O is suitable and available are designed around direct I/O, which significantly increases their design and maintenance costs.

If indirect I/O is used, a next decision should be which parallel ports to use for address, data and control signals for the device to be connected. If directly coupled, the question is similar, which of the 5 parallel ports in the MMC2001 should be selected, or else what external port should be built. In the latter case, we need to consider whether an output port should be basic or readable or shadowed, a set or clear port, or whether an address trigger or address register output is indicated. The availability of existing ports and the I/O port latency is the most important criterion. If the port is available on the MMC2001 and is not needed by other devices, the first choice is to use one of these ports. If there is competition for their use, then external devices need to be considered.

If an MMC2001 port is to be used, note that some pins are generally used for ISPI and other devices, so their main attraction is for contrasting the operation using a parallel port to the same operation for using the ISPI or UART, and they are thus less desirable for general parallel port use. KCD and KRD, suit a 16-bit port in a single-chip MMC2001. KCD and KRD suit 1-bit or 8-bit ports since these ports are in the MMC2001. The choice of a first I/O port follows from the discussion above, but the selection of a fourth or fifth device usually becomes less clear. We can use less desirable ports for less critical devices, but fortunately the MMC2001 has severals useful ports.

For external ports other than the serial 'three-wire interface', a basic output port is cheapest but the software cannot read the data in it, so many *structs* having bit fields, and explicit software equivalent to bit fields won't work with basic output ports. A readable output port is most general, but also about twice as expensive as the basic output port. However, a RAM shadow, or software keeping a duplicate copy of the output data, can be used to avoid building a readable output port.

The hardware for a basic output, readable output, input, or other external port, can be implemented using simple TTL MSI chips. This chapter showed how to use SSI gate chips and decoders to implement the address decoder, and the popular 74HC244 and 74HC374 medium-scale integrated circuits to implement these ports. While other chips can be used, these chips are often used in printed circuit card microcomputers that are mass-produced and intended for a wide variety of applications.

Besides hardware cost which govern the choices above, software cost are more often critical. Programming decisions generally affect clarity, which is reflected in the cost to write, maintain, and debug, programs, and often affect static or dynamic efficiency. We now consider the designer's choice of software techniques to be used with I/O device.

A fundamental question is which language to use. Use assembly language for code that must operate quickly, or with precise timing, and to use microcontroller features not available through the compiler. However, programming in C or C++ is about an order of magnitude cheaper to design and maintain than assembly language. Because of its simplicity and wide acceptance, C is the language of choice for most I/O interfacing code. C provides operators that directly specify pointer usage, shifting, or masking, whereas other languages like PASCAL, BASIC, or JAVA miss some key parts that C provides. However, C++ offers object-oriented programming which we discuss later.

I/O devices can be accessed with constant pointers, as in *d = (char*)0x2d020000;* with variable pointers, as in char **ptr = (char*)0x2d020000; d = *ptr;* using a global variable positioned with embedded assembly language origin statements, with vector indexing, as in char **ptr = (char*)0x2d020000; dp[0];* and with pointers to structures, as in *spiPort->bd = 2;* We have shown examples of each. Constant pointers appear useful where an I/O port is accessed once or twice. Global variables work well for most I/O devices. Vectors are clearer when a device has a number of same-sized ports. Structure pointer access is very useful for devices with many ports of assorted sizes. Finally, any long program can likely benefit from the intelligent use of *#define* statements to rename these constructs to be meaningful port names to provide self-documenting code.

Objects provide a mechanism to efficiently implement the capabilities of device drivers such as device independence and I/O redirection. They achieve a major fraction of the capabilities of operating system I/O device drivers, but with a small fraction of the overhead. They provide I/O independence, which permits compile-time substitution, without changing the body of the code outside the I/O device objects, and I/O redirection, which permits the same flexibility at run time. They provide this independence and redirection by protecting functions and data so that there are no subtle interactions among I/O devices and the main procedure and its subroutines, other than the directly stated interactions in public function member arguments and data members. Also this ability to provide protection enhances documentation and maintainability, and thus reduces software design cost. One is less likely to confuse or abuse a variable that is bound to the functions that use it. If C++ is unavailable, its mechanisms can be emulated in C by enforcing appropriate conventions in symbolic names and function arguments.

Objects described in this chapter provide more than just a way to show students who have had a course in C++ how object-oriented programming can be used in interface design. The fundamental ideas of object-oriented programming can elevate the programmer from mere software to full system design. By considering the object as encapsulating the function and data member, as well as the I/O port and external hardware, the designer can have predesigned and tested objects (functions, data, and external hardware) that can be inserted into an application as a unit, 'plug-and-play,' or be available in a library, to significantly reduce the design cost of a microcontroller-based system. Further uses of objects will appear in the next and following chapters. An outstanding article, 'Object-Oriented Development' by Grady Booch in the IEEE

Computer Society Press tutorial *Object-Oriented Computing*, Gerald Peterson, Ed., vol 2. led us to appreciate the use of objects in the design of embedded microcomputer systems. You might consult it for additional insights on this approach.

Device drivers provide additional capabilities, expeically for multi-tasking systems. However, they require considerable more care, because the programmer has to adhere to rather firm rules regarding the use of structures and data blocks. The example device drivers shown in §5.3.7 show a good technique for writing these device drivers. Get the basic procedures to work, as in §5.3.1, §5.3.2, and §5.3.4, and merge them with a moderately compatible driver such as was shown in §5.1.6. Device drivers should be used where their additional power is useful, but not where I/O classes provide enough power.

5.5   Conclusions

The interfacing of a microcomputer to almost any I/O system has been shown to be simple and flexible, using parallel and serial I/O devices. We studied different ways data can be passed through a port, into or out of a microcontroller. We saw some I/O software that moved data through a microcomputer, moved data into a buffer, and implemented a traffic light controller and IC tester using the simple I/O devices. Because timing is important to them, we studied the timing of such program segments. We studied indirect and serial I/O, which are especially attractive to the MMC2001 and other microcontroller systems. Next, we considered how the ISPI can assist in serial I/O. We can use the same approach to designing an IC or an I/O system as we can for studying it, and thus develop an understanding of why it was designed as it was and how it might be used. In the remaining chapters, these techniques are extended to analog interfacing, counters, communications interfacing, and display and magnetic recording chips.

Do You Know These Terms?

See page 30 for instructions.

synchronous	basic output port	address register	unit control data
I/O device	readable output	output	(UCD)
port	port	direction port	stub
input port	shadowed output	control port	driver-dependent
output port	address trigger	initialization ritual	parameters
isolated I/O	read address trigger	configure	(DDP)
input instruction	write address	device-independent	unit control block
output instruction	trigger	I/O redirection	(UCB)
memory-mapped	address trigger	object driver	logic-timer control
I/O	sequence	function member	light pattern
lock		checking	sequence

cycle | present internal | block | Interval (Mode)
delay loop | state | template element | Serial Peripheral
interpreter | input state | nasty problem | Interface (ISPI)
Mealy sequential | output state | direct I/O | master
machine | next internal state | indirect I/O | slave
 | linked-list structure | |

Problems

Problem 1 is a paragraph correction problem. See the problems at the end of Chapter 1 for guidelines. Guidelines for software problems are given in the problems at the end of Chapter 2, and for hardware problems, at the end of Chapter 3. When modifying Ariel Device Drivers, show only those procedures, functions, or declarations that are modified.

1*. A port is a subsystem that handles I/O. Memory-mapped I/O is used on the MMC2001 and is popular even on microcomputers that have isolated I/O, because it can use instructions that operate directly on memory and is more reliable in the face of a runaway stack than is isolated I/O. However, if a program error writes over I/O devices, a lock can be used to prevent the calamity. A basic output port is a tristate driver and a decoder; the decoder needs only to look at the address and the RW line to see if the device is to be written into. An input port is a tristate driver and a decoder; the decoder needs only to look at the address and the RW line to see if the device is to be read. A basic output port is a read-only port that cannot be written by the program. Therefore, the program should keep an extra copy of a word in such an output port if it wants to know what is in it. The data can be recorded automatically in memory by using an address trigger. Such output devices are commonly used because they are cheaper than readable output devices. An address output line uses a register to capture the low-order address bits when the high-order address bits match the decoder pattern.

2. A group of eight 1-bit input ports is to be addressed at locations 0x2C30 to 0x2C3E so they will be read in the sign bit position. Show the logic diagram of such a port, inputting with a 74HC251 (Figure 5.23), completely decoding the low-order 16 bits using only a 74HC4078 and 74HC30 (Figure 4.10).

3. A group of eight 1-bit output ports is to be addressed at locations 0x73A0 to 0x73AE so they will write the sign bit of these words. Show the logic diagram of such a port, whose output latches are in the 74HC259 addressable latch (Figure 5.23), Use only the 74HC4078, and 74HC30 (Figure 4.10) to fully decode the low-order 16 address bits. (Note: This group of output ports can have the same address as a read-write memory using a shadow, so that when words are written in the memory, the sign bits appear in the outputs of the corresponding latch to be used in the outside world.)

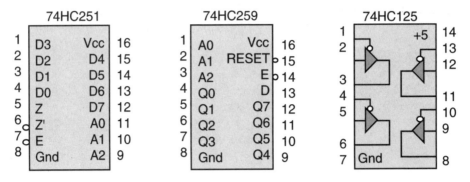

Figure 5.23. Some ICs for I/O

4. Suppose a 1-bit input device using a 74HC125 (see Figure 5.23) inputs a signal A in the sign bit of location 0x2000.

a. Show a logic diagram of the input device using incomplete decoding, using chips from Figure 4.10 to implement the decoder, decoding all necessary address and control signals (CLK, RW, etc.) but not using chip selects, assuming that the program uses only addresses 0 to 0xFFFF (for RAM), 0x2000 (for this device), 0x6000 to 0x7FFF (for internal I/O), and 0xF000 to 0xFFFF (for ROM). Full credit is given to the design using the least number of chips.

b. A wave form is initially low, when your program begins sampling it, then it goes high and then low for the rest of the time. Write a C or C++ function *short pwdth( )* to return the width of this positive pulse (in microseconds) as accurately as you can, assuming the CLK clock is 32 MHz.

c. What is part b's worst-case pulse-width measuring error, in microseconds?

5. Suppose a 1-bit output device using a 74HC74 (see Figure 4.12) outputs the least significant bit of location 0x2000. The output is to be a square wave.

a. Show a logic diagram of the output device completely decoding the low-order 16 bits, using chips from Figure 4.10 to implement the decoder, decoding all 16 bits of the address and appropriate control signals (E, RW, etc.) but not using chip selects. Show all lines connected to +5 or ground.

b. Write a self-initializing procedure *void squr(short n)* to generate a square wave with frequency in Hz. given in *n* using delay loops, assuming E is 8 MHz.

c. What is the lowest and highest frequency that part b can generate?

6. An output device having 16 output bits is addressed at location 0xD3A2. If a number *2n + 1* is written into this location, the *n*th 1-bit latch is set, $0 \le n < 16$ and if a number *2n* is written into this location, the *n*th 1-bit latch is cleared. Show a logic diagram of such a device, whose address decoder fully decodes the low-order 16 bits and whose latches are in two 74HC259 addressable latches. Use only two 74HC259s, a 74HC04, 74HC32, 74HC4078, and 74HC30. Show all chips and pin numbers.

7. Show the logical design of the decoder for a readable output port in Figure 5.3 that can be read from or written in at location 0x2d020000. The decoder shown therein should be implemented with a minimal number of available gates using the chips in Figure 4.10. Show only the decoder, and not the components shown in Figure 5.3.

8. Show the logic diagram of a 16-bit readable output port at 0x2d000200. Use CS0 and other control signals. The program will always read or write 16 bits, and never read or write 8 bits, using this port. However, do not show pin numbers on the data bus and gates or flip-flops connected to the data bus (use 'vector' notation to indicate a bus). Give assembly language instructions to read this port, write data into it, and increment its value. Give C or C++ statements to do these operations.

9. Show the logic diagram of two ports, which are both addressed at location 0x2d008020. Use CS0 and other control signals. In one, a readable output port, the 7 least significant bits of the output word are readable; in the other port, an input port, the most significant bit read is input from the outside world. The address decoder is to be completely specified using a minimum of chips. Use a 74HC244 for the input port and a 74HC374 for the output port, and gates from Figure 4.10 for the decoder.

10. Show the logic diagram of a readable set and a clear port, which are both addressed at location 0x2d0FFF22. If a 1 is written in bits 7 to 0, the corresponding port bit is set, otherwise the data is unchanged; if a 0 is written in bits 15 to 8, port bits 7 to 0 are cleared, otherwise the data is unchanged. Data stored in the port must be available to the outside world. Reading 16 bits from 0xFF22 reads the port bits to data bus lines 7 to 0 only. The program will never read or write 8 bits using this port. The address decoder is to be completely specified using a minimum of chips. . Use CS0. Use a 74HC244 for the input port, 74HC08 and 74HC32 for the set-clear logic, a 74HC374 for the output port, and 74HC133, 74HC04, 74HC32, and 74HC4078 for the decoder. Show all chips and pin numbers and M.CORE address and data bus lines and control signals used.

11. A 16-bit input port at location 0x2d006F3A is connected to a 16-bit serial-in parallel-out shift register to input into a 16-bit wide buffer 1-bit serial data shifted into the shift register. Bits are stored, in each 16-bit word in the buffer, msb first. Use CS0.

> a. Show the logic diagram for this pair of ports using two 74HC244s and a pair of 8-bit serial in parallel out shift registers that use 74HC164s. The address decoder is to be completely specified using a minimum of 74HC133s and 74HC4078s.

> b. Write a (fastest) assembly language program for a 32-MHz MMC2001 to store these data into a buffer. How many bits/second can it collect?

12. Write a templated class *Port* with function member checking. In the constructor, if CHECK is defined and direction is specified, the address must be a valid direction port address, otherwise *errors* is asserted to warn the user. If CHECK is not defined, the code is identical to *Port*'s in 5.2.2.

13. Write templated class *BitPort* to access any bit of any parallel port, derived from *Port*. Its constructor's arguments are *p*, *b*, *d* where *p* is the 32-bit parallel port address, *b* is a bit number, and *d* is F (0) for input and T (1) for output.

14. Write a device driver and main routine calling it, to output the message 'Hi There', for a basic output port (§5.1.1) that is similar to the device driver in §5.1.7. Assume only that the port address is passed in the device-specific parameters.

15. Write a device driver and main routine calling it, to input 10 characters, for an input port (§5.1.1) that is similar to the device driver in §5.1.7. Assume only that the port address is passed in the device-specific parameters.

16. Write the fastest self-initializing procedure *inbuf(char *a, char *p, short n)* using assembly language embedded in C, to input *n* data bytes from port *p* to a buffer at location *a*. Give the rate at which words can be input to the buffer for this procedure, assuming a 32-MHz CLK clock.

17. Write an array interpreter (like a linked-list interpreter) and an array char *a[ ][ ][ ]* as in the first *main* procedure in §5.3.4, to control Figure 5.9's traffic light. For internal state *s* and input state *x*, the array *a[s,x][ ]* entries are next state, output state sent to the keypad data port, and time delay. Use a procedure *delay()* that delays 1 second. The main sequence is shown in Table 5.2. A late night sequence has north-south lanes red and east-west lanes yellow, both on for 1 second and off for 1 second. A fire truck emergency sequence is the north-south lanes red while the others are green for 20 seconds, then the east-west yellow for 2 seconds while north-south is still red, after which the north-south are green and the others are red for 10 seconds, and then the main or late-night sequence is resumed with its first line. Transitions to new sequences exhibit no time delay and output the output state of the sequence being begun. *KRD*'s input state is either 0 if neither late night nor emergency occur, 1 if an emergency occurs, and 2 if it is late at night and no emergency occurs. The emergency input state can last only one state transition to begin the emergency sequence. The other input states last as long as the sequences are to be executed. *enum* the main sequence states as *M0*, *M1*, *M2*, and *M3*, having values 0 to 3, the late night sequence as *N0* and *N1*, having values 4 and 5, and the emergency sequence as *E0*, *E1*, and *E2*, having values 6, 7, and 8.

18. Consider a vending machine controller. Its input port at location 0x1003 has value 0 if no coins are put into the machine, 1 if a nickel, 2 if a dime, 3 if a quarter, and 4 if the coin return button is pressed. The output port, at location 0x1003, will dispense a bottle of pop if the number 1 is output, a nickel if 2 is output, a dime if 3 is output, and a quarter if 4 is output. This vending machine will dispense a bottle of pop if 30 cents have been entered, will return the amount entered if the coin return button is pressed, and otherwise will keep track of the remaining amount of money entered (i.e., it will not return change if greater than 30 cents are put in). All control is done in software.

a. Show the logical design of the I/O hardware, assuming incompletely specified decoding if the program uses only addresses 0 to 0x80, 0x1003 (for these ports), and 0xFF00 to 0xFFFF, and the ports are only 3-bits wide. Use a minimum of 74HC04s and 74HC20s, and 74HC244 and 74HC374 for input and output port chips.

b. Show this controller's tabular and graphical sequential machine. Internal states S = {zero, five, ten, fifteen, twenty, twentyfive} will be the total accumulated money. Input states I = {B, N, D, Q, R} represents that no (blank) inputs are given; that a nickel, a dime, a quarter are given; or that the coin return button has been pressed. Output states O = {b,p,n,d,q} will represent the fact that nothing (blank) is done, or a bottle of pop, a nickel, a dime, or a quarter respectively, is to be returned. If multiple outputs are indicated, output a pop first, and go to the state that represents the amount of changes left in the machine, and output the larger coin first. Assume the coin return button is pressed repeatedly to return all the coins.

c. Show a self-initializing procedure *void seqMch( )* to implement this sequential machine by a linked-list interpreter, and show the linked-list as a three dimensional array, using an index as the link. Activate the solenoids for 0.1 seconds (assume procedure *void delay( )* causes a 0.1 second delay), and then release the solenoids that dispense the bottles and the money. Guard against responding to an input and then checking it again before it has been removed. (Hint: Respond to an input only when it changes from that input back to the blank input.)

d. As an alternative to part c, show a self-initializing procedure *void algor( )* to implement this sequential machine using arithmetic and conditional expressions in an 'algorithm'. Use the same assumptions as in part c. Note the ease or difficulty of modifying part c's state machine, or the code in part d, when the cost of a soda is changed to 35 cents (don't jump to conclusions).

19. Design a keyless entry module (KEM) for a car.

a. The KEM has five SPST switches (A to E); when any is pressed (only one can be pressed at a time) its corresponding binary number (A is 1, B is 2, C is 3, D is 4, E is 5) is input in negative logic to KRD, so, for instance if switch E is pressed, KRD bits 0 and 2 are asserted low (representing TLT or binary number 5). Show a logic diagram of an input state encoder that uses a 74HC08 to OR the switch signals into port input bits (in negative logic) and switches with pullup resistors.

b. The hardware state decoder of part a was found more costly than a software solution that inputs each key's signal directly to a KRD pin (switch A to bit 0, B to 1, C to 2, D to 3, E to 4). (Do not draw the logic diagram of this new hardware circuit.) Write *char getkey( )* that returns the currently pressed key's value (noKey returns 0, A returns 1, B - 2, C - 3, D - 4, E - 5) using this new hardware. (Hint: table-lookup may provide the shortest program, or you can give a shorter one.)

c. The KEM recognizes pressing A* C* D* B*, where A* means *char getkey( )* returns the value 1 (key A) one or more times, followed by one or more 0s (noKey). Give the Mealy model sequential machine table that recognizes when the code is entered. Let the input state **A** represent that *char getkey( )* returns the value A (1) one or more times, followed by one or more values noKey (0), etc. Input states **B**, **C**, **D**, and **E** are similarly defined. Let the internal state **a** represent that no keys have been recognized, that an illegal sequence has been input, or that an output pulse has been sent, **b** represent that the code A* is recognized, c represent that the code A* C* is recognized, and **d** represent that the code A* C* D* is recognized. Output state 1 indicates the whole code is recognized and a pulse should be generated, otherwise the output state is 0. Show the (Mealy model) sequential machine table.

d. Write a self-initializing procedure *void validate( )* that produces a 500 msec. positive pulse on KCD bit 0 when the sequence is recognized, then continues to check for the sequence, forever. You must use the sequential machine interpreter model, rather than if-then or case statements. The procedure must store the state machine in a compact global *char* vector *v[4,5,2]*, which you must show initialized in C, where the left index applies to the present internal state, the middle to the input state and the right to the output and next internal state. Use the *#define* statements to make your declaration more readable. Use a for-loop delay, assuming that a constant *N* in it will be defined later to provide the correct pulse length.

20. Using §5.3.5's procedure *short check( )*, write *#define* statements to generate patterns *A, B,..., Z* and test vectors *unsigned short v[ ]* to test the:

a. 74HC32 OR gate. b. 74HC04. c. 74HC10 NAND gate (see Figure 4.10).
d. 74HC138 decoder (see Figure 4.10). Follow this test procedure: for E1, E2 and E3 asserted, check for all combinations of A2, A1, A0; then for A2, A1, A0 all L (0) check for all combinations of E1, E2 and E3, but you do not check one of the patterns already checked in the first sequence (15 tests, rather than 64).
e. 74HC74 (see Figure 4.12). Follow this test procedure:

1. Assert S, with clock and D low. Check that Q and Q-bar are set.
2. Clock a 0 into both flip-flops. Check if Q and Q-bar are clear.
3. Drop clock with data high. Check to see if both flip-flops remain cleared.
4. Clock a 1 into both flip-flops. Check to see if both flip-flops are set.
5. Clear both flip-flops asserting only R, Check to see if both flip-flops are clear.

21. Write a self-initializing procedure *short ina(a)* for Figure 5.15d, similar to *out-a(a, d)* in §5.4.1, to return the 6818A's time of day, using indirect I/O. It should emulate Figure 5.15c's read cycle's timing diagram. However, if multiple outputs change at the same time, change *ds* first, *as* second, *rw* third, *cs* last. (Disregard a possible timing problem, in which you may be reading the 6818A's time when it is changing.)

22. An LCD display connects to KCD as in Figure 5.16, but the LCD's Rw signal is connected to KCD bit 6 instead of ground. When Rw and CLK are T (1) and Rs is F (0) the most significant bit outputs a busy bit which is T (1) when the display is not ready for a new command, changing to F (0) when a new command can be given. Write the initialization and output procedures, and procedures to read the busy bit to synchronize writing to the LCD: *char get4(char a)* returns a nibble, *char get(char a)* returns a byte, *put4(char d, char a)* outputs a nibble, and *put(char d, char a)* outputs a byte, where *d* is data and *a* is the LCD register accessed (if *a* is zero, Rs is 0, otherwise Rs is 1).

23. Write self-initializing procedures for the other configurations in Figure 5.18, using port S, but not using the ISPI. Assume address $a = 0$ selects the leftmost '589 or '595, 1 selects the middle '589 or '595, and 2 selects the right '589 or '595.

   a. For Figure 5.18a, *char serialIn(char a)* returns the byte input from the *a*th '589.

   b. For Figure 5.18a, *void serialIn(char *s)* puts three bytes input from the '589s into *s*.

   c. For Figure 5.18b, *void serialIn(char *s)* puts three bytes input from the '589s into *s*.

   d. For Figure 5.18c, *void serialOut(char c, char a)* outputs c to the *a*th '595.

   e. For Figure 5.18d, *void serialOut(char c, char a)* outputs c to the *a*th '595.

   f. For Figure 5.18d, *void serialOut(char *s)* outputs three bytes from *s* to the '595s.

24. The program at the end of §5.4.3 outputs a command to and inputs the measured temperature from the Dallas Semiconductor 1620. This chip has an output Tcom on pin 5, which can be used to control a heater or air conditioner, and registers TH for high and TL for low-temperature limits. Tcom becomes asserted when the measured temperature exceeds TH and becomes negated when the measured temperature goes below TL. Use §5.4.3's procedures *send* and *receive,* in both parts below:

   a. TH is written with the command 0x01 and read with the command 0xa1. Write a procedure *short putHi(short t)* that outputs t to TH and reads it back, returning 1 if the value read back is not the same as the value output, and 0 otherwise.

b. TL is written with the command 0x02 and read with the command 0xa2. Write a procedure *short putLow(short t)* that outputs t to TL and reads it back, returning 1 if the value read back is not the same as the value output, and 0 otherwise.

25. Repeat problem 24 using the ISPI to shift the data in and out, using §5.4.4's global variable ports and *enum*erated symbolic names. Data are shifted on the rising edge of SCLK, at 4 MHz, and SCLK is initially high.

26. Write C or C++ procedures *void main( )* to initialize the ISPI, for the MMC2001's 32-MHz CLK clock and output on SCLK, MOSI, and SS, as follows:

a. Use §5.3.4's global variable port names and *enum*erated symbolic names to shift on the falling edge of SCLK, at 4 MHz, with SCLK initially high. The ISPI outputs are open collector.

b. Use a vector *char spi[4];* to access all ISPI ports, to shift on the rising edge of SCLK, at 1 MHz, with SCLK initially high.

c. Use vector *char spi[4];* to access all ispi ports and give *#defines* equating *SPCR, SPICR, SPDR, SPSR, KCD* and *KCDD;* to *spi*'s elements, to shift on the falling edge of SCLK (initially low), at 250 KHz.

d. Use §5.3.4's struct *ISPI* to access ISPI control ports 1 and 2, to shift on the rising edge of SCLK, with SCLK initially high.

e. Use a pointer *SpiPtr* to §5.3.5.5's struct *ISPI* to access control ports 1 and 2, to shift on the rising edge of SCLK, with SCLK initially low.

27. In §5.3.6, The ISPI shifts most significant bit first. Rewrite its `put` and `serialXfr` function members to shift data least significant bit first.

28. Write a device driver and main routine calling it, to output the message 'Hi There', for a LCD (§5.3.2) that is similar to the device driver in §5.3.7, but using an 8-bit directly addressed LCD. Assume only that the port address is passed in the device-specific parameters.

29. Write a device driver and main routine calling it, to output the message 'Hi There', for an ISPI (§5.3.2) that is similar to the device driver in §5.3.7, but passing 8-bit data rather than 16-bit data through the ISPI.

6

Interrupts and Alternatives

The computer has to be synchronized with the fast or slow I/O device. The two main areas of concern are the amount of data that will be input or output and the type of error conditions that will arise in the I/O system. Given varying amounts of data and different I/O error conditions, we need to decide the appropriate action to be taken by the microcomputer program. This is studied in this chapter.

One of the most important problems in the design of I/O systems is timing. In §5.3.2, we saw how data can be put into a buffer from an input device. However, we ignored the problem of synchronizing with the source of the data, so we get a word from the source when it has a word to give us. I/O systems are often quite a bit slower, and are occasionally a bit faster, than the computer. A typewriter may type a fast 30 characters per second, but the MPU can send a character to be typed only once every 266,667 memory cycles. So the computer often waits a long time between outputting successive characters to be typed. Behold the mighty computer, able to invert a matrix in a single bound, waiting patiently to complete some tedious I/O operation. On the other hand, some I/O systems, such as hard disks, are so fast that a microcomputer may not take data from them fast enough. Recall that the time from when an I/O system requests service (such as to output a word) until it gets this service (such as having the word removed) is the latency. If the service is not completed within a maximum latency time, the data may be overwritten by new data and be lost before the computer can store them.

Synchronization is the technique used to get the computer to supply data to an output device when the device needs data, get data from an input device when the device has some data available, or to respond promptly to an error if ever it occurs. Over ten techniques are used to match the speed of the I/O device to that of the microprocessor. Real-time synchronization is conceptually quite simple; in fact we have already written a real-time program in the previous chapter to synchronize to a traffic light. Gadfly synchronization requires a bit more hardware, but has advantages in speed and software simplicity. Gadfly was actually used in synchronizing the ISPI. More powerful interrupt synchronization techniques — polled and vectored — require more hardware. Direct memory access and context switching are faster synchronization mechanisms. Shuttle, indirect, time-multiplexed, and video memories can be used for very fast I/O devices. These synchronization techniques are also discussed in this chapter. The next chapter continues this theme, to synchronize to very slow devices.

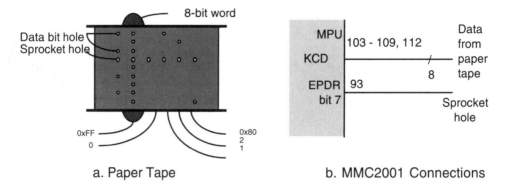

a. Paper Tape   b. MMC2001 Connections

Figure 6.1. Paper Tape Hardware

We first study the synchronization problem from the I/O device viewpoint, introducing busy/done states and terminology. An example of a paper tape reader, illustrating the collection of data from a tape into a buffer will be used to illustrate the different approaches to I/O synchronization. Paper tape (See Figure 6.1a.) is used in environments like machine shops, whose dust and fumes are hostile to floppy disks.

Data from the data port can be read just as in the previous chapter and will be put into a buffer, as we now discuss. The pattern of holes across a one-inch-wide paper tape corresponds to a byte of data; in each position, a hole is a true value, and the absence of a hole is a false value. Optical or mechanical sensors over each hole position connected to the port pins signal an H (T) if the sensor is over a hole. We will read data from MMC2001's *KCD*, to realize this collection. At any time, the values of such a pattern of holes under the paper tape head can be read by an instruction like *d = KCD*. It can be put in a buffer by a statement like: `for (pnt=buffer; pnt<buffer+0x100; ) *(pnt++) = KCD`; However, in the last chapter we ignored, and in this chapter we focus on, the problem of getting the data at the right time, when the hardware presents it. The user can advance the paper manually. In this example, we have to read one byte of data from the pattern of holes when the sprocket hole sensor finds a sprocket hole.

A simple but general model (a Mealy sequential machine) of the device describes how a computer synchronizes with an I/O device so it can take data from it or send data to it. (See Figure 6.2a.) In this model, the device (or equivalently, its object) has three states: the *IDLE*, *BUSY*, and *DONE states*. The device is in the IDLE state when no program is using it. When a program begins to use the device, the program puts it in the BUSY state. If the device is IDLE, it is free to be used, and, if BUSY, it is still busy doing its operation. When the device completely its operation, it enters the DONE state. Often, DONE implies the device has some data in an output port that must be read by the program. The state transition from BUSY to DONE is associated with the availability of output from the device to the program. When the program reads this data, it puts the device into IDLE if it does not want to do any more operations, or into BUSY if it wants more operations done. An error condition may also put the device into DONE and should provide some way for the program to distinguish between a successfully completed operation and an error condition. If

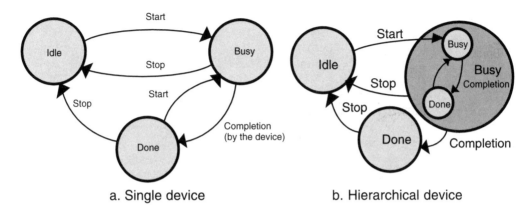

a. Single device   b. Hierarchical device

Figure 6.2. State Diagram for I/O Devices

the program puts the device into BUSY or IDLE; this is called, respectively, *starting* or *stopping* the device. The device enters DONE by itself. This is called *completing* the requested action. When the device is in DONE, the program can get the results of an operation and/or check to see if an error has occurred.

The IDLE state indicates the paper tape reader is not in use. The user starts the paper tape reader, putting it into the BUSY state. In that state, he or she pulls the paper tape until the next pattern is under the read head that reads a word from the tape. When the sprocket hole is under the tape reader or no paper is left, the reader reads a byte and enters the DONE state. The computer recognizes that when the reader is in the DONE state, data from the pattern should be read through the data port and put into the buffer in the next available location, or that an error condition might exist. Once read, if the program intends to read the next pattern because more words are needed to fill the buffer, it puts the reader back into the BUSY state to restart it. If another pattern should not be read because the buffer is full, the device goes to the IDLE state to stop or to ignore it. If an error condition is signaled, the device is left in the DONE state so it will not be used until the error is read and possibly fixed or reported. To read three bytes, the tape reader might pass through the following states: IDLE, BUSY, DONE, BUSY, DONE, BUSY, DONE, IDLE. (See Figure 6.2a.) Data are read each time the paper tape reader goes into the DONE state. Note that there is a difference between the IDLE state and the DONE state. In the DONE state, some data in the input port are ready to be read, and the I/O device is requesting the computer to read them, or an error has rendered the device unusable; while in the IDLE state, nothing is happening, and nothing need be done.

In some I/O systems which do not return values or error messages back to the computer, however, DONE is indistinguishable from IDLE, so only two states are required. Consider a paper tape punch. IDLE corresponds to when it is not in use. BUSY corresponds to when the program has put a byte out to it but the byte has not yet been completely punched. DONE corresponds to when the holes are punched, and another byte of data can be sent out. In this case, with no error conditions to examine in DONE, the DONE and IDLE states are indistinguishable, and we can say the device has just two states: IDLE and BUSY.

States can be hierarchically defined (Figure 6.2b). If the IDLE/BUSY/DONE states of Figure 6.2a synchronize the arrival of a byte in a serial input device, the BUSY state indicates that within a byte, bits have been requested but not all have arrived. As a bit is requested, a lower-level BUSY state (Figure 6.2b) is entered; when the bit arrives, a lower-level DONE state is entered. The lower-level state machine ping-pongs back and forth between the lower-level BUSY and DONE states until the whole byte has been serially input, while the higher-level state machine remains BUSY. Similarly, for the arrival of bytes described by the state machine in Figure 6.2a, there can be a higher-level state machine associated with the buffer into which the bytes are placed. This state machine is BUSY as long as the buffer is dedicated for input but not filled with data. This state machine is DONE when the entire buffer is filled and is ready to be used.

A typical microcomputer having several I/O devices has as many BUSY/DONE sequential machines, one for each device, and possibly a BUSY/DONE sequential machine for every buffer being emptied and filled. There exist product machines that can be used to describe interactions between devices. For instance, if a devices has states i1, b1, and d1, and another device has states i2, b2, and d2, then the product machine has states i1i2, i1b2, i1d2, b1i2, b1b2, b1d2, d1i2, d1b2, d1d2, where for instance i1b2 means that state machine 1 is in the IDLE and state machine 2 is in the BUSY state. In general, if there are n devices or buffers, there are 3^n states in their product machine. While many synchronization problems can be clearly defined for the simple state machine rather than the product machine, the product machine can define problems that occur when two or more devices or buffers interact, such that a condition arises only when each machine, and therefore the product machine, is in a specific state.

The general synchronization problem is to attend to a device or buffer when it enters its DONE state, in order to get input data, check for and respond to errors, or initiate an activity for future work. In this section, we consider realtime and gadfly synchronization techniques written in simple C and C++ object-oriented procedures.

6.1   Programmed Synchronization

Two ways a microcomputer can synchronize with a slower I/O device, real-time and gadfly synchronization, are programmed in the *main* procedure or in procedures it calls, and are studied below, as discussed in this section. These correspond to the two most fundamental synchronization principles: wait for (a time) and wait until (an event).

6.1.1   Real-time Synchronization

Real-time synchronization uses program timing delays to synchronize with the delays in the I/O system, to either equal or exceed the time a device is in BUSY. See Figure 6.3a. Two cases are: (1) using a procedure's inherent delays, and (2) using a wait loop

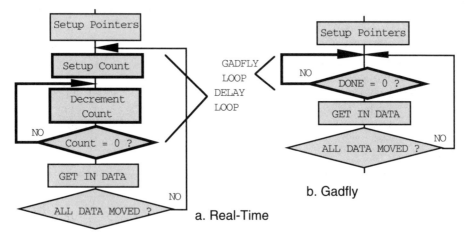

Figure 6.3. Flow Charts for Programmed I/O

or other time-consuming statements to "pad" the processor's delay to match or exceed the I/O device's BUSY state time. If that happens to be the rate at which the paper tape is pulled, and the procedure is started so as to pick up the first pattern of holes, putting it into *buffer[0]*, (a preposterous idea) the processor is synchronized with the device.

```
void main() {char buffer[0x100]; int i;
    for(i=0;i<0x100;i++) buffer[i] = KCD;
}
```

To illustrate the second case, we insert a delay loop as we did in the traffic light application in §5.3.3. We can modify the procedure as follows:

```
void main() {char buffer[0x100]; int i, j;
    for(i = 0; i < 0x100; i++) { j = N; while(−−j); buffer[i] = KCD;}
}
```

The statically efficient delay loop, $j = N$; *while(−−j)*; is empirically adjusted, by defining constant N, so that the loop executes during the BUSY state which is related to the rate at which the paper tape is pulled. A longer delay can be implemented using nested loops: $i = N1$; $j = N2$; *while(−−j) while(−−i)*; If the procedure is started so as to pick up the first pattern of holes, putting it into *buffer[0]*, (a nontrivial challenge) the processor is synchronized with the device. Because required program statement timing delays already synchronize the processor to the tape reader, we are also using real-time synchronization. This program's assembler language is shown below.

```
00000000   24F0    SUBI    R0,$10          ; allocate for saving registers
00000002   007D    STM     R13-R15,(R0)    ; save registers used in subroutine
00000004   3281    BGENI   R1,$08          ; generate 0x100 for allocation
00000006   0510    SUBU    R0,R1           ; allocate vector buffer
```

```
00000008  600E    MOVI    R14,$00     ; initialize i to 0
0000000A  664D    MOVI    R13,$64     ; delay loop counter
0000000C  240D    SUBI    R13,$01     ; count down
0000000E  2A0D    CMPNEI  R13,$00     ; until zero
00000010  E7FD    BT      *-4         ; abs = $0000000C loop
00000012  1207    MOV     R7,R0       ; get address of buffer
00000014  12E6    MOV     R6,R14      ; get i
00000016  1C76    ADDU    R6,R7       ; get address of buffer[i]
00000018  7506    LRW     R5,[*+24]   ; abs = $00000030 get addr. of KCD
0000001A  A705    LD.B    R7,(R5,0)   ; get data
0000001C  B706    ST.B    R7,(R6,0)   ; put into buffer[i]
0000001E  200E    ADDI    R14,$01     ; increment i
00000020  3287    BGENI   R7,$08      ; generate final count
00000022  0D7E    CMPLT   R14,R7      ; compare i to final count
00000024  E7F2    BT      *-26        ; abs = $0000000A loop back
00000026  3281    BGENI   R1,$08      ; generate 0x100 for deallocation
00000028  1C10    ADDU    R0,R1       ; deallocate vector buffer
0000002A  006D    LDM     R13-R15,(R0); restore registers used in subroutine
0000002C  20F0    ADDI    R0,$10      ; deallocate saved registers
0000002E  00CF    JMP     R15         ; return to caller
00000030  1000    #KCD                ; address of KCD
00000032  3006
```

In real-time synchronization, the device has IDLE, BUSY, and DONE states, but the computer may have no way of reading them from the I/O device. Instead, it starts operations and keeps track of the time in which it expects the device to complete the operation. Busy-done states can be recognized by the program segment being executed in synchronization with the device state. IDLE is any time before we begin reading the tape; the delay loop corresponds to BUSY; and DONE is when *buffer[i]* = *KCD;* is executed. The device is started, and the time it takes to complete its BUSY state is matched by the time a program takes before it assumes the device is in DONE. While an exact match in timing is occasionally needed, usually the microcomputer must wait longer than the I/O device takes to complete its BUSY state. In fact, the program is usually timed for the longest possible time to complete an I/O operation.

Real-time synchronization is considered bad programming by almost all computer scientists. It is rarely used in M·CORE microcontroller systems. Dynamic memories can require refresh cycles, interrupts, and DMA cycles discussed in §5.4.1, which can occur at unpredictable times. A cache memory supplying instructions or data can speed up the execution of instructions, thus affecting delays based on their execution. If timing delays are implemented in high-level languages such as a *while* loop in C, the delay time can change when a later version of the compiler or operating system is used. It is difficult to provide a delay of a fixed time by means of delays inherent in instruction execution. The effort in writing the program may be the highest because of the difficulty of precisely tailoring the program to provide the required time delay, as well as being logically correct. This approach is sensitive to errors in the speed of the I/O system. If some mechanical components are not oiled, the I/O may be slower than what the program is made to handle. The program is therefore often timed to handle the worst possible situation and is the least responsive synchronization technique.

However, real-time synchronization requires the least hardware; it can be used with a basic input or output device, without the need for further hardware. It is a practical alternative in applications such as the traffic light controller discussed in 5.2.2 or a microcomputer that is dedicated to control a printer.

6.1.2   Gadfly Synchronization

The sprocket hole input can be sensed in the *gadfly* synchronization technique to pick up the data exactly when they are available. The program continually "asks" one or more devices what they are doing (such as by continually testing the sprocket hole sensor). This technique is named after the great philosopher, Socrates, who, in the Socratic method of teaching, kept asking the same question until he got the answer he wanted. Socrates was called the "gadfly of Athens" because he kept pestering the local politicians like a pesky little fly until they gave him the answer he wanted (regrettably, they also gave him some poison to drink). This bothering is usually implemented in a loop, called a *gadfly loop*, in which the microcomputer continually inputs the device state of one or more I/O systems until it detects DONE or an error condition in one of the systems. (See Figure 6.3b.) Gadfly synchronization is often called polled synchronization. However, polling means sampling different people with the same question — not bothering the same person with the same question. Polling is used in interrupt handlers discussed later; in this text, we distinguish between a polling sequence and a gadfly loop.

The MMC2001 has a very simple edge port *EPDR* that is very useful for gadfly and interrupt synchronization. (See Figure 6.4.) Its bits can also be used in monitoring control switches and position sensors. Edge ports detect falling, rising, or both edges on its inputs, depending on a pair of consecutive bits in *EPPAR*. Consider input (*EPDDR* bit 0 is F) *EPDR* bit 0; if *EPPAR* bit 1 is T, then *EPFR* bit 0 sets upon a falling edge on *EPDR* bit 0 (pin 102), and if *EPPAR* bit 0 is T, then *EPFR* bit 0 sets upon a rising edge on *EPDR* bit 0 (pin 102). If both *EPPAR* bits 1 and 0 are T, *EPFR* bit 0 sets upon either edge. *EPFR* bit 0 is cleared by writing a 1 into it (it is a clear

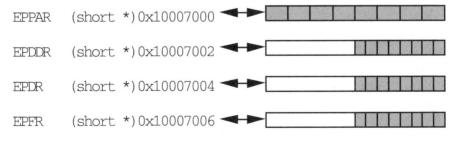

Figure 6.4. M·CORE Edge Ports

port). Similarly, each bit of *EPFR* is controlled by a pair of bits of *EPPAR*. *EPPAR* bits 15 and 14 control *EPFR* bit 7. A gadfly C procedure using bit 7, and its assembler language, follows:

```
volatile short EPPAR@0x10007000, EPDDR@0x10007002, EPDR@0x10007004,
    EPFR@0x10007006;
void main() {char buffer[0x100]; int i;
    for(KCDD = i = 0, EPPAR = 0x8000; i < 0x100; i++)
        {EPFR = 0x80; while ((EPFR & 0x80) == 0); buffer[i] = KCD; }
}
```

This program's assembly language is listed below:

```
000010B0   2470   SUBI    R0,$08          ; allocate for saving registers
000010B2   007E   STM     R14-R15,(R0)    ; save registers used in subroutine
000010B4   3281   BGENI   R1,$08          ; generate 0x100 for allocation
000010B6   0510   SUBU    R0,R1           ; allocate vector buffer
000010B8   6007   MOVI    R7,$00          ; initialize i to 0
000010BA   7610   LRW     R6,[*+64]       ; abs = $000010FC addr. direction reg.
000010BC   B706   ST.B    R7,(R6,0)       ; make direction input
000010BE   32F6   BGENI   R6,$0F          ; make 0x8000
000010C0   750E   LRW     R5,[*+56]       ; abs = $000010F8 addr. assign reg.
000010C2   D605   ST.H    R6,(R5,0)       ; make falling edge
000010C4   3276   BGENI   R6,$07          ; make 0x80
000010C6   750B   LRW     R5,[*+44]       ; abs = $000010F4 addr. flag register
000010C8   D605   ST.H    R6,(R5,0)       ; clear flag (clear port)
000010CA   3276   BGENI   R6,$07          ; make 0x80
000010CC   7409   LRW     R4,[*+36]       ; abs = $000010F0 addr. flag register
000010CE   A504   LD.B    R5,(R4,0)       ; get status register data
000010D0   0E65   TST     R5,R6           ; check bit 7
000010D2   EFFB   BF      *-8             ; abs = $000010CA gadfly loop
000010D4   1205   MOV     R5,R0           ; get address of buffer
000010D6   1C75   ADDU    R5,R7           ; get address of buffer[i]
000010D8   7305   LRW     R3,[*+20]       ; abs = $000010EC get addr. of KCD
000010DA   A403   LD.B    R4,(R3,0)       ; get data
000010DC   B405   ST.B    R4,(R5,0)       ; put into buffer[i]
000010DE   2007   ADDI    R7,$01          ; increment i
000010E0   3C16   LSLI    R6,$01          ; generate final count
000010E2   0D67   CMPLT   R7,R6           ; compare i to final count
000010E4   E7EF   BT      *-32            ; abs = $000010C4 loop back
000010E6   3281   BGENI   R1,$08          ; generate 0x100 for deallocation
000010E8   1C10   ADDU    R0,R1           ; deallocate vector buffer
000010EA   00CF   JMP     R15             ; return to caller
000010EC   1000   #KCD                    ; address of data register
000010F0   1000   #EPFR                   ; address of flag register
000010F4   1000   #EPFR                   ; address of flag register
000010F8   1000   #EPPAR                  ; address of assignment register
000010FC   1000   #KCD                    ; address of direction register
```

The gadfly loop, `while ((EPFR & 0x80) = = 0);` waits until *EPDR* bit 7 input falls, which is when a sprocket hole is about to pass from under its sensor. Data are picked up when this loop test fails, when the sprocket hole is still under its sensor.

The programs above exhibits the IDLE, BUSY, and DONE states. IDLE is any time before we execute the procedure and DONE is when `buffer[i] = KCD;` is executed, the same as in real-time synchronization. The device is BUSY when executing a gadfly loop. Gadfly exhibits lower latency than real-time synchronization, because the gadfly loop terminates exactly when the device is DONE, while the delay loop must delay for the worst-case time the device is BUSY. Real-time synchronization requires no hardware in addition to the basic I/O port, while gadfly synchronization requires hardware such as a edge flag bit. Both techniques generally loop while the device is in BUSY, and that can be a long time for slow I/O devices. Interrupts, discussed later, can free the processor to do other things while it waits for DONE.

6.1.3   Handshaking

In addition to synchronization, a device may require a *handshake* signal to control the hardware. For the paper tape reader, a handshake output bit R might be used to engage the motor pulling the paper tape; R would be asserted whenever the program decides to input the next byte. R can be a positive or negative logic level signal, asserted until the data are input, or a a positive or negative logic pulse asserted for, say, a microsecond. We now add handshake mechanisms to the paper tape readers using real-time and gadfly synchronization.

First, consider an M·CORE real-time paper tape reader having data input on a basic input port (figure 5.1), when *EPDR* bit 7 falls, using handshake signal on *EPDR* bit 6, which is pulsed high momentarily to request data.

```
void main() {char buffer[0x100]; int i, *ptr = (char *) 0x2d200000;
    EPPAR = 0x8000; // set flag on falling edge of bit 7
    EPDDR = 0x40;    // arrange to output port bit 6 as a handshake
    EPDR = 0; // initialize it low
    for(i=0; i < 0x100; i++) {
        EPDR = 0x40; EPDR = 0;   // pulse the handshake signal
        EPFR = 0x80; while( ! (EPFR & 0x80)); // gadfly sync
        buffer[i] = *ptr; // read port
    }
}
```

Note that, each time data are received, the handshake signal is high for about a 100 ns before the real-time delay, and the subsequent input operation.

Consider an MMC2001 gadfly paper tape punch that outputs data on *KCD* using real-time synchronization, and a handshake signal on *EPDR* bit 1, which becomes low to punch data, remains low for 1 msec., and returns high after the data is punched. We assume the MMC2001 uses a 32-MHz CLK and the while loop takes 3 memory cycles.

```
void main() {char buffer[0x100]; short i, j;
   KCDD = 0xff; EPPAR = 2; EPDR = EPDR = 2;   // hndshk out bit 1 initialized high
   for(i = 0; i < 0x100; i++) {
   EPDR = 0;    // begin handshake by asserting the signal low
   j = 32000/3; while( j-- ) ;   // real-time synchronization
   KCD = buffer[i];   // write data
   EPDR = 2;    // end handshake by negating the signal
   }
}
```

6.1.4   Some Examples of Programmed I/O

We have already shown real-time programming, with a traffic light controller, and
gadfly synchronization, with the ISPI. Here we will provide three more illustrative
examples of programmed I/O: generating infrared remote control signals, inputting
magnetic card reader signals, and generating BSR X-10 signals.

The familiar infrared remote control controls TV sets and other home electronics. A rather large number of deliberately different control formats are used so that
commands given to one unit will not be inadvertently obeyed by another. A typical
format sends data least significant bit first; for an F (0), KCD bit 0 is H (1) for 350
μsec.; for T (1) it is H (1) for 700 μsec. See Figure 6.5. A H (1) causes a 555 to
generate a 38 KHz square-wave. See Figure 6.5b. The procedure *sendIr* sends 11 bits
to the infrared LED.

```
void sendIr(int data) { char i;
   for(i = 0, KCDD = 1; i < 11; i++)
      {KCD | = 1; if(data & 1) wait(); wait();
      KCD &= ~1; wait(); data >>=1; }
}
```

```
void wait() {int i = N; while(--i) ;}    /* N is 350 * 8 / 3 to wait 350 μs. */
```

The also familiar magnetic credit card reader generates a falling-edge clock as it
inputs data bits serially, most significant bit first. See Figure 6.6. The procedure
receiveMagCard receives 16 bits serially, through *EPDR* bit 0.

```
int receiveMagCard (){ int i, data;
   for(i = 0, EPDR = 0; i < 16; i++)
      {while(!(EPDR&2)) ; while(EPDR&2) ; data=(data << 1) | (EPDR & 1);}
   return data;
}
```

An *X-10* controller (Figure 6.7a) and module (Figure 6.7b), originally designed
by BSR Inc., now distributed by Radio Shack, Sears, and other mail-order houses,
controls lamps and appliances in the home via a signal sent over household 110 volt
60-Hz "power wiring". One or more controllers and modules, each costing about

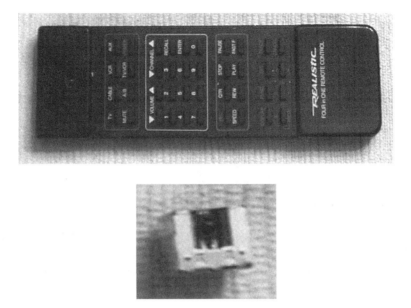

a. Remote Control and Receiver module

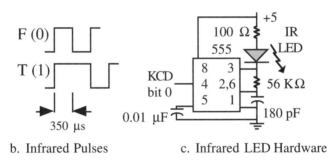

b. Infrared Pulses   c. Infrared LED Hardware

Figure 6.5. Infrared Control

$15, are plugged into power wiring sockets, without the need for any other wiring. A controller sends commands as bursts of 100-KHz signals over the power wiring to the modules (Figure 6.7d). One bit is sent each half-cycle of the *60Hz* signal. An F (0) bit is no burst, and a T (1) bit is three 100-KHz bursts evenly spaced in the half cycle. The TW523 module (Figure 6.7b), uses three optical isolators (*opto-isolators*) described in Figure 6.7e. One inputs a squared-up *60 Hz* wave and the other outputs a *signal*. Each side of Figure 6.7e's opto-isolators have separate ground and +5 V supplies, so 110 V power is not applied to the MMC2001. The TW523 safely isolates power from the computer and user. It provides a 60 Hz square wave and decoded signal input, and a microcontroller can drive it to produce the 100-KHz waveforms to output the signal.

a. Pin Connections

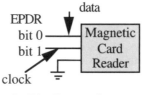

b. Pin Connections

Figure 6.6. Magnetic Card Reader

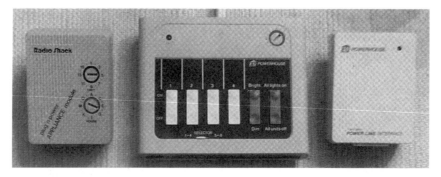

a. Bsr Receiver   b. BSR Controller   c. TW523 Module

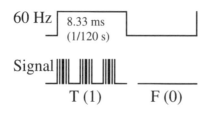

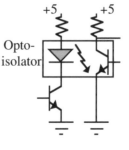

d. BSR X-10 Pulses   e. Bsr X10 Transmitter

Figure 6.7. BSR X-10

sendBsr sends 16 bits through *EPDR* bit 0, synchronized to *EPDR* bit 1.

```
#pragma NO_ENTRY
#pragma NO_EXIT
#pragma NO_RETURN
#pragma NO_FRAME
void pulse(void){ asm{

        subi    r0,16                   ; allocate space for saved registers
        stm     r12-r15,(r0)            ; save registers used in subroutine
        movi    r14,10                  ; count set for ten pulses
        lrw     r13,[L5]                ; get address of output port
L1:     movi    r12,1                   ; generate a 1
        st.h    r12,(r13,0)             ; write it to the output port
        lrw     r12,[L6]                ; get count for 100 KHz square wave
        setc                            ; preset carry for loopt
L2:     loopt   r12,L2                  ; loop until C clears, when r12 is 0
        movi    r12,0                   ; generate a 0
        st.h    r12,(r13,0)             ; write it into the output port
        lrw     r12,[L6]                ; get count for 100 KHz square wave
        setc                            ; preset carry for loopt
L3:     loopt   r12,L3                  ; loop until C clears, when r12 is 0
        subi    r14,1                   ; decrement outer loop counter
        cmpnei  r14,0                   ; until it becomes zero
        bt      L1                      ; loop until then
        lrw     r12,[L7]                ; count for waiting rest of 1/6 of cycle
        setc                            ; preset carry for loopt
L4:     loopt   r12,L4                  ; loop until C clears, when r12 is 0
        ldm     r12-r15,(r0)            ; restore registers used in subroutine
        addi    r0,16                   ; deallocate space for saved registers
        jmp     ;r15                    ; return to caller
L5:     dc.w    0x10007004              ; address of output port
L6:     dc.w    * 32                    ; count for 100 KHz square wave
L7:     dc.w    32000000/360 - 20 * ( 5 * 32) }}

void sendBsr (int data) { int i;
    while(EPDR & 2) ;   /* make sure we start on an edge */
    for(i = 0, EPDR = 1; i < 16; i++) {
        while( ! (EPDR & 2)) ;   /* wait for next 1/2 cycle */
        if(data & 0x8000) { pulse(); pulse(); pulse(); }
        data <<= 1;   /* next bit */
        while(EPDR & 2) ;   /* wait for next 1/2 cycle */
        if(data & 0x8000) { pulse(); pulse(); pulse(); }
        data <<= 1; /* next bit */
    }
}
void main(){ sendBsr (0x55aa); }
```

sendBsr illustrates some interesting techniques. We wait for either edge of the
60 Hz square wave because bits are sent each half cycle. The top half and bottom half
of *sendBsr* differ only in which edge the gadfly loop is waiting for. If a T (1) is sent,
then the procedure *pulse* is executed three times; If a F (0) is sent, then the procedure *pulse* is not executed. *pulse* illustrates the use of embedded assembly language to produce 10 cycles of 100 KHz square wave and a small delay.

6.1.5   Object-Oriented Classes for Programmed I/O

An object-oriented class *SyncPort* for real-time and gadfly synchronization mechanisms, derived from §5.1.2's generic *Port* class, is presented below. It is generalized
to accommodate the handshake functions of the Motorola 6811 microcontroller's
StrA and StrB pins and associated control logic, which are not implemented in
hardware in the MMC2001. This class accommodates the synchronization techniques
that use programmed I/O (i.e. real-time and gadfly synchronization).

There are two parts to this general synchronization mechanism, a first part in
which an input signal edge or delay that indicates a device's transition from BUSY to
DONE, and a second part which is an output signal that can provide a pulse or a
level handshake signal that prods the device to do something.

To accommodate various synchronization and handshake options, we pass, to
SyncPort's constructor, a pointer to a *struct* of type *SyncCtl*. The constructor
saves this pointer. This *struct* 's *Obit* is an output bit pattern, in which exactly
one bit is 1 (T), which is used for the handshake output on *EPDR*. *Ibit* is an intput
bit pattern, in which exactly one bit is 1 (T), which indicates which bit of *EPFR* is
used in gadfly synchronzation, or is all zeros, which indicates real-time synchronization is used. If gadfly is selected, *DlyEdge* is the pattern ORed into *EPPAR* to
select the edge used; if real-time is selected, *DlyEdge* is the real-time delay. Bits *E1*,
E2, *E3*, *E4*, and *E5*, if set, cause the handshake output bit to rise or fall at
different times. *E1* causes the input to rise first, *E2* causes the input to fall next, *E3*
causes the input to rise next, *E4* causes the input to rise after the gadfly or real-time
loop is done, *E5* causes the input to fall last. For instance, to pulse the output high
just before the loop is entered, assert *E1* and *E2*, to hold the output low while the
loop is executed, assert *E2* and *E4*.

```
typedef struct
    { long DlyEdge, char Obit, Ibit, E1:1, E2:1, E3:1, E4:1, E5:1; } SyncCtl;

template <class T> class SyncPort : public Port<T> {

    protected: SyncCtl *ctl;

    void preSync (void) {
        if(ctl->Ibit) EPFR = ~ctl->Ibit; if(ctl->E1) EPDR | = ctl->Obit;
        if(ctl->E2) EPDR &= ~ctl->Obit; if(ctl->E3) EPDR | = ctl->Obit;
        if(ctl->Ibit) while(EPFR & ctl->Ibit);  /*gadfly loop*/
        else {long count = ctl->DlyEdge; while(count--) ; }   /*delay loop*/
}
```

```
void postSync(void)
   {if(ctl->E4) EPDR | = ctl->Obit; if(ctl->E5) EPDR &= ~ctl->Obit;}

public : SyncPort(SyncCtl *ctl,long a,char attr=0, T dir=0) :
     Port(a, attr, dir) { this->ctl = ctl;
     if(ctl->Ibit) EPPAR | = ctl->DlyEdge;    /*set rise/fall*/ EPDR = Obit;
     if(ctl->E1)EPDR &= ~ctl->Obit; else EPDR | = ctl->Obit;
     }
   }

   virtual T get(void){T c; preSync(); c=Port<T>::get(); postSync();}

   virtual void put(T c) { preSync(); Port<T>::put(c); postSync(); }
};
```

We now redo the examples in §6.1.3 to use the class *SyncPort*. First, consider the gadfly paper tape reader having data input on an input port at 0x2d020000 (Figure 5.1), syncronized on the falling edge of *EPDR* bit 1, using handshake signal on *EPDR* bit 0, which is pulsed high momentarily to request data. *SyncCtl*'s *Obit* is set to 0x01, *Ibit* is set to 0x02, *DlyEdge* is set to 8L, and *E1* and *E2* are set. The object's second argument is the address of the data port which is 0x2d020000. It uses no output control or direction port, so the third and fourth parameters are zero (so they are omitted).

```
SyncCtl ctl = {0x00000008L, 0x01, 0x02, 1, 1, 0, 0, 0 );
SyncPort<char> p(&ctl, 0x2d020000};

void main() {char buffer[0x100]; p.get(b, 0x100) }
```

Note that since the class's function members do all the work, the program itself is trivially simple. Now consider our gadfly paper tape punch whose data are outputs on *KCD*, using *EPDR* bit 3 for synchronization, which pulses low before data are punched, and returned high after the data is read. The punch synchronized using a real-time delay of 1 msec. which we assume takes a count of 4000. The constructor's first argument is the synchronization parameter. In it, *EPDR* bit 3 is to be output, which is initially high, becoming low before synchronization and going high after data are punched. The constructor's second argument is the address of the data port which is *0x10003006*. Its direction register is two locations lower than its data port address, and its direction is output. Note that this *main* procedure for punching tape using real-time synchronization is very similar to the *main* procedure for the paper tape reader using gadfly synchronization because the class's function members do all the work. In each case, we merely declare the object, which invokes the constructor, and use vector put and get member functions.

```
SyncCtl ctl = {4000L, 0x04, 0, 0, 1, 0, 1, 0 );
SyncPort <char> p(&ctl, 0x10003006, offset2, 0xff};

void main() {char buffer[0x100]; p.put(b, 0x100); }
```

These examples suggest that a library of devices and their associated classes can be written and debugged in advance, and copied into the application late in the design. Object-oriented programming separates the I/O software from the main procedure to make such a substitution cleanly and correctly, without side effects.

6.1.6   Device Drivers for Programmed I/O

We now develop a device driver to use real-time or gadfly synchronization. Essentially, we put §6.1.5's *preSync()* and *postSync()* procedures into §5.1.7.2 device driver's *svcFun()*.

Global variables, *#includes*, *#defines*, *structs ucd0*, *charDev*, *function0*, and procedures *put()*, *get()*, and *main()* are the same as in §5.1.7.1. However, the device-specific parameter *struct* is augmented to provide for the synchronization options.

```
typedef struct param0_s { char *port, *dir, initdirection;
    long DlyEdge; short Obit, Ibit, E1:1, E2:1, E3:1, E4:1, E5:1;
} param0_t;
```

We first present the paper tape reader's device-specific parameter. Then we list the device driver procedures. Finally we show the paper tape punch's device-specific parameter. We list the driver functions in the middle because we want to keep them on one page.

This paper tape reader has data input on *KRD* at *0x10003007*; it is synchronized on the falling edge of *EPDR* bit 1, and uses *EPDR* bit 0 to handshake, which is pulsed high momentarily to request data. We initialize the device-specific parameters thus:

```
param0_t param0 = { (char *)0x10003006, (char *)0x10003004, 0,
    0x00000008L, 0, 0x08, 1, 1, 0, 0, 0};
```

The device driver procedures are listed below. Note that they are very similar to the driver in §5.1.6.2, with §6.1.5's member functions inserted to provide synchronization.

```
extern int32 initFun(void);extern void svcFun(ucd_t *, int, cdev_t *);
#define _PIO_DIRECTION ((u_int32) 0x0000000a) /* comnd to set dir of port */

param0_t param0=
    { (char*)0x10003007,(char  *)0x10003005,0,4000L,0x01,
    0x02,1,1,0,0,0};
void preSync(param0_t *pptr) {
    if(pptr->Ibit) EPFR = pptr->Ibit; if(pptr->E1) EPDR | = pptr->Obit;
    if(pptr->E2) EPDR &= ~pptr->Obit; if(pptr->E3) EPDR | = pptr->Obit;
      if(pptr->Ibit) while(EPFR & pptr->Ibit);   /*gadfly loop*/
      else {long count = pptr->DlyEdge; while(count--) ; }   /*delay loop*/
}
void postSync(param0_t *pptr)
    {if(pptr->E4) EPDR = pptr->Obit; if(pptr->E5) EPDR &= ~pptr->Obit;}
```

```
int32 initFun( ucd_t *ucp ) {/* arg. pointer to unit control data */
    param0_t *ddptr = (param0_t *)ucp->ddp;  /* -> device-specific parameters */
    if(ddptr->dir) *(ddptr->dir) = ddptr->initdirection;
    if(ddptr->Ibit) EPPAR | = ddptr->DlyEdge; EPDR=ddptr->Obit;
    if(ddptr->E1)EPDR &= ~ddptr->Obit; else EPDR | = ddptr->Obit;
    return SUCCESS;
}


void svcFun( ucb_t *ucb, int fun, cdev_t *dparams ) {
    ucd_t *ucd = &ucb->ucd;  /* find initFun's argument */
    param0_t *pptr = (param0_t *)(ucd->ddp);  /* get device-specific args */
    switch(fun){
      case _PIO_READ:
          while(dparams->lub--) {
              preSync(pptr);
              *dparams->aub++ = *pptr->port;
              postSync(pptr) ;
          }
          break;
      case _PIO_WRITE:
          while(dparams->lub--) {
              preSync(pptr);
              *pptr->port = *dparams->aub++;
              postSync(pptr);
          }
        break;
      case _PIO_DIRECTION: *pptr->dir = *(dparams->aub); break;
      default: _drv_pio_done(ucb, EOS_BAD_PARAM, 0); return;
    }
    _drv_pio_done(ucb, SUCCESS, 0);   /* special call to end service */
}
```

The paper tape punch has data output on *KCD*, whose direction *KCDD* is initialized to 0cff; it pulses *EPDR* bit 3 low before data are punched, and restores it
after 1 ms.

```
param0_t param0 = {(char *)0x10003006, (char *)0x10003004, 0xff,
    0x00000008L, 0, 0x08, 0, 1, 0, 1, 0};
```

This device driver illustrates programmed I/O synchronization. It is a realistic
device driver, suitable for paper tape readers and punches, but it does not take
advantage of timer and interrupt signal synchronization avaliable in Ariel. Using
these synchronization mechanisms, we can avoid tying up the computer in a delay
loop or a wait loop. We discuss these improved synchronization techniques in later
sections.

6.2   Interrupt Synchronization

In this section, we consider interrupt hardware and software. We first look at the
MMC2001 MPU and interrupt controller. Then we go through the steps in an
interrupt. Then we consider interrupt handlers, and the accommodation of critical
sections. Finally, multiple interrupts are discussed, using two techniques called
polling and vectored interrupts, which are simple extensions of the single interrupt
case.

Interrupt techniques can be used to let the I/O system interrupt the processor
when it is DONE, so the processor can be doing useful work until it is interrupted.
Also, latency times resulting from interrupts can be less than latency times resulting
from a large gadfly loop that tests many I/O devices, or a variation of the gadfly
approach, whereby the computer executes a procedure, checks the I/O device, next
executes another procedure, and then checks the devices, and so on — and that can
be an important factor for fast I/O devices. Recall the basic idea of an interrupt from
§1.2.3: that a program P currently being executed can be stopped at any point, then a
"handler" program H is executed to carry out some task requested by the device and
the program P is resumed. The device must have some logic to determine when it
needs to have the processor execute the H program. P must execute the same way
whenever and regardless whether H is executed. All the information that P needs to
resume without error must be saved.

The MMC2001 interrupt controller can request *normal* or *fast* interrupts. While
both mechanisms have the same latency, fast interrupts can be handled while
normal interrupts are disabled or are being handled, so they can be reserved for
devices that require lower latency. When the MPU gets a normal interrupt, the
32-bit word at address 0x28 + the contents of the vector base register (VBR or
CR1), is put in PC. Fast interrupts load the 32-bit word at address VBR + 0x2c
into the PC.

Upon reset, the exception vector base register VBR (cr1) is cleared to read the
reset vector, and the low end of the memory map is in ROM, so the interrupt vectors
are initially in ROM and cannot be modified. But the VBR can be loaded with a base
address that is in RAM, so that the interrupt handler address can be written at run
time. Other interrupt and exception vectors should be copied from the old vector in
ROM to the new vector in RAM, to be able to trace, or handle program exceptions
such as divide by zero.

The interrupt handler can be executed using the alternative set of registers
(Figure 2.1). In order to do this, the alternate stack pointer, r0', needs to be set up to
point to a stack buffer area in RAM, which is to be used by interrupt handlers.

```
#define BASE r2
#define STACK r3
#define HANDLER r4
#define OFFSET r5
#pragma NO_ENTRY
#pragma NO_FRAME
#pragma NO_RETURN
```

```
void initInt(long b,long s,void (*h)(),unsigned char o){ asm{
  mfcr       r1,cr1          /* get previous exception vector */
  mtcr       BASE,cr1        /* put new exception vector */
  movi       r7,31           /* use 127 to move vectored interrupt vectors as well */
  setc                       /* set condition so first loopt will branch back */
Loop: ld.w   r6,(r1,0)       /* get copy of a vector */
  st.w       r6,(r2,0)       * put copy into new vector */
  addi       r1,4            /* next source vector */
  addi       r2,4            /* next destination vector */
  loopt      r7,Loop         /* until all 128 vectors are copied */
  mfcr       r1,cr1          /* get exception vector */
  st.w       STACK,(r1,0)    /* put auxiliary stack pointer into reset vector */
  mfcr       r7,cr0          /* get PSW */
  bseti      r7,1            /* set A to access auxiliary register set */
  mtcr       r7,cr0          /* put back PSW */
  mfcr       r6,cr1          /* get current exception vector again */
  ld.w       r0,(r6,0)       /* get copy of stack pointer */
  mfcr       r7,cr0          /* get PSW */
  bclri      r7,1            /* clear A to access normal register set */
  mtcr       r7,cr0          /* put back PSW */
  mfcr       r6,cr1          /* get current exception vector */
  lsli       OFFSET,2        /* convert fourth operand, interrupt number, to offset */
  addu       OFFSET,r6       /* add vector base address, get addr of interrupt vector */
  addi       HANDLER,1       /* set lsb to handler address, to use auxiliary registers */
  st.w       HANDLER,        /* insert handler address as interrupt address */
             (OFFSET,0)
  movi       r7,0x4C         /* generate relative offset for trap #3 */
  addu       r6,r7           /* combined address of trap #3 vector */
  lrw        r7,[aTrap]      /* get address of trap handler */
  st.w       r7,(r6,0)       /* put in exception vector */
  jmp        r15             /* return from subroutine */
trap3:mfcr   r1,cr4          /* get previous return address */
  mfcr       r7,cr2          /* get previous saved PSW */
  bseti      r7,31           /* set S to put in supervisor mode */
  mtcr       r7,cr2          /* put new saved PSW */
  addi       r1,2            /* move past trap opcode */
  mtcr       r1,cr4          /* put new return address */
  rte
aTrap: dc.w trap3
}}
```

Additionally, we may wish to execute I/O routines in supervisor mode in order to get at some registers that are only accessible in the supervisor mode. A short trap #3 handler will be used to switch into the supervisor mode, as discussed shortly.

The procedure *initInt()* above loads the VBR with its first argument, copies the previous interrupt vectors to their new location, installs a new interrupt handler, the third argument, into the new interrupt vector at an offset, which is the fourth argument, and initializes the alternate stack vector to the second argument. It will also install a trap #3 handler to permit switching to the supervisor mode. For example,

initInt(0x30000000, 0x30000200, handler, 10);

will put *0x30000000* into the VBR and copy 32 vectors that were pointed to by the VBR into the new vector at *0x30000000,* put the address of the interrupt handler *handler* at the *10*th vector (location *0x30000028*), and initialize the auxiliary stack to *0x30000200.* This prepares normal interrupts to be handled by *handler()* using the auxiliary set of registers, and the execution of trap #3 to enter the supervisor mode. We will consistently use this procedure before any normal interrupts are used. Fast interrupts can similarly be initialized by making the last argument *11* instead of *10.* Two or more executions of *initInt()* can be used to set up two or more interrupts. Alternatively, the second interrupt can be directly installed in the new interrupt vector. For instance, **(long *)0x3000002C=(long)fastHanlder+1;* can be used, after *initInt()* has been executed, to set up the fast interrupt handler.

In *initInt()* *#pragma NO_ENTRY* prevents the setup of entry code that saves the registers used in the subroutine, *#pragma NO_FRAME* prevents the setting of the stack frame, and *#pragma NO_RETURN* prevents putting the JMP 15 (RTS) instruction at the end of the procedure, since it is explicitly put in the body of the procedure.

The procedure above must be executed in supervisor mode. While it is possible to use supervisor mode for all code execution, it is not difficult to use supervisor mode only for I/O code execution, and use user mode for other program execution. To switch to user mode, you can execute the procedure *toUser()* shown below.

```
#pragma NO_ENTRY
#pragma NO_FRAME
void toUser(void) { asm{
    MFCR R1,CR0    /* get PSW */
    BCLRI R1,31    /* clear S bit */
    MTCR R1,CR0 /* put back PSW */
}}
```

To enter supervisor mode, you can execute asm *trap #3.* This will execute the trap handler that is located in the *initInt()* procedure, which sets the S bit in the saved PSW, which becomes the PSW after *trap #3* is executed. (Note that this embedded assembler language directive must be at the end of a line, with no text following it.) At the beginning of each I/O routine, execute asm *trap #3,* and at the end of each such routine, call *toUser()* to let the rest of the code execute in user mode.

6.2.1   MMC2001 Interrupt Controller

Figure 6.8 shows MMC2001 Interrupt Controller Ports. Normal and fast interrupts each have their own sets of ports. Figure 6.8a gives the five 32-bit control and status ports, *INTSRC, NIER, FIER, NIPND,* and *FIPND.* and Figure 6.8b shows how the bit positions in each port are assigned to each interrupt source. A T bit read from *INTSRC* indicates an interrupt request. *NIER* indicates the normal interrupt enable, and *NIPND* indicates the normal interrupt pending. *FIER* indicates the fast interrupt enable, and *FIPND* indicates the fast interrupt pending. The ports can be declared as shown below; these declarations can be included in each program that uses these ports.

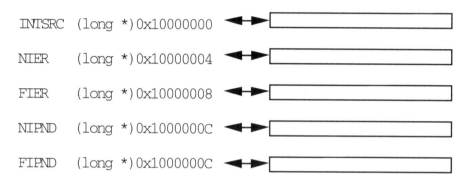

```
INTSRC   (long *) 0x10000000
NIER     (long *) 0x10000004
FIER     (long *) 0x10000008
NIPND    (long *) 0x1000000C
FIPND    (long *) 0x1000000C
```

a. Port Locations

b. Port Bits

Figure 6.8. MMC2001 Interrupt Controller Ports

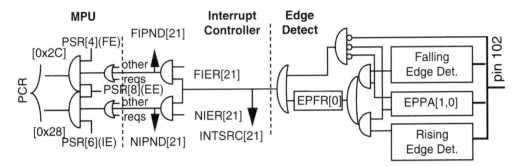

Figure 6.9. INTO Hardware

```
volatile long INTSRC@0x10000000, NIER@0x10000004,
   FIER@0x10000008, NIPND@0x1000000C, FIPND@0x1000000C;
```

We shall use *INTO*, which is bit 21, and later use the nearly identical *INT1*, which is bit 22, in the following examples. *INTO*'s hardware path is shown in Figure 6.9. As discussed in §6.1.2, depending on the bits *EPPA[1,0]*, a rising or falling edge on pin 102 (*INTO* input pin) will set *EPFR[0]*, which can be read, and cleared by writing a T into it (it is a clear port). If *EPPA[1,0]* is not FF, this flip-flop's value is also read as *INTSRC[21]*; if *NIER[21]* is T, it is the value read as *NIPND[21]*, and if *FIER[21]* is T, it is the value read as *FIPND[21]*. (If *EPPA[1,0]* is FF, *INTSRC[21]* is the complement of pin 102.) If *NIPND[21]* is T, and processor status register *PSR* bits 6 and 8 are T, then a normal interrupt will occur. This causes the long word at VBR + 0x28 to be loaded into the program counter, and the number 21 to be put into *PSR[22 to 16]*. If *FIPND[21]* is T, and processor status register *PSR* bits 4 and 8 are T, then a fast interrupt will occur. This causes the long word at VBR + 0x2C to be loaded into the program counter, and the number 21 to be put into *PSR[22 to 16]*.

6.2.2   Steps in an Interrupt

We now consider an M·CORE microcontroller that has just one interrupt, due to a falling edge on the edge port bit 0. We will assume that *EPPA[1,0]* is set to FT to detect falling edges, that *NIER[21]* is set to enable edge port bit 0 interrupts, and processor status word *PSR* bits 6 and 8 are set to permit normal interrupts. Figure 6.9's hardware simplifies to that shown in Figure 6.10. If, in our example, the paper tape were pulled though the reader, the sprocket hole signal, attached to pin 102, would be a train of 256 pulses, which falls and then rises each time a hole is under the sensor. A falling edge causes an edge interrupt. The six-step sequence of actions, leading to an interrupt and servicing it, are outlined below.

1. When the external hardware determines it needs service either to move some data into it or out of it or to report an error, we say that *the device requests an*

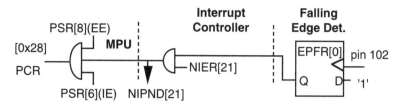

Figure 6.10. Simplified Edge Interrupt Request Path

interrupt. This occurs in the paper tape reader when the signal *EPDR* bit 0 (pin 102) falls.

2. If *EPDR* bit 0 are configured to sense falling edges by putting TF in *EPPA* bits 1 and 0, and *NIER* bit 21 is T, *edge* port *device* 0 is *enabled* for falling edge interrupts.

3. If the MPU's program status word (*PSW*) bits 6 (*IE*) and 8 (*EE*) are T, we say the *MPU is enabled for normal interrupts*, otherwise the *MPU is masked (or disabled) for normal interrupts*. *IE* and *EE* are also controlled by hardware in the next step. If a signal from (any) device is sent to the controller, we say the *MPU sees a request*, or *a request is pending*, and an interrupt will occur, as described below.

4. The MPU cannot stop in the middle of an instruction, whenever *PSW* bit 7 (*IC*) is clear. Therefore, if the MPU recognizes an interrupt, it *honors a normal interrupt* at the end of the current instruction; when it does so, it saves the PSW in special register *CR2* and the program counter in special register *CR4*. It then clears *PSW* bit 8 (*EE*) and *PSW* bit 6 (*IE*) to disable exceptions and normal interrupts, sets *PSW* bit 31 (*S*) to put the MPU in supervisor mode, clears *PSW* bits 15 and 14 (*TM*) to prevent tracing and *PSW* bit 12 (*TC*) to inhibit virtual memory address translation, and loads the 32-bit word at VBR + 0x28 into the program counter to process this interrupt. Importantly, *PSW* bits are changed <u>after</u> the former *PSW* value was saved in *CR2*.

5. Location 0x28 (plus the contents of control register *CR1*, which is the Vector Base Register VBR). contains the address of the normal interrupt *handler*. The handler is like a "subroutine" which performs the work requested by the device. It may move a word, or it may report or fix up an error. One of a handler's critically important but easy to overlook functions is that it must explicitly remove the cause of the interrupt (by negating the interrupt request) unless the hardware does that for you automatically. This is done in our case by writing 1 into bit 0 of the *EPFR* port.

6. When it is completed, the handler executes an RTE instruction; this restores the processor status register and program counter to resume the program where it left off.

Each interrupt handler's address must be put in a 32-bit word in memory where the interrupt mechanism reads it. Different interrupts have different interrupt numbers (fully described in Table 6.1). The number for a normal interrupt is 10. As described in step 5 above, when an interrupt having number n occurs, the interrupt mechanism adds $4*n$ to the vector base register VBR to read four bytes. These bytes are the address of a handler; they are read into the program counter. These bytes should be in RAM, so that the program can write them with the address of its handler, rather than in ROM, where the debugger program usually puts them. `initInt()` listed at the beginning of §6.2, initializes the VBR and copies previously defined interrupt handler addresses; it can, and should, be used to insert the first interrupt. For example, `initInt(0x30000000, 0x30000200, handler, 10);` puts the address of the interrupt handler `handler` at the 10th vector (location `0x30000028`). Additional interrupts can be written in a RAM location set up by `initInt()`, by simply writing the addresses at the offset corresponding to the interrupt number. For instance, `*(long *)0x3000002C = (long)fastHanlder+1;` can be used, after `initInt()` has been executed, to set up the fast interrupt handler, `fastHanlder`, which is interrupt number 11. In most examples in this and later chapters, we have only one interrupt, which is written using `initInt()`.

We note that whatever means are used to put the handler's address in the interrupt vector, the keyword `interrupt` should be written before the handler procedure name `handler` in order to replace all `rts` (`jmp r15`) instructions with `rte` instructions.

We stress that as soon as it honors an interrupt seen on a line, the MMC2001, like most computers, clears the `IE` and `EE` bits to prevent honoring another interrupt from the same device, or any kind of an exception. If it did not inhibit interrupts, the handler's first instruction would be promptly interrupted — an infinite loop that will fill up the stack. You do not have to worry about setting `IE`, because the preceding sequence's step 6 restores the condition code register, including the `IE` and `EE` bits, to the values they had before the interrupt was honored. The M·CORE edge interrupt, and most other device interrupts need to clear the source of the interrupt before RTE is executed. For these devices, if the RTE is executed and the source of the interrupt is not cleared, this same device will promptly interrupt the processor again and again — hanging up the machine. Before the handler executes RTE or changes `IE`, it *must* remove the interrupt source!

An interrupt request takes two steps. A flip-flop triggers if an edge occurs on its input. A switch in series with an input that can set this flag is called an *arm*; if it is closed the device is *armed*, and if opened, the device is *disarmed*. Any switch between the flip-flop and the M·CORE controller is called an *enable*; if all such switches are closed the device is *enabled*, and if any is opened, the device is *disabled*. Arming a device lets it record a request to request an interrupt, either immediately if it is enabled, or later if it is disabled. Disarming a device prevents honoring an interrupt now or later. Disable an interrupt to postpone it, if you can not honor it now, but you may honor it later when interrupts are enabled at a later time.

The programs above exhibits the IDLE, BUSY, and DONE states. IDLE is any time before we execute the procedure and DONE is when an interrupt occurs. The device is BUSY when waiting for an interrupt. Interrupt exhibits longer (worst case) latency than gadfly synchronization, because the instruction must complete, the registers must be saved on the stack, and a few instructions in the handler must be executed before the device gets served (by reading data from the port or writing data to the port).

Normally, if *PSW* bit 7 (*IC*) is clear, upon execution of long instruction might have just begun execution when an interrupt occurs. However, if *PSW* bit 7 (*IC*) is set, these instructions can be interrupted within their execution cycles. Further, if other devices use interrupts, a device's interrupt can be disabled for a long time while executing the other device's handler.

Interrupt synchronization requires more hardware such as a line (and pin) for the interrupt request, and a gate (and flip-flop) to enable the interrupt. However interrupt synchronization can avoid the loop while the device is in BUSY, and that can be a long time for slow I/O devices. This can free the processor to do other things while it waits for DONE.

6.2.3   Interrupt Handlers and Critical Sections

Three techniques are generally used in interrupt handlers: altering a global variable, writing into a buffer, and pushing into a queue. These are illustrated with the paper tape example. We then discuss the problem of critical sections, and how to manage them.

6.2.3.1   A Handler That Changes a Global Variable

First, we use a global variable *char flag*, which the *main* program gadflies upon. main inhibits interrupts using procedure *disableInt();. main* enables and clears the edge interrupt. Interrupts are then enabled, using procedure *enableInt();.* These procedures, shown below, take care of a critical section, as discussed at the end of this subsection. flag is cleared. Upon each falling edge, indicating the arrival of another pattern, an interrupt occurs and *flag* is incremented. Each time *main* sees *flag* nonzero, input data is transferred into the buffer. Then the edge interrupt is disabled and we exit.

In this example, we use the function *enableInt* to enable interrupts and the function *disableInt* to disable interrupts. These two functions are specified with pragmas so that they do not save and restore registers; these procedures will be used in subsequent examples whenever interrupts are to be enabled or disabled.

```
#pragma NO_ENTRY
#pragma NO_FRAME
void enableInt(void) { asm{
    MFCR R1,CR0 /* get PSW */
    BSETI R1,6 /* set IE */
    BSETI R1,8 /* set EE */
    MTCR R1,CR0 /* put back PSW */
}}
```

```
#pragma NO_ENTRY
#pragma NO_FRAME
void disableInt(void) { asm{
    MFCR R1,CR0 /* get PSW */
    BCLRI R1,6 /* clear IE */
    MTCR R1,CR0 /* put back PSW */
}}
```

The first subsection shows a program whose interrupt handler sets a flag variable
in an M·CORE microcontroller. The interrupt handler is coded in assembler language as shown below the C++ program. Note the RTE instruction at the end.
Also, note that the handler does not save the registers it uses. The alternate set of
registers is used in the handler, while the regular set of registers is used in the
program that was interrupted. The procedure initInt(); generates an odd interrupt vector, which, when this vector is read after an interrupt, causes the handler
to use the alternate set of registers.

```
char flag;
interrupt void keyInt(){EPFR = 1;    /*clear interrupt edge flip-flop*/ flag++; };
void main() { char buffer[10], i;
    disableInt(); NIER |= 1 << 21;   /* enable edge interrupt for bit 0 */
    EPPAR = 1;    /* interrupt on bit 0 rising edge */ EPFR = 1;    /* clear flag */
    initInt(0x30000000, 0x30000200, keyInt, 10);   /* set up handler's addr */
    KCDD = 0;    /* make keyboard column inputs */
    enableInt();    /* enable normal interrupts and exceptions */
    for(KCDD = i = 0; i < 10; i++) {
        flag = 0;    /* prepare for next wait for interrupt */
        while(flag == 0) ;    /* wait for interrupt */
        buffer[i] = KCD;    /* store data */
    }
    NIER &= ~(1 << 21);    /* disable edge interrupt for bit 0 */
}
```

The interrupt handler is compiled into assembly language as shown below.

```
300010B4 6017  MOVI   R7,$01       ; generate a 1
300010B6 7603  LRW    R6,[*+12]    ; abs = $300010C4 get addr. EPFR
300010B8 D706  ST.H   R7,(R6,0)    ; write 1 to clear EPFR
300010BA 7403  LRW    R4,[*+12     ; abs = $ address of flag
300010BC A504  LD.B   R5,(R4,0)    ; get flag
300010BE 2005  ADDI   R5,$01       ; increment it
300010C0 B504  ST.B   R5,(R4,0)    ; put flag back
300010C2 0002  RTE                 ; return to caller
300010C4 1000  #EPFR               ; address of EPFR register
00000012 3006                      ;
300010C8 3000  #flag               ; address of flag variable
00000016 0018
```

6.2.3.2   A Handler That Fills or Empties a Buffer

This first technique shown above really just transfers the synchronization from interrupt-based hardware, to gadfly-based software. In another common technique, the interrupt handler stores data in the buffer. This second technique is illustrated in this subsection.

In the example below, main sets up the index to the buffer and a edge request interrupt as in the previous example. It then waits for the buffer's index to reach its end. EPDR bit 0's falling edge occurs each time another paper tape sprocket hole comes under the reader. This causes the interrupt handler to be entered, which removes the source of the interrupt and reads the pattern into the next buffer element. When all data has arrived, the edge interrupt is disabled. This gadfly loop in main waits for the buffer to be DONE, rather than for the arrival of each byte.

```
unsigned char buffer[10], index;

interrupt void keyInt(void){EPFR = 1; buffer[index++] = KCD;}
void main() {
    disableInt();             /* prevent int. while setting up */
    index = 0;                /* set buffer pointer to beginning */
    EPPAR = 1; EPFR = 1;      /* enable rise edge interrupt, clear flag */
    initInt(0x30000000, 0x30000200, keyInt, 10);   /* set up handler */
    enableInt();              /* enable interrupts */
    while(index < 10) ;       /* wait for buffer to be done */
    disableInt();             /* prevent int. when all data is collected */
}
```

This interrupt handler is coded in assembler language as shown below:

```
300010B4  6017    MOVI    R7,$01        ; generate a 1
300010B6  7606    LRW     R6,[*+24]     ; abs = $300010D0 get addr. EPFR
300010B8  D706    ST.H    R7,(R6,0)     ; write 1 to clear EPFR
300010BA  7406    LRW     R4,[*+24]     ; abs = $300010D4 get addr. index
300010BC  A504    LD.B    R5,(R4,0)     ; get index
300010BE  1253    MOV     R3,R5         ; copy index
300010C0  2005    ADDI    R5,$01        ; increment index
300010C2  B504    ST.B    R5,(R4,0)     ; save index
300010C4  7505    LRW     R5,[*+20]     ; abs = $300010D8 get addr. buffer
300010C6  1C35    ADDU    R5,R3         ; get addr. buffer[index]
300010C8  7305    LRW     R3,[*+20]     ; abs = $300010DC addr. KCD
300010CA  A403    LD.B    R4,(R3,0)     ; get input data
300010CC  B405    ST.B    R4,(R5,0)     ; save it in buffer[index]
300010CE  0002    RTE                   ; return to interrupted routine
```

Chapter 6   Interrupts and Alternatives

6.2.3.3 A Handler That Uses a Queue for Input

The third technique's interrupt handler pushes data acquired from the input port into a queue until *main* pops the data to use it. Do not be confused between the queue, which is a temporary storage place to hold data from the handler to the main program, and the buffer which eventually stores the data. *main* just stores data that is popped from the queue into the buffer, but a typical program analyzes data, or uses it in some way, as it pops it from the queue. This example uses a queue whose length is a power of 2, to generate more efficient code for the M·CORE instruction set.

```
char index, buffer[10], queue[8], top, bot, error;
void push(char data){queue[top=(top+1)&7]=data; if(bot==top)
    error=1;}
char pull(void){if(bot==top)error=1; return queue[bot=(bot+1)&7];}
char get(void){ while(bot == top) ; return pull(); }
interrupt void keyInt(void){EPFR=1; push(KCD);   /* input and push data */}
void main() { char i;
    disableInt(); NIER |= 1 << 21;      /* enable edge interrupt for bit 0 */
    EPPAR = 1;   /* interrupt on bit 0 rising edge */
    EPFR = 1;   /* clear flag */
    initInt(0x30000000, 0x30000200, keyInt, 10);   /* set up handler's addr */
    KCDD = 0;   /* make keyboard column inputs */ enableInt();
    for(i = 0; i < 10; i++)
        buffer[index++] = get() ;   /* fill buffer */
    NIER &= ~(1 << 21);   /* disable edge interrupt for bit 0 */
}
```

The interrupt handle is coded in assembler language as:

```
3000114C    24F0    SUBI    R0,$10          ; make room for registers used
3000114E    007D    STM     R13-R15,(R0)    ; save register used, return address
30001150    601E    MOVI    R14,$01         ; put 1
30001152    7D05    LRW     R13,[*+20]      ; abs = $30001168 address of EPFR
30001154    DE0D    ST.H    R14,(R13,0)     ; clear source of interrupt
30001156    7D03    LRW     R13,[*+12]      ; abs = $30001164 address of KCD
30001158    A20D    LD.B    R2,(R13,0)      ; get data from port
3000115A    FFAC    BSR     *-166           ; call push
3000115C    006D    LDM     R13-R15,(R0)    ; restore regs used, PC
3000115E    20F0    ADDI    R0,$10          ; deallocate
30001160    0002    RTE                     ; return to caller
30001162    0000    BKPT                    ; align port addresses
30001164    1000    MFCR    R0,CR0          ; address of EPFR
30001166    3006    BCLRI   R6,$00
30001168    1000    MFCR    R0,CR0          ; address of KCD
3000116A    7006    JMPI    [*+24]
```

While this handler appears to be about as simple as the previous example's handler, it calls the rather long push procedure, which is coded in assembler language as:

```
300010B6    B200    ST.B      R2,(R0,0)      ; save data to be pushed
300010B8    760F    LRW       R6,[*+60]      ; abs = $300010F4 address of size
300010BA    A706    LD.B      R7,(R6,0)      ; get size
300010BC    1275    MOV       R5,R7          ; make copy of size
300010BE    2007    ADDI      R7,$01         ; increment size
300010C0    B706    ST.B      R7,(R6,0)      ; write back size
300010C2    2295    CMPLTI    R5,$0A         ; if size less than 10
300010C4    E003    BT        *+8            ; abs = $300010CC skip next seg.
300010C6    6017    MOVI      R7,$01         ; generate 1 for error
300010C8    760A    LRW       R6,[*+40]      ; get addr. of error
300010CA    B706    ST.B      R7,(R6,0)      ; set error to one
300010CC    7608    LRW       R6,[*+32]      ; abs = $300010EC get addr. of top
300010CE    A706    LD.B      R7,(R6,0)      ; get top
300010D0    1275    MOV       R5,R7          ; copy top
300010D2    2007    ADDI      R7,$01         ; increment top
300010D4    B706    ST.B      R7,(R6,0)      ; replace top
300010D6    2E75    ANDI      R5,$07         ; wraparound top
300010D8    B506    ST.B      R5,(R6,0)      ; replace top
300010DA    7703    LRW       R7,[*+12]      ; abs = $300010E8 addr. of queue
300010DC    1C57    ADDU      R7,R5          ; addr. of queue[top]
300010DE    A600    LD.B      R6,(R0,0)      ; get input data
300010E0    B607    ST.B      R6,(R7,0)      ; store into queue[top]
300010E2    2070    ADDI      R0,$08         ; deallocate subroutine's temporary
300010E4    00CF    JMP       R15            ; return to caller
300010E6    0000    BKPT                     ;
300010E8    3000    BCLRI     R0,$00         ; address of queue
300010EA    0110    XTRB2     R1,R0          ;
300010EC    3000    BCLRI     R0,$00         ; address of top
300010EE    0118    XTRB2     R1, R8         ;
300010F0    3000    BCLRI     R0,$00         ; address of error
300010F2    011B    XTRB2     R1, R11        ;
300010F4    3000    BCLRIo    R0,$00         ; address of size
300010F6    011A    XTRB2     R1, R10        ;
```

As in the two previous examples, main prepares the same edge interrupt. Then it calls the input procedure to get a byte at a time, and writes each byte into the buffer. However, in this example, the interrupt handler pushed the data obtained from the input port onto the queue, using the push procedure. The get procedure gadflies as long as the input queue is empty; when the queue is nonempty, get pops an item, returning it to the caller. main then puts this data into its buffer. When the buffer is filled, the handler inhibits edge interrupts and the main procedure exits. Observe how long the handler takes to run. More complex software in the handler degrades latency.

As in previous discussions of synchronization methods, we discuss interrupt synchronization BUSY/DONE states. The paper tape reader is IDLE until after initialization sets it up, and after *main* disables interrupts from the paper tape reader just before it exits. It becomes BUSY after initialization. The DONE state is entered when an interrupt occurs and data are moved from *KCD* to the input queue. After the data are moved, the device generally returns to the BUSY state.

This example actually accommodates a problem that is more thoroughly discussed in §6.2.2.5. The problem is the possibility of an interrupt when a variable in a register is changed, and the variable is changed by the handler. This problem is made more common in a RISC computer like the M·CORE which maintains data in registers.

6.2.3.4   A Handler That Uses a Queue for Output

The output operation is quite similar to input in all but the last case involving a queue. We consider specifically our running paper tape punch example because it illustrates two important points about interrupt software.

The device generally causes edge port bit 0 pin signal to fall if it is prepared to accept output data. The output interrupt should only be enabled when there is data to be output. Otherwise, an interrupt will occur, the handler will be entered, but there will not be anything to do because there will not be any data to output. Frustration!

In interrupt handlers using queues, output queue size indicates whether or not there is data to be output. If the queue is empty, there is no data to be output and the output interrupt should be disabled, but if the queue is nonempty, there is data to be output, and the output interrupt should be enabled. The output interrupt is initially not enabled, but is enabled in *put* and disabled in the interrupt handler to follow this rule. (Observe, by contrast, that input interrupts are always left enabled.) Also a handshake, *EPDR* bit 1, is asserted exactly when the queue is nonempty, to advance the tape. We also have to terminate device use. The termination procedure should wait until the output queue is empty before it disables interrupts and terminates use of the device. When the output queue is empty, the interrupt handler clears *NIER[21]* and the *main* procedure gadflies on this bit, only exiting main when the bit is clear (and therefore the queue is empty).

```
char index, buffer[10], queue[8], top, bot, error;

void push(char data)
    {queue[top = (top + 1) & 7] = data; if(bot == top) error = 1;}

char pull(void)
    {if(bot == top) error = 1; return queue[bot = (bot + 1) & 7];}

void put(char data) {
    char oldTop = top; while(top == ((bot + 7) & 7)) ; push( data );
    if(oldTop == bot) NIER |= 1 << 21;
}
```

```
interrupt void keyInt(void)
   {EPFR=1; KCD = pull(); /*store data*/ if(bot==top) NIER &= ~(1<<21);}

void main() { char i;
   disableInt(); EPPAR = 1; EPFR = 1; KCDD = 0xff;
   initInt(0x30000000, 0x30000200, keyInt, 10);
   for(i = 0; i < 10; i++)
     put(buffer[index++]);     /* fill buffer */
   while(NIER & (1 << 21));     /* wait for edge interrupt bit 0 to be disabled */
}
```

These procedures show some interrupt handler mechanisms, but these mechanisms are inefficient because gadflying on a global variable, such as a buffer index or a queue length, wastes time. Unless the microprocessor can do something else in the meantime, gadfly synchronization would be simpler, have lower latency, and be easier to debug.

6.2.3.5   Critical Sections

The correct management of critical sections is a very important aspect of interrupt software. A *critical section* is part of a procedure, which, if interrupted, can cause incorrect results. An example, in §2.2.2's queue for an input device like a paper tape reader, $size$ is incremented in the handler when an item is pushed, and decremented in a procedure called by $main$, when an element is pulled from the queue. $size$ may be copied from its memory location to a register like GPR r14, decremented there, and put back. If an interrupt occurs between the time $size$ is read until it is rewritten and the interrupt handler's $push$ procedure increments $size$, then the interrupted program will cancel the change made by the handler. For instance, if $size$ were initially 3, indicating there are 3 words in the output queue, and the get procedure pulled a word, it would decrement $size$ to 2. But if, while $size$ is effectively moved to GPR r14, an interrupt occurs and the interrupt handler saw a request to output a word for that interrupt request, it would push a word. It would increment the value of $size$ that was in memory, changing it from 3 to 4. When the interrupted program that was decrementing size is resumed, it will write the number 2 in the memory variable $size$. But there are now 3 words in the queue, not 2. Subsequent queue size checking will be faulty.

The chances of a critical section fault happening are actually very small. But if you believe in Murphy's law, such an error will occur at the worst possible time. Therefore, to write correct programs, you must avoid any possibility of such a critical section error.

In §6.2.2.1, reading $flag$ does not cause a critical section. The main procedure reads $flag$ in a gadfly loop. If $flag$ is put into a register such as GPR r14 when an interrupt occurs, the $main$ program will not see a change in $flag$. However, the next time through the loop, the changed $flag$ is seen. The effect of an interrupt is merely delayed, but is correctly handled, so there is no critical section. Similarly, in §6.2.2.3 and §6.2.2.4, top and bot are compared to detect a queue overflow, rather than checking a variable *size* as in §2.2.2. This approach does not cause a critical section.

When interrupt synchronization is used, the initialization of interrupt hardware and the variables, for instance counters and pointers for a queue, is often a critical section, although some study and considerable experience might be required to diagnose this problem. For instance, in the examples above, the initialization of edge ports might accidentally set a flag flip-flop by causing an edge to occur. Or the flip-flop might have been left set from prior use, and not cleared, before this program uses it. The interrupt will then occur as soon as interrupts are enabled. But there is no data over the heads of the paper tape reader, and garbage is read and entered into the buffer. This is why the initialization procedure writes 1 into the edge flag register, to clear interrupt requests before interrupts are enabled. (Clearing a flag register with interrupt disabled is equivalent to having disarmed the device, as discussed earlier. This alternative is therefore called *software disarming* of the device.) Because this analysis is usually quite difficult, for any device using interrupts, we strongly advocate always disabling interrupts before, and enabling them after, an I/O device is initialized, and software disarming the device.

Critical section errors can be correctly avoided by inhibiting interrupts in main program segments that change variables that are changed, read, or tested in an interrupt handler. Before a potential critical section is being executed, *disableInt();* prevents interrupts from happening while the data is in a register, and *enableInt();* permits interrupts when the critical section is left.

6.2.4   Polled Interrupts

For multiple (polled) interrupts, the *interrupt handler* just finds out which interrupt request needs service. In this case, when the interrupt occurs, its handler is executed. It *polls* the possible interrupts to see which one caused the interrupt. The polling program checks each possible interrupt in *priority* order, highest priority interrupt request first, until it finds a device that requested service, executes that request, and clears that interrupt request.

Consider an M·CORE system having two paper tape readers. The first reader's data and sprocket holes are on *KCD* and *EPDR* bit 0, respectively. The second reader's data and sprocket holes are on *KRD* and *EPDR* bit 1, respectively. See Figure 6.11. Either interrupt causes execution of *handler* whose address is in VBR + 0x28. This interrupt handler, written in embedded assembler language, reads the *NIPND* port, and uses the *ff1* (find first one) to determine the leftmost request. Note that *ff1* returns 31 - *n* if bit *n* is the leftmost bit that is set. It then computes an offset to vector which it uses to go to a specific handler. Each different interrupt source generally has its own *handler* that actually services the interrupt. This handler is a C or C++ procedure that ends in an *rte* instruction, which may have local variables and may call other procedures, to satisfy the needs of the interrupting device. Each handler, such as *handler1()*, *handler2()*, etc., will handle one source of interrupt on a device. Once there, it removes the source by clearing the flag bit and

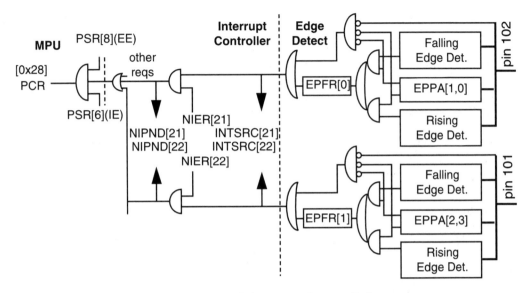

Figure 6.11. Polled Interrupt Request Path

writes data into the associated buffer. To make the program work sensibly, *main* terminates when either buffer is full.

```
unsigned char bufferA[10], bufferB[10], indexA, indexB;

interrupt void handler1(void){ EPFR = 1; bufferA[indexA++] = KCD;}

interrupt void handler2(void){ EPFR = 2; bufferB[indexB++] = KRD;}

const long Jvector[33] = {0,0,0,0,0,0,0,0,0,0,0,0,0,0,0,0,0,0,0,0,0,
     (long)handler1,(long)handler2 };   /* Jvector[21], [22] have addresses */
```

```
#pragma NO_ENTRY
#pragma NO_EXIT
#pragma NO_RETURN
#pragma NO_FRAME
interrupt handler() { asm{
          LRW      r1,L0          ; address of Jvector[33]
          LD.W     r1,(r1,0)      ; get NIPND set if both pending and enabled
          FF1      r1             ; identify postion of leftmost bit set
          RSUBI    r1,31          ; convert so NIPND bit # -> Jvector index
          LRW      r2,L1          ; address of vector
          IXW      r2,r1          ; combine with vector base
          LD.W     r2,(r2,0)      ; get handler address from Jvector[33]
          JMP      r2             ; go to handler
L0:       DC.w     0x1000000C     ; address of NIPND
L1:       DC.w     Jvector        ; address of jump vector
     }}
```

```
void main(){ disableInt(); indexA=indexB=0;  /*set buffer pointer to begin-
    ning*/
    NIER |= (( 1 << 21) | ( 1 << 22)); EPFR = 3; EPPAR = 5;
    initInt(0x30000000, 0x30000200, keyInt, 10); enableInt();
    while((indexA<10) && (indexB<10));   /* wait for either buffer to be done */
    NIER &= ~(( 1 << 21) | ( 1 << 22));   /* disable edge interrupts only */
}
```

Note that a 1 (T) appears in *NIPND* if the edge flag has set, and the enable is also set. In the handler, the edge flag is cleared and data are written into the appropriate buffer. If you only read *EPFR*, going to a handler if a bit in it is T, but this bit's enable bit is not T, something else caused the interrupt, and will continue to request the interrupt after the handler is exited with RTE. The flag bit that does not have its enable bit in *NIER* set could not cause the interrupt and should not be serviced. There are even situations where this failing to service the interrupt making the request can hang up the machine, if the flag, for the interrupt being handled, remains set. Therefore interrupt polling checks for both the flag and the enable by reading *NIPND*. This differs from a gadfly loop, which tests only the flag *EPFR*. Polling checks different bits than gadfly loops check.

The device corresponding to the leftmost 1 in *NIPND* is called the highest priority device. This fixed priority scheme has the disadvantage that a high priority device, that frequently requests interrupts, can hog the use of the machine. Other devices with lower priority might therefore be starved, or may have excessively long latency, causing errors.

An alternative scheme is called a *round-robin* priority scheme; generally, the polling is effectively done in an infinite program loop (see Figure 6.12). When the *i*th interrupt request in the priority order gets an interrupt and is serviced, the i + 1*th*

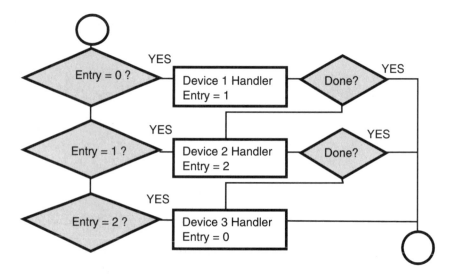

Figure 6.12. General Round-Robin Polling Process

interrupt request assumes the highest priority; so whenever the next interrupt occurs, the polling program starts checking the $i + 1th$ interrupt request first. If no handler clears the interrupt request, the handler exits and is immediately reentered to check for interrupts. An embedded assembler language M·CORE round-robin polling handler is shown below.

```
#pragma NO_ENTRY
#pragma NO_EXIT
#pragma NO_RETURN
#pragma NO_FRAME
interrupt handler() { asm{
          LRW       r1,L0         ; address of NIPND
          LD.W      r1,(r1,0)     ; get NIPND, set if both pending and enabled
          LRW       r3,L1         ; address of entry
          LD.B      r2,(r3,0)     ; contents of entry
          BGENR     r2,r2         ; produce a mask having T where last interrupt
          SUBI      r2,1          ; produce a mask having T to right of last int.
          AND       r2,r1         ; bit pattern of requests to right of last int.
          FF1       r2            ; identify postion of leftmost bit set
          CMPLTI    r2,32         ; if less than 32, it is a bit number, so go too
          BT        L             ; the routine that goes to the vector entry
          FF1       r1            ; else identify postion of any leftmost bit set
          MOV       r2,r1         ; put number of interrupt in r2
L:        RSUBI     r2,31         ; convert so NIPND bit # -> Jvector index
          ST.B      r2,(r3,0)     ; save number of interrupt in entry
          LRW       r2,L2         ; address of Jvector
          IXW       r2,r1         ; combine interrupt number with vector base
          LD.W      r2,(r2,0)     ; get handler address from Jvector[33]
          JMP       r2            ; go to handler
L0:       DC.w      0x1000000C    ; address of NIPND
L1:       DC.w      entry         ; address of global variable entry
L2:       DC.w      Jvector       ; address of global jump vector Jvector
}}
```

In the round-robin interrupt handler that takes best advantage of M·CORE's architecture and the MMC2001 interrupt controller module, the *NIPND* port is read, and the *ff1* instruction is used, to poll up to 32 requests in one step, as shown in the fixed priority polling scheme just discussed. But in round-robin priority, a global variable *entry* remembers the last interrupt that was serviced, to determine which interrupt now has the highest priority. It generates a mask that eliminates devices having the same or higher priority than the last interrupt had, in a first evaluation of the prior request; if that fails to select an interrupt, the program segment reevaluates all pending interrupts without masking any requests. As in the fixed priority mechanism, the address of the handler is read from the jump vector. The handler, which is then executed, ends in *rte*.

The polling method and priority ordering affect latency. Priority polling provides shortest latency to the prior device, and longest latency to the least prior device, while round-robin priority provides average latency to each device. Aside from this, the method and ordering only affect performance when, in the product state machine, several devices change from busy to done in a short time (the latency of the first device to change state).

6.2.5 Vectored Interrupts

The previous example shows how multiple interrupts are polled in the interrupt handler. Polling may take too much time for some interrupt requests that need service quickly. A *vectored interrupt* technique replaces the interrupt handler software by a hardware mechanism, so that the interrupt request handler is entered almost as soon as the device requests an interrupt. We illustrate this technique by using normal and fast interrupts.

Consider two paper tape readers attached to ports exactly as in the previous section, but one of them uses the normal interrupt hardware, as in previous sections, while the other uses the fast interrupt. See Figure 6.13. While the first interrupt causes execution of the handler whose address is in 0x30000028, the second interrupt causes execution of the handler whose address is in 0x3000002C. The C program for the MMC2001 is shown on the next page.

```
char bufferA[10], bufferB[10], indexA, indexB;

interrupt void handler1(void) { EPFR = 1; bufferA[indexA++] = KCD; }

interrupt void handler2(void) { EPFR = 2; bufferB[indexB++] = KRD; }

void main() { disableInt(); indexA = indexB = 0;
    NIER |= 1 << 21; FIER |= 1 << 22; EPFR |= 3; EPPAR |= 5;
    initInt(0x30000000, 0x30000200, handler1, 10); /* set up normal int */
```

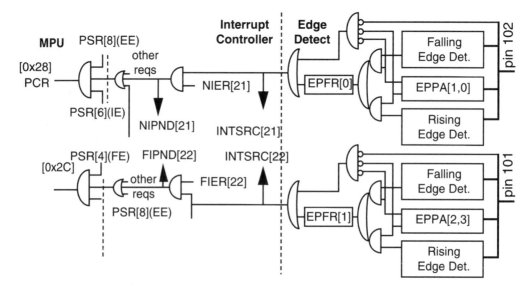

Figure 6.13. Vector Interrupt Request Path

```
*(long *)0x3000002C=(long)handler2+1; /* set up fast int */ enableInt();
while((indexA<10) && (indexB<10)) ; /* wait for either buffer to be done */
NIER &= ~( 1 << 21); FIER &= ~( 1 << 22); /* disable edge interrupts only */
}
```

When an interrupt is honored in step three (§6.2.1), the MMC2001 goes directly to the handler. A falling edge on *EPDR* pin 0 makes *handler1* execute immediately. A falling edge on *EPDR* pin 1 makses *handler2* execute immediately. There is no need for polling, even using the efficient *ff1* instruction, thereby reducing latency time.

6.2.6 Vectored Interrupts for Other Devices

As Table 6.1 shows, the M·CORE architecture has a lot of hardware interrupt vectors. For instance the normal interrupt studied in earlier sections is executed whenever any bit in *NIPND* is set, and *PSW* bits 6 and 8 are set. Also, as discussed in the last section, the fast interrupt is executed whenever any bit in *FIPND* is set, and *PSW* bits 4 and 8 are set. The number on the left is the *interrupt number* used at the begining of a HIWARE interrupt handler. For instance, the handler for normal interrupts is designated *interrupt 10,* and the handler for fast interrupts is designated *interrupt 11*. The number in the middle is the *interrupt offset* where the hardware gets the 32-bit number which is put into the program counter. For instance, the address of the handler for normal interrupts is at VBR + 0x28 and for

Table 6.1. M·CORE Interrupt Vectors

Number	Offset	Assignment
0	00	Reset
1	04	Misaligned Access
2	08	Access Error
3	0c	Divide by Zero
4	10	Illegal Instruction
5	14	Priviledge Violation
6	18	Trace Exception
7	1c	Breakpoint Exception
8	20	Unercoverable Error
9	24	Soft Reset
10	28	INT Autovector
11	2c	FINT Autovector
12	30	Hardware Accelerator
13	34	Reserved
14	38	Reserved
15	3c	Reserved
16–19	40 – 4c	Trap #0 – 3
20–31	50 – 7c	Reserved
32–127	80 – 1fc	Reserved for Vector Int

fast interrupts, at VBR + 0x2c. Using vectored interrupts, the polling routine is executed very quickly in hardware, and the winning handler is jumped to, or vectored to, right after the PC and PSW registers are saved.

Additional vectors are for MPU error conditions, such as misaligned access, which handles word and half-word accesses using odd addresses, access error which handles accesses to nonexistent memory locations, and divide by zero. The illegal instruction and privilege violation vectors are used when an attempt is made to execute such an illegal instruction, or priveleged instruction while in user mode, and trace and breakpoint vectors are used by debuggers. Unrecoverable errors handle cases where an exception is attempted when PSR bit 8 (EE) is clear, and soft reset is executed when a pin, $SRST$, is asserted. The hardware accelerator vector handles attempts to use extra hardware procesors when they are not attached. Trap vectors handle trap instruction executions. Remaining vectors (32 to 127) handle interrupts in which the I/O device provides an interrupt vector number. These I/O device vectors are unimplemented in the MMC2001, but they permit future M·CORE microcontrollers to explicitly vector more I/O interrupts.

6.2.7   Examples of Interrupt Synchronization

The interrupt is useful in managing asynchronous requests and repetitive tasks. This section illustrates the use of interrupts in in responding to key requests and inputting and outputting ISPI data.

6.2.7.1   Keyboard Handling

The M·CORE's edge port, $EPDR$, and key pad ports KCD and KRD, are ideally suited to responding to sensors and push buttons, to begin procedures whenever such sensor detects a situation needing correction, or a button is pressed requesting an action. Key contacts occur asynchronously, at random and unpredictable times, and are therefore handled by interrupts. Mechanical switches "bounce". A common debouncing technique is shown here for the edge port, but the key pad ports have a hardware debounce circuits. Finally, multiple keys have to be scanned; a simple scanning technique is given. The fact that several sensors or push buttons might request actions in short order, while the microcontroller is taking care of one of them, is resolved in the next chapter.

When mechanical switches or sensors close, a metal contact rebounds when it hits a metal plate, causing multiple closed-open events. If each transition from open to closed causes an edge, each edge causes an interrupt, and each interrupt initiates an action, then one physical closure of a switch or sensor may initiate multiple actions. This problem of *contact bounce* is mechanically addressed by using bounceless contacts, such as opto-electronic sensors or mercury-wetted contacts, by electrically debouncing the switch using analog techniques, such as putting a capacitor across it, using digital techniques, such as using a set-clear flip-flop, or by computationally accounting for multiple bounces. The most common technique used in microcontrollers, a computationally accounting technique, is, upon detecting an apparent switch closure, seeing if it remains closed for 5 ms before recognizing the closure. This time permits most switches time to stop bouncing. This *"wait-and-see"* technique has some disadvantages, but provides more than adequate performance at the least cost and complexity of all the techniques discussed above. An example of this technique is illustrated for a single switch, a linear select, and a coincident select or matrix keyboard.

Figure 6.15a illustrates a single switch or sensor connected to *EPDR* bit 0. An external pull-up is attached so that if the switch is open, the pin voltage is high. When the switch is closed, the pin voltage falls, and a falling edge can generate an interrupt. In this short program that follows the single interrupt technique shown in §6.2.2.1, when *EPDR* bit 0 sees a falling edge, it generates a normal interrupt, and after 5 ms, if the switch remains closed and the input is low, the handler puts a T (1) on *EPDR* bit 1. Constant N is selected to provide a 5-ms delay, in this and other keyboard examples.

```
interrupt void keyInt()
    {short i; for(i=0; i<N; i++); if(EPD&1) EPD |= 2; EPFR = 1; };

void main() { char buffer[10], i;
    disableInt();NIER |= 1<<21; EPPAR=1; EPFR=1;
    initInt(0x30000000, 0x30000200, keyInt, 10); KCDD=0; enableInt();
}
```

In an MMC2001, a *linear select keyboard* has (say, 8) switches, each connected between a *EPD* pin, configured as an input with external pull-ups, and ground. See Figure 6.15b. When any switch is closed, a falling edge can generate an interrupt. The normal interrupt polling mechanism of §6.2.3 is used for each edge port interrupt. The handler for each edge port can be like the handler *keyInt* described above.

An MMC2001 *matrix*, or *coincident select, keyboard* is connected as (say, 8) rows of wires connected to *KRD* pins and (say, 8) columns of wires connected to *KCD* pins. (See Figure 6.15c.) Although electrically connected in a two-dimensional array, they can be physically positioned suited to an application such as a terminal keyboard, for instance. In the MMC2001, rows are connected to input *KRD*, having pull-ups. Columns are connected to output *KCD*. The main program initially makes *KCD* an output, whose pins are all low, and configures the keyboard control ports shown in Figure 6.14. Then, if any one key is pressed in any row and in any column, a KRD input falls. Hardware debouncing waits 1/64th s, and then if the key remains low, an interrupt occurs. After thus being debounced by hardware, the key's row and column number are catenated into a key code, which is pushed into a queue, to be pulled later.

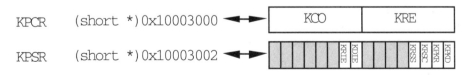

Figure 6.14. Keyboard Control and Status Ports

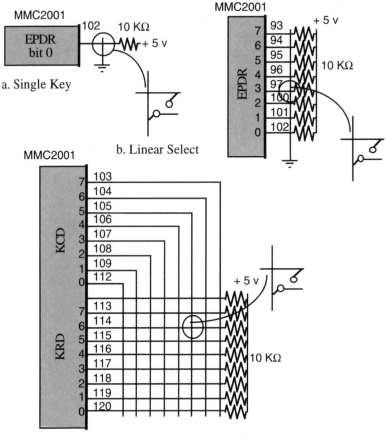

Figure 6.15. Keys and Keyboards

Initialization begins by setting up the interrupt and preparing the keyboard ports. As indicated in the comments, the row and column directions are set up as in §5.1.3. Then the rows are made open drain, and columns are connected to the keypad interrupt hardware. The key depress interrupt is enabled on the keyboard hardware device, and the keyboard interrupt is enabled in the interrupt hardware device.

```
volatile short KPCR@0x10003000, KPSR@0x10003002;

void main() { char buffer[10], i;
    disableInt(); KCDD = 0xff;   /* column direction output*/
    KRDD = 0;   /* row direction input */
    KPCR = 0xffff;   /* columns are open drain, rows connect to interrupts */
    KPSR = 0x10c;   /* enable key depress interrupt */
    NIER |= 1 << 6;   /* enable normal interrupt from keyboard */
    initInt(0x30000000, 0x30000200, handler, 10); enableInt();
}
```

```
void interrupt handler() { char r, c; short i, j;
    KPSR = 1;    /* clear status bit by writing one into it */
    for(i = 0; i < 8; i++){    /* scan 8 columns */
        KCD = ~(1 << i); c = ~KRD;    /* get port data for that row *
        for(j = 0, r = 1; j < 8; j++, r <<= 1;) {
            if ( c & r ) {    /* if a bit pattern in r intersects with the input in d */
                push( ( i << 3) + j );    /* assemble key code, push it */
                i = j = 8;    /* terminate loops */
            }
        }
    }
    KCD = 0;    /* all columns low to interrupt if any key is pressed */
}
```

6.2.7.2   ISPI Interrupt Synchronization

The ISPI can input and output serial data, as has been discussed in §5.3.4. In this example (see Figure 6.16), the ISPI repetitively inputs and outputs multiple bytes of data using interrupts. The 6812 MOSI pin feeds a series of '595s, *MISO* collects from a series of '589s, SCLK feeds each chip's shift clock, and SS feeds each chip's RCLK and LD, as discussed in §4.4.3. Global vectors *inBuffer* and *outBuffer* hold data received from the '589s and sent to the '595s respectively, and a global *index* is used to access the vector elements in the interrupt handler. *main* initializes the ISPI as before, but the ISPI interrupt enable, *IRQ_EN,* is set to permit interrupts. *SPDR* is written to start the ISPI. The shift clock rate is set low to give time for *main* to do other work. Each time eight bits have been shifted in and out, an ISPI interrupt occurs. An address register sequence, reading *SPSR* and *SPDR,* clears the interrupt and writes *SPDR*'s data into *inBuffer.* Data from *outBuffer* is written into *SPDR,* which also starts the ISPI again. If *index* attains its maximum, it is cleared and the *SPIGP* pin is pulsed low twice to transfer each 589's data between shift register and output register, and each 595's data between input register and shift

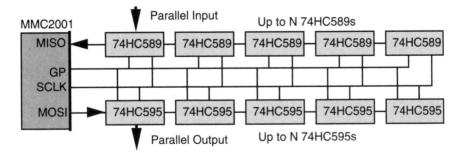

Figure 6.16. ISPI Network

register. *main* can now read from *inBuffer* and write data into *outBuffer;* the ISPI moves this data in and out of the computer.

```
interrupt void handler() { char dummy;
    dummy = SPSR; inBuffer[index] = SPDR; SPDR = outBuffer[index++];
    if(index == 5)
        {index=0;  SPCR|=SPIGP;  SPCR&=~SPIGP;  SPCR|=SPIGP;
        SPCR&=~SPIGP;}
}
void main() { disableInt();
    NIER |= 1 << 20;   /* enable ISPI normal interrupt */
    SPCR = SPI_EN + MSTR + IRQ_EN + 0x70 + 7; SPDR = 0;
    initInt(0x30000000, 0x30000200, handler, 10);
    enableInt();   /*enable int */ while(index) ; SPCR = 0;   /* disable int */
}
```

The ISPI can be repetitively read from and written into a buffer, and data sent to the ISPI can be read from another buffer. The interrupt handler can be used to continuously exchange the external data with the buffer data. Then the external data can appear to be in the MMC2001 RAM memory in *inBuffer[],* so that input data can be read from the input buffer and output data need only be put into memory in *outBuffer[].* It is for this reason that the ISPI and external "three-wire" interface chips are widely used, and why, because of this mechanism, the MMC2001 has need of only a few parallel ports.

6.2.8   Object-Oriented Classes for Parallel Port Interrupts

Interrupts can be very nicely handled using objects, especially when queues are used to pass the data. We first discuss for input, the class *IQFPort* (input queue flag port) below is derived from *SyncPort* and uses class *Queue* from §2.3.3 and the normal interrupt handlaer from §6.2.3. We then discuss for output, the class *OQFPort*. Both assume that *initInt()* has been executed to set up the vector at 0x30000000.

6.2.8.1   An Input Queued Flag Port Class

IQFPort's constructor, its arguments, and handshake mechanisms are very similar to those of the *SyncPort* class. Both classes are derived classes of *Port*. Handshaking is not ususally used with input interrupts, because data is provided to the device asynchronously, but is provided here such that if *get* is executed when the queue is empty, a handshake signal can be pulsed, or asserted until the interrupt handler responds to put data in the queue, to request data. A *struct* of type *SyncCtl* specifies the size of the queue, whether to use normal interrupts (if it is 0) or fast interrupts (if it is 1). A bit of *EPDR* is specified in *Obit* for outputting the handshake signal, and a bit is specified in *Ibit* for inputting the edge that indicates entry into the done state. The polling routine for normal interrupts uses *normVector* as *Jvector* was used in §6.2.3, using the normal interrupt handler that is described there. The polling routine for fast interrupts uses a fast interrupt handler in an analogous manner, substituting *FIPND* for *NIPND* and using *fastVector* in place of *Jvector*. The interrupt handler reads the port, pushing the result on the queue. The *get* function pulls the word.

```
long normVector[33], fastVector[33];    // handler addresses

const char mapBit[8] = { 0x80, 0x40, 0x20, 0x10, 8, 4, 2, 1 };
typedef struct
{longDlyEdge,charObit,Ibit,Fast:1,E1:1,E2:1,E3:1,E4:1,E5:1;}SyncCtl;

interrupt void handler(void) { device->service();}

template <class T> class IQFPort : public Port<T> {
    friend void handler(void);
    protected: SyncCtl *ctl; Queue<T> *Q; char Ibit;

    void preSync(void) {
        if(ctl->E1) EPDR|= ctl->Obit; if(ctl->E2) EPDR &= ~ctl->Obit;
        if(ctl->E3) EPDR|= ctl->)Obit;
    }

    void postSync(void) {
        if(ctl->E4) EPDR |= ctl->Obit; if(ctl->E5) EPDR &= ~ctl->Obit;
    }

    public : IQFPort
        (unsigend char QSIZE,SyncCtl *ctl,long a,char attr=0, T dir=0):
        Port(a, attr, dir) : { char i;
            this->ctl = ctl; Q = new Queue<T>(QSIZE); disableInt();
            EPPAR |= ctl->DlyEdge; EPDR = |ctl->Obit;
            if(ctl->E1) EPDR &= ~ctl->Obit; else EPDR |= ctl->Obit;
            for(i = 0; i < 8; i++) if(ctl->Ibit == mapBit[i]) break;
            if(ctl->Fast)    { /* fast interrupt */
              fastVector[i + 21] = (long)handler;
              FIER |= 1 << (i + 21);
            }
            else {    /* normal interrupt */
              normVector[i + 21] = (long)handler;
              NIER |= 1 << (i + 21);
            }
            enableInt();
    }
    virtual T get(void) { T data;
        if( ! Q->Qlen ) preSync(); while( ! Q->Qlen ) ;
        disableInt(); data = Q->pull(); enableInt();
        attr |= Q->errors; return data;
    }
```

```
    void service(void) {
        EPFR = ctl->Ibit; Q->push(Port<T>::get()); attr |= Q->errors;
        if( Q->Qlen == Q->Qsize ) postSync(); return 1;
    }
} *device;
```

A paper tape reader is a straight-forward application of the class *IQFPort* without the need for a derived class; it is shown below. The declaration of the object sets the data port to be *KCD,* using edge port bit 0 for fast interrupt on the falling edge, and edge port bit 1 for handshake, which is asserted low when *get* is entered, and negates high after the interrupt brings in the data.

```
SyncCtl ctl = {0x0002, 0x01, 0x02, 1, 0, 1, 0, 1, 0};
IQFPort<char> p(10, *ctl, 0x10003006, offset2);
void main(){ char buffer[0x100]; p.get(buffer, 0x100); }
```

The constructor's first parameter specifies the queue size. The next parameter controls synchronization and handshaking. Its high 16 bits are put into *EPPAR,* and specify that the falling edge sets the flag. The next 8 bits indicate that the handshake will fall when input is requested and rise when it is recieved. The next four bits indicates that the handshake signal is on pin 2, the next bit (which is absent) indicates that normal interrupts are used, and the last three bits (which are absent) designate that *EPD* bit 0 is selected for edge detection. The constructor's next parameter gives the port address. The remaining missing parameters are zero, for the attributes and direction.

When an interrupt occurs, *handler* reads data from this object's port and pushes it onto its queue. The main procedure *gets* data by pulling it from the object's queue, but if the queue is empty, the *get* procedure gadflies until the queue is nonempty.

6.2.8.2   An Output Queued Flag Port Class

For output, the *class OQFPort* (output queue flag port) below is derived from *Port* and uses class *Queue* from §2.3.3 and handlers from §6.2.3. *OQFPort*'s constructor arguments are essentially identical to those of *IQFPort:* its first argument gives the size of the queue, and a bit of the next argument's *struct, bitPos,* specifies whether to use normal interrupts (if it is 0) or fast interrupts (if it is 1). Other bits of this *struct* supply handshake parameters in the same manner as in *IQFPort. put* pushes a word. The interrupt handler pulls the result from the queue, which it writes to the port.

Interrupts must be enabled only when the queue is nonempty. When *put* is called, it pushes its argument onto the output queue and enables interrupts if the queue was empty, becoming nonempty. When an interrupt occurs. *handler* pulls data from the object's output queue and output to its port. If the output queue empties, the output interrupt is disabled.

Handshaking is often needed, such that if *put* is executed when the queue becomes non-empty, a handshake signal can be pulsed, or asserted until the interrupt handler responds to pull data in the queue. Handshaking is done using member functions *postSync()* and *postSync()*.

The termination of an object of this class, done in the destructor, waits for the output queue to empty before shutting off interrupts. If it failed to do this, then when the program terminates the object, some data having been pushed into the output queue would be lost. To remove the interrupt enable, saved data member *intInfo* indicates which edge bit was used, and both normal and fast interrupt requests from it are disabled.

```
long normVector[33], fastVector[33];    // handler addresses
const char mapBit[8] = { 0x80, 0x40, 0x20, 0x10, 8, 4, 2, 1};
typedef struct
{long DlyEdge, char Obit, Ibit, Fast:1, E1:1, E2:1, E3:1, E4:1, E5:1;} SyncCtl;

interrupt void handler(void) { device->service(); }

template <class T> class OQFPort : public Port<T> {
    friend void handler(void);
    protected: SyncCtl *ctl; Queue<T> *Q; char Ibit; char intInfo;

    void preSync(void) {
        if(ctl->E1) EPDR|= ctl->Obit;
        if(ctl->E2) EPDR &= ~ctl->Obit;
        if(ctl->E3) EPDR|= ctl->)Obit;
    }

    void postSync(void)
        {if(ctl->E4) EPDR |= ctl->Obit; if(ctl->E5) EPDR &= ~ctl->Obit;}

    public : OQFPort
        (unsigend char QSIZE, SyncCtl *ctl, long a, char attr=0, T dir=0):
        Port(a, attr, dir) : { char i;
            this->ctl = ctl; Q = new Queue<T>(QSIZE); disableInt();
            EPPAR |= ctl->DlyEdge; EPDR = |ctl->Obit;
            if(ctl->E1) EPDR &= ~ctl->Obit; else EPDR |= ctl->Obit;
            for(i = 0; i < 8; i++) if(ctl->Ibit == mapBit[i]) break;
            if(ctl->Fast)      /* fast interrupt */
              {fastVector[i+21]=(long)handler; FIER|=1<<(i+21);}
            else   /* normal interrupt */
              {normVector[i+21]=(long)handler; NIER|=1<<(i+21);}
            intInfo = i; enableInt();
    }

    virtual void put(T data) {
        while(Q->Qlen >= Q->Size) ;
        disableInt(); Q->push(value = data); enableInt();
        attr |= Q->errors; if(Q->Qlen == 1) preSync();
    }
```

```
void service(void) {
   EPFR = Ibit; Port<T>::put(Q->pull()); attr |= Q->errors;
   if( ! Q->Qsize ) postSync();
}

~OQFPort(void) { char i;
   i = Q->size; while(Q->size) ; if(i) postSync();
   FIER &= ~(1 << (21 + (intInfo & 7)));
   NIER &= ~(1 << (21 + (intInfo & 7)));
}
} *device;
```

A paper tape punch, shown below, is a straightforward direct application of the class OQFPort. The declaration of the object sets the data port to be KRD, using edge port bit 0 for synchronization on the falling edge, and edge port bit 1 for handshake, which is asserted low as long as the queue is nonempty. Note that the destructor is explicitly called at the end of main to flush the queue and disable interrupts.

```
SyncCtl ctl = {0x0002, 0x01, 0x02, 1, 0, 1, 0, 1, 0};
IQFPort<char> p(10, *ctl, 0x10003007, offset2, 0xff);
void main(){ char buffer[0x100]; p.get(buffer, 0x100); ~IQFPort(); }
```

6.2.9   Interrupt Device Drivers for Parallel I/O

We implement interrupt examples as device drivers similar to those of §6.2.3. We first show a driver that only handles interrupts to increment a global counter, then we illustrate filling a buffer, and finally, we discuss a driver to fill and empty a queue.

6.2.9.1   An Interrupt Device Driver Accessing a Global Variable

Ariel facilitates the insertion and servicing of interrupts. In the drv_t struct, the fourth element, if nonzero, is the address of an interrupt service procedure which is called up by the interrupt handler, and its sixth element is the interrupt priority. Higher priority interrupts are polled before lower priority interrupts. The ucd struct's last parameter is a vector whose first element is this driver's interrupt vector number (for normal interrupts this is 10). Execution of _pio_create automatically installs the interrupt. The interrupt service procedure, a standard C procedure, is passed the address of the driver's Unit Control Block UCB, and a pointer to an optional local control block LCB. It should return a true (1) if the interrupt is handled, or false (0) if not.

Whenever the edge port bit 7 sees a falling edge, the normal interrupt handler, intFun(), merely increments a global variable in the manner of §6.2.3.1. This example can be used to verify that interrupts are handled by the device driver. Note that a call to _pio_create takes care of installing an interrupt whose address is in function0's fourth element.

Here is a simple device driver that increments a global variable.

```
uxid_t id; rv_t status; unsigned char index;
const drv_t function0 = {initFun, 0, 0, intFun, 0, 100, 0};
ucd_t ucd0 = {_EXTNAME('C','M','P','O'), &function0,0,0,0,0,{10,0}};
main() { id = _pio_create(&ucd0); while(index < 10) ; }
int32 initFun( ucd_t *ucp ) { EPPAR = 0x8000; EPFR = 0x80; return 1; }
int32 intFun(ucb_t *ucb, lcl_t *lcl) { EPFR = 0x80; index++; return 1;}
```

6.2.9.2   An Interrupt Device Driver Filling a Buffer

This subsection's driver fills a buffer as was done in §6.2.3.2. This driver needs to execute service calls. Only service calls in Table 6.2 can be called from within an interrupt handler (in the E, or exception, state). But the interrupt handler can call service procedure _drv_entsst() whose first parameter is the name of a procedure that executes as part of the operating system (S, or serialized system, state). There the procedure can call services in Table 6.3. _drv_pio_done() terminates the task's use of the driver, and _drv_sig_send() sends a signal to the task doing I/O. The service calls in Table 6.2 and Table 6.3 are Ariel subroutines called with JSR rather than TRAP instruction; the latter services, listed in other chapters, can't be called in E or S/SE state.

The full driver and its calling routines are listed below. Procedures running in task (T) state are shown first, ending in main(), and then the device driver procedures are shown.

```
const drv_t function0={initFun,svcFun,ctlFun,intFun,0,10,0};
ucd_t ucd0 = {_EXTNAME('P','T','R','O'),&function0,0,0,0,0,{100,0}};
rv_t status; uxid_t i_o_Id   /* device drvr id */ , tskId   /* current task's id */;
void get(char *data, u_int32 n) { cdev_t charDev;
    charDev.aub = data; charDev.lub = n; charDev.spl = charDev.axb = 0;
    _pio_cmd(i_o_Id, _PIO_READ, 0L, &charDev, &status, _CONT_NOCOOR);
    _sig_wait(_TIME_UNLIMITED);   /* wait for signal */
}
```

Table 6.2. Ariel Exception Service Routines

_drv_chkmem()	check for an address that can be written into
_drv_del_irq()	remove interrupt handler (E_state entry)
_drv_minalign()	return minimum processor data alignment
_drv_conv_ms()	translate time parameters to MS
_drv_tod_get()	return address of time-of-day
_drv_seteva()	set address of exception vector area
_drv_geteva()>	set address of exception vector area
_drv_splu()	set processor level to unit, return level
_drv_splx()	set processor level to X, return level
_drv_entsst()	enter S-State processing

```
void close() { _pio_control(i_o_Id, 0, 0); }    /* like C++ destructor */
void main(){char s[10];
    if(!ISSVCERR(i_o_Id=_pio_create(&ucd0)))    /* if no error upon creation */
        {get(s, 10); close();}    /* input some data, then close driver */
}

int32 initFun( ucd_t *ucp ) {EPPAR = 1; KCDD = 0; return SUCCESS;}

void svcFun(ucb_t *ucb,int fun,cdev_t *dparams)
    {EPFR = 1;    /* clr flag */ NIER |= (1 << 21);    /*enable edge port 0 int. */}

void ctlFun( ucb_t *ucb, int fun, cdev_t *dparams)    // remove interrupt
    {if(fun == 0)_drv_del_irq(10,intFun,100);}    // from Ariel's poll table

void intSvcFun(ucb_t *ucb, int funNum, cdev_t *params) {    /* do in S/SE state */
    _drv_sig_send(ucb->tcb,0); /* send signal to resume the task doing I/O */
    _drv_pio_done(ucb,SUCCESS,0);    /* permit driver to be used by other tasks */
}

int32 intFun(ucb_t *ucb, cdev_t *params) {    /* interrupt service in E state */
    *params->aub++ = KCD;    /* get data */
    EPFR = 1;    /* remove source of int. */
    if(params->lub != 0) params->lub--;    /* count of input bytes */
    else {    /* when count reaches zero */
        NIER &= ~(1 << 21);    /* disable this device's interrupts */
        _drv_entsst(intSvcFun, ''pp'', ucb, params);    /* intSvcFun() exits */
    }
    return SUCCESS;    /* signify ok */
}
```

Table 6.3. Ariel Internal Service Routines

_drv_add_irq()	add interrupt handler (E-state entry)
_drv_waitstd()	start (standard) timer pre-assigned to the unit
_drv_cancelstd()	cancel (standard) timer
_drv_waitaux()	start (auxiliary) timer
_drv_cancelaux()	cancel (auxiliary) timer
_drv_cmp_alloc()	allocate memory from Ariel's memory buffer
_drv_cmp_free()	deallocate memory taken from Ariel's memory buffer
_drv_getid()	find id with given name
_drv_pio_done()	end processing on current request
_drv_getsr()	get processor status register, disable interrupts
_drv_setsr()	sets the processor status register
_drv_fbp_free()	de-allocate fixed block from Fixed Block pool
_drv_fbp_alloc()	allocates block from Fixed Block Pool
_drv_entsch()	direct return to scheduler
_drv_msb_put()	send message to the end of message buffer
_drv_mbx_send()	send message to mailbox
_drv_sig_send()	send signal to task
_drv_efg_global()	set global event flags
_drv_efg_local()	set local event flags of given task
_drv_tsk_start()	start task
_drv_tsk_startq()	start task (queue this request if task is not dormant)

Creating the driver sets up the interrupt linkage as in §6.2.9.1, and executes *initFun ()* which initializes *KCD* and the edge port bit 0, with interrupts disabled. Calling *get ()* puts the receive buffer's address, count, and command into *charDev* and executes *svcFun ()* which enables interrupts. *get ()'s _sig_wait ()* blocks the task, waiting for a signal. When an interrupt occurs, *intFun ()* is entered and data from *KCD* is put in the next location of the buffer. When the buffer's count decrements to zero, this device's interrupts are disabled and *intSvcFun ()* is called, using *_drv_entsst ()*, to send a signal to the task, putting the blocked task back in active state, and to permit other tasks to access the device. When the driver is no longer needed, *close ()* is called, which uses service call *_pio_control ()* to immediately execute a control procedure *ctlFun ()* to remove this device's interrupt from the polling table. *_pio_control ()* is called to terminate use of a driver, rather than *_pio_cmd ()*, because its request is not queued, but immediately causes execution of *ctlFun ()*.

6.2.9.3   An Interrupt Device Driver Using a Queue

The last subsection's handler requires the task to call *get ()* before any input can be received through the port; prior input data (*unsolicited data*) are discarded. This subsection's driver uses a queue as was done in §6.2.3.3. The queue permits data to be input before the task requests it. This simple queue is implemented with C procedures *pstop ()* and *plbot ()* from §2.2.2, which are called from the interrupt handler *intFun ()* and the *get ()* procedure, repectively.

Queues should be kept in the driver's local data. Recall from §5.3.7 that 46 bytes of local data can be put in the *ucb struct* at the element *gpr*. The data stored there is declared as a device-dependent *struct ugpr_t* as shown below. The queue, its indexes and counts, and a flag called *w* that indicates waiting, are stored in this *struct*.

```
typedef struct { unsigned char d[10]   /*queue*/, s, e, t, b, w;} ugpr_t;
```

A pointer to the *struct* is passed as a last argument to driver procedures *pstop ()* and *plbot ()* and its elements are used essentially as in §2.3.3. *struct* element *w* is used as a flag, being set when the queue is empty and the task waits for an interrupt to return to the active state. The driver procedures for pushing and pulling data are shown below.

```
void pstop (int item_to_push, ugpr_t *p) { if ((++p->s) > 10) p->e = 1;
    else { if (p->t ==10) p->t = 0; p->d[p->t++] = item_to_push; }
}

int plbot (ugpr_t *p) { if ((--p->s) < 0) p->e = 1;
    if (p->b == 10) p->b = 0; return p->d[p->b++];
}
```

Gobal variables and the *main ()* procedure and *structs* needed by its *_pio_create ()* are listed below.

```
uxid_t i_o_Id; rv_t status; imst_t imst = {0xffffffff, 0xffffffff};

const drv_t function0={initFun, svcFun, 0, intFun, 0, 100, 0};

ucd_t ucd0 = { _EXTNAME('P','T','R','O'),&function0,0,0,0,0,{10,0}};

main(){char s[10];if(!ISSVCERR(i_o_Id=_pio_create(&ucd0)))get(s,10);}
```

The _pio_create() service call, executed in main(), executes initFun().

```
int32 initFun( ucd_t *ucp )
    {NIER |= 1 << 21; EPPAR = 1; EPFR = 1; KCDD = 0; return SUCCESS;}
```

get() calls _pio_cmd() which calls svcFun(), as ususal. Basically, when a data item arrives, an interrupt causes execution of intFun() which pushes input data into the queue. Then svcFun() pulls data from the queue, moving it to the output buffer. However, if this input queue is empty, svcFun() returns to get() with a count (dparams->spl) of fewer than the requested number of bytes (dparams->lub), which causes _pio_cmd() to wait for a signal. Waiting must be done in the T (task) state, because neither T nor S/SE state is able to give up control of the MPU. When an interrupt executes intFun() to push data into a previously empty queue, a signal is sent. This causes _pio_cmd() to be reentered, to move more data into its output buffer.

```
void get(char *data, short n){ cdev_t charDev;
    charDev.aub=data; charDev.lub=n; charDev.axb=0; charDev.spl=0;
    do{/* repeat this loop until all data is returned (should also exit if an error occurs) */
        _pio_cmd(i_o_Id,_PIO_READ,0L,&charDev,&status,_CONT_NOCOOR);
        if(dparams->spl == dparams->lub) break;    /* if all data received, exit */
        _sig_wait(_TIME_UNLIMITED);  /* otherwise, wait for a signal, to get more */
    } while(1);
}

void svcFun( ucb_t *ucb, int fun, cdev_t *dparams) {
    while(dparams->spl < dparams->lub) {/* repeat while data are being received */
        if(((ugpr_t *) &ucb->gpr)->s==0)   /* if empty queue */
            {((ugpr_t *)&ucb->gpr)->w=1; return;}   /* set waiting flag, return */
        else dparams->aub[dparams->spl++] = plbot(&ucb->gpr);
    }
    _drv_pio_done(ucb,SUCCESS,0);   /* permit other tasks to use device driver */
}

intFun(ucb_t *ucb, cdev_t *params) {
    EPFR = 1;   /* remove source of int. */ pstop(KCD, &ucb->gpr);   /* push data */
    _drv_entsst(intSvcFun, ''pp'', ucb, (ugpr_t *)&ucb->gpr);
}

void intSvcFun(ucb_t *ucb, ugpr_t *gpr)    // if waiting flag set, send signal
    { if(gpr->w) _drv_sig_send(ucb->tcb,0); gpr->w = 0;}   // & clear flag
```

6.3   Fast Synchronization Mechanisms

Previously, we discussed synchronization mechanisms used for slower I/O devices. There are six mechanisms used for faster devices. These are direct memory access, context switching, coprocessing, and shuttle, indirect, time-multiplexed, and video memory.

6.3.1   Direct Memory Access

Direct memory access (DMA) is a well-known technique, whereby an I/O device gets access to memory directly without having the microprocessor in between. By this direct path, a word input through a device can be stored in memory, or a word from memory can be output through a device, on the device's request. It is also possible for a word in memory to be moved to another place in memory using direct memory access. Another technique, *context switching*, is actually a more general type of DMA. The *context* of a processor is its set of MPU registers (as Texas Instruments uses the term) and the instruction set of the processor. To switch context means to logically disconnect the existing set of registers — bringing in a new set to be used in their place — or to use a different instruction set, or both. Context switching is discussed in the next section.

One of the fastest ways to input data to a buffer is direct memory access. Compared to techniques discussed earlier, this technique requires considerably more hardware and is considerably faster. A *DMA channel* is the additional logic needed to move data to or from an I/O device. In DMA, a word is moved from the device to a memory in a *DMA transfer*. The device requests transferring a word to or from memory; the microprocessor CPU, which may be in the middle of an operation, simply stops what it is doing for one to five memory cycles and *releases control* of the address and data bus to its memory by disabling its tristate drivers that drive these buses; the I/O system including the DMA channel is then expected to use those cycles to transfer words from its input port to a memory location. Successive words are moved this way into or from a buffer.

The DMA device has a DESTINATION address, a COUNT, and a DONE status bit. The DESTINATION and COUNT registers are initialized before DMA begins, and are incremented, and decremented, respectively, as each word is moved.

Two DMA techniques are generally available. An internal I/O device or an external I/O device can use DMA. If an external I/O device wishes to input or output data, it causes an edge on a pin control signal, which *steals a memory cycle* to transfer one word in *cycle steal mode* or it asserts a level which halts the microprocessor to transfer one or more words in *burst mode*, as long as the level remains asserted. This sequence of events happens when a DMA request is made to input a byte using cycle stealing:

1.   A falling edge occurs on an input pin, signaling the availability of data.
2.   In the DMA controller, a request is made and granted, and a memory cycle is stolen from the processor.

3. The controller signals the I/O device to read a byte from a port.

4. The byte is written into memory using a DESTINATION address.

5. The COUNT value is decremented. If it becomes 0, the DONE status bit is set. The program gadflies or interrupts on this DONE bit, and resumes when it is set.

There is a two-level busy-done state associated with DMA, as with any I/O transfer that fill or empties a buffer, as discussed at the beginning of this chapter. The low-level busy-done state is associated with the transfer of single words. BUSY is when a word is requested from an input device and has not been input, or is sent to an output device and has not been fully output (the hardware is punching the paper, in the paper tape example). The high-level busy-done state is associated with the transfer of the buffer. BUSY is when the buffer is being written into from an input device and has not been completely filled, or the buffer is being read from into an input device and has not been completely emptied. The DMA channel synchronizes to the low-level busy-done state to move words into or out of the I/O device. The computer can synchronize with the high-level busy-done state in the ways discussed so far. A real-time synchronization would have the processor do some program or execute a wait loop until enough time has elapsed for the buffer to be filled or emptied. Gadfly synchronization was used in the example given above. An interrupt could be used to indicate that the buffer is full.

I/O used in high-level languages is often *buffered* or *cached*. For input, a buffer or *cache* is maintained, and filled with more data than are needed. In *lazy* buffer management, the buffer is filled with data only when some data input is requested, but more data are put into the buffer than is requested in order to take data from the buffer, rather than from the input device, when some more data are needed later. In *eager* buffer management, the buffer is filled with data before some data input is requested, so that it will be in the buffer when it is requested. This technique makes the I/O device faster.

Finally, DMA can be used to synchronize to the high-level busy-done state; a kind of DMA^2. In larger computers such as an IBM mainframe, such a pair of DMA channels is called an *I/O Channel*. In an I/O channel, a second DMA that synchronizes the high-level busy-done state of the first DMA channel will refill the COUNT, SOURCE, DESTINATION, and CONTROL of the first DMA channel that moves words synchronizing to the low-level busy-done state. Thus, after one buffer is filled or emptied by the first DMA channel, the second DMA channel sets up the first DMA channel so the next buffer is set up to be filled or emptied. The second DMA channel's buffer is conceptually a program called the *I/O channel program*. This channel itself has busy-done states. BUSY occurs when some, but not all, buffers have been moved, and DONE occurs when all buffers have been moved. How can this busy-done state synchronizing the high-level busy-done states be synchronized? Here we go again. It can be synchronized using real-time, gadfly, interrupt, or DMA synchronization. However, DMA^3 is not very useful; DMA would not be used to reload the COUNT, SOURCE, DESTINATION, and CONTROL of the second DMA channel.

Direct memory access requires more hardware and may restrict the choice of some hardware used in I/O systems. The DMA channel must be added to the system, and the other I/O chips should be selected to cooperate with it. However, the amount of software can be less than with other techniques because all the software does is initialize some of the ports and then wait for the data to be moved. The main attraction of DMA is that the data can be moved during a memory cycle or two anytime, without waiting for the MPU to use software to move the data.

Although DMA is not available on the MMC2001, it is implemented on the Motorola 68340 and HC08 microcntrollers. As we discuss in Chapter 8, there is room for additional devices in the MMC2001, so a DMA controller might be added to it.

6.3.2   Context Switching

An interesting variation to DMA, uniquely attractive because it is inexpensive, is to use two or more microprocessors on the same address and data bus. See Figure 6.17. One runs the main program. This one stops when a device requests service, as if a DMA request were being honored, and another microprocessor starts. When the first stops, it releases control over the address and data buses, which are common to all the microprocessors and to memory and I/O, so the second can use them. The second microprocessor, which then can execute the interrupt request handler, is started more quickly because the registers in the first are saved merely by freezing them in place rather than saving them on a stack. The registers in the second could already contain the values needed by the interrupt request handler, so they would not need to be initialized. DMA using a DMA chip is restricted to just inputting a word into, or outputting a word from, a buffer; whereas the second microprocessor can execute any software routine after obtaining direct memory access from the first microprocessor.

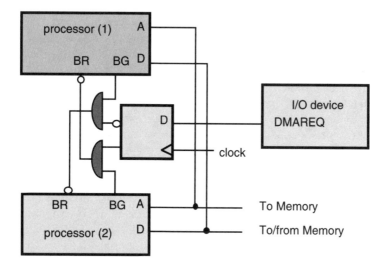

Figure 6.17. Connections for Context Switching

A complex operation is easy to do with *context switching*. While ordinary DMA cannot do this operation, context switching is faster than interrupt synchronization because, not only are the registers not saved, but also they remain in the second processor, so they usually don't have to be initialized each time an interrupt occurs.

Finally, any set of microcomputers having DMA capability can be used in this manner; the one operating the main program need not be the same model as the one handling a device. This means you can put a new microprocessor in your old microcomputer. The old microprocessor is turned on to run programs left over from earlier days, and the new microprocessor is turned on to execute the new and better programs. This is an alternative to simulation or emulation in microprogramming. It is better because the best machine to emulate itself is usually itself. And putting two microprocessors in the same microcomputer hardly has an impact on the system's cost.

A coprocessor, such as the floating-point 68881, essentially uses the same concept as context switching. Whenever the main processor detects a floating-point add that should be executed in the 68881, it gives up the bus to the coprocessor which does one instruction, and then the main processor resumes decoding the next instruction. A one-instruction context switch makes the main processor and coprocessor appear to be part of a processor having the main processor's instructions plus the coprocessor's instructions. While coprocessors can be designed to handle I/O, they usually handle data computation.

Though this technique is not used often by designers because they are not familiar with it, it is useful for microcomputers because the added cost for a microprocessor is so small and the speed and flexibility gained are the equivalent of somewhere between those attained by true DMA and vectored interrupt, a quality that is often just what is required.

6.3.3 Memory Buffer Synchronization

The last techniques we will consider that synchronize fast I/O devices involve their use of memory, which is not restricted by memory conflicts with the microprocessor. One technique uses a completely separate and possibly faster memory, called a *shuttle memory*. A variant of it uses an I/O device to access memory, like indirect I/O, and is called an *indirect memory*. Another uses the same memory as the microprocessor, but this memory is fast and can be *time-multiplexed*, giving time slices to the I/O device. In a sense, these techniques solve the synchronization problem by avoiding it, by decoupling the microprocessor from the I/O by means of a memory that can be completely controlled by the I/O device.

Figure 6.18a shows a shuttle memory. The multiplexer connects the 16 address lines that go into the shuttle buffer and the 16 data lines that go to or from the buffer to the microprocessor or the I/O device. The buffer memory is shuttled between the microprocessor and the I/O device. When the buffer is connected to the I/O device, it has total and unrestricted use of the shuttle memory buffer whenever I/O operations take place. The microprocessor can access its primary memory and I/O at this time without conflict with the I/O's access to its shuttle memory because they are separate

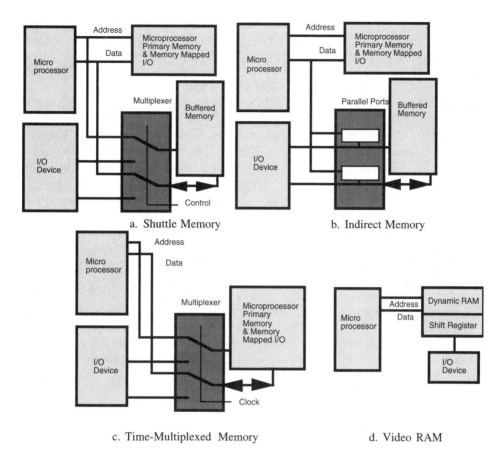

Figure 6.18. Fast Synchronization Mechanisms Using Memory Organizations

from the shuttle memory used by the I/O device. The multiplexer switches are both in the lower position at that time. Then when the microprocessor wishes to get the data in the shuttle memory, the multiplexer switches are put in the upper position, and the microprocessor has access to the shuttle memory just as it has to its own primary memory. The buffer appears in the memory address space of the microprocessor. The microprocessor can load and store data in the shuttle memory. The synchronization problem is solved by avoiding it. Synchronization is required as data are moved to and from the I/O device from and to the shuttle memory; but the buffer memory is wholly controlled by the I/O device, so that it is not too difficult. The microprocessor can move data to and from the shuttle memory at leisure. It can even tolerate the delays that result from handling an interrupt at any time, when it moves data from one location in its memory to another location in it. There is no need for synchronization in that operation.

We built a parallel computer called TRAC, which used shuttle memories. The shuttle memories were connected to one processor or another processor. Once connected to a processor, the shuttle memory behaved like local memory and did not

experience memory contention. Caches were simple to use with this variation of a
shared memory. In I/O devices discussed in this book, the shuttle memory similarly
removes the problem of memory contention from the synchronization problem.

A variation of a shuttle memory uses a parallel I/O device like MMC2001 ports
in place of the multiplexer. (See Figure 6.18b.) The external port pins of an I/O
device connect to the address and data buses of the memory. The processor writes an
address to a parallel output port and then reads (or writes) the data to (or from)
another port to access the memory. The only way to read or write in the buffer is to
send addresses and data to or from the I/O device, just as we accessed an indirect I/O
device in §5.4.1, so we call it *indirect memory*. The M6818A's RAM from addresses
0xE to 0x3F is an indirect memory. Indirect buffer memory is completely separate
from the microprocessor primary memory. When the (fast) memory is not controlled
by the microprocessor through the I/O ports, it can be completely controlled by the
I/O device, so it can synchronize to fast I/O devices. Only the memory-mapped
parallel I/O device takes up memory space in the primary memory, whereas the
shuttle memory technique has the whole shuttle memory in the primary memory
address space when the processor accesses it. But to access the buffer memory, you
use slow subroutines as you do in indirect I/O.

Indirect memory using MCM6264D-45 8 K by 8 chip can be implemented on the
MMC2001. (See Figure 6.19.) The 12-bit memory chip address pins, and three
control pins E1, G and W are connected to keyboard *KPDR* and its 8-bit data pins are
connected to edge port *EPDR*. Tristate drivers (74HC244's) connect the external
device to the memory when the MMC2001 is not accessing it, so when not being
used, all MMC2001 port bits are made inputs to allow the 74HC244s to access the
memory signals. The MMC2001 reads the memory, following the timing diagram in
Figure 6.18a, by making *KCD* and *EPDR* outputs, making *E1*, *G*, and *W* high, outputting the high byte of the address to *KCD*, and the low byte to *EPDR*, asserting *E1*
low, asserting *G* low, reading the data from *PORTC*, negating *E1* high, and negating
G high. Writing, following the timing diagram in Figure 6.18b, is done by making
KPDR output, making *E1*, *G*, and *W* high, outputting the address to *KPDR*, asserting *W*
low, asserting *E1* low, making *EPDR* output, writing the data to *EPDR*, negating *E1*
high, and negating *W* high. The hardware is connected as shown in Figure 6.18c. Only
connections to the MMC2001 are shown.

```
enum {E1 = 0x8000, G = 0x4000, W = 0x2000};

char get(int a) { char data;
    KCDD = 0xffff; EPDDR = 0; KPDR = a | E1 | G | W;;
    KPDR &= ~E1; KPDR &= ~G; data = EPDR; KPDR |= E1; KPDR |= G;
    KCDD = EPDR = 0; return data;
}

void put(int a, char data) {
    KCDD = 0xffff; KPDR = a | E1 | G | W;
    KPDR &= ~W; KPDR &= ~E1; EPDDR = 0xffff; EPDR = data;
    KPDR |= E1; KPDR |= W; KCDD = EPDR = 0;
}
```

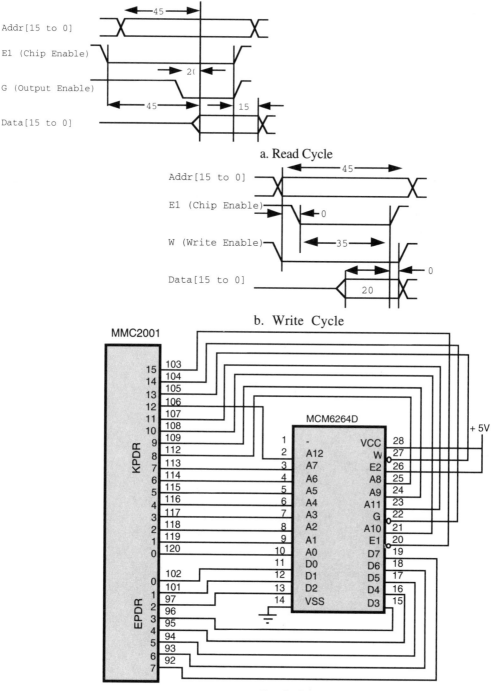

a. Read Cycle

b. Write Cycle

c. Logic Diagram

Figure 6.19. Indirect Memory Using an MCM6264D-45

A very similar mechanism uses the same memory for the primary memory and the buffer memory, but that memory is twice as fast as is necessary for the processor. (See Figure 6.18c.) In one processor memory cycle, the memory executes two memory cycles — one for the processor and one for the I/O device. The multiplexer is switched to the I/O device (for the first half of the memory cycle) and to the processor (for the last half of the memory cycle) to time-multiplex the memory. The I/O device always gets one memory cycle all to itself because the processor only uses the other memory cycle.

The time-multiplexed memory uses the same memory as the microprocessor, but this memory is twice as fast, and the processor gets one time slice, then the I/O device gets one time slice, in an endless cycle. It is obviously less costly than the shuttle and indirect memories because a single large memory is used rather than two smaller memories. Its operation is very similar to DMA. In fact, it is sometimes called *transparent DMA*. However, the memory must be twice as fast as the processor, and the I/O device must synchronize to the processor (CLK) clock in this technique.

The shuttle and indirect memories are more costly, but a very fast (40-nanosecond cycle time) memory can be used in the buffer and run asynchronously at full speed when accessed by the I/O device, but run about the speed of the CLK clock when the processor accesses it. All three techniques provide for faster synchronization to the I/O device than the techniques discussed in the previous subsection. They can transfer data on every memory cycle, without handshaking with the processor, to acquire it or memory. They find considerable use in CRT, hard disk, and fast communication devices.

Finally a video RAM (Figure 6.18d) is a dynamic memory in which a row can be read from DRAM into a shift register, or written into DRAM from a shift register, in one memory cycle. The shift register can then shift data into or out of an I/O device. The TMS48C121 is a (128K, 8) DRAM with a (512,8) shift register, which can shift data into or out of an I/O device at 30 nanoseconds per byte.

One of the main points of this section is that extra hardware can be added to meet greater synchronization demands met in fast I/O devices. While DMA is popular, it is actually not the fastest technique because handshaking with the microprocessor and the cycle time of the main memory slow it down. Shuttle or indirect memories that use fast static RAMs can be significantly faster than DMA. Moreover, for all of these techniques, the controlling software can usually be quite slow, and thus can be coded in C without loss of performance compared to programs coded in assembler language.

6.4   A Designer's Selection of Synchronization Mechanisms

The designer should select the synchronization mechanism and the software that implements that mechanism, from the techniques mentioned in this chapter. We first review the synchronization mechanisms. Then we summarize the software used with these mechanisms.

The designer should first determine whether synchronization is fundamentally synchronous or asynchronous, and whether the external world is slower, or faster, than the computer. Figure 6.20 summarizes the techniques presented in this chapter.

Real-time is a synchronous synchronization technique, in that the I/O device's timing is synchronized to the processor's timing to determine when data are to be transferred. It uses the least hardware and is practical if an inexpensive micro-computer has nothing to do but time out an I/O operation. However, real-time synchronization, using just the delay of the execution of instructions needed to carry out the algorithm being implemented, are difficult to program. Real-time synchro-nization using delay loops are inefficient because a computer generally cannot do anything else when it is in a delay loop; this mechanism should not be used in M·CORE systems. The next chapter presents counter-timer hardware, which better provides for synchronous synchronization.

Gadfly is an asynchronous synchronization technique, in that the I/O device's timing is not synchronized to the processor's timing but the processor locks onto its timing when data are to be transferred. Gadfly programs are easier to write than real-time programs, but require that the hardware provide an indication of DONE. Also, a computer generally cannot do anything else when it is in a gadfly loop, so this is as inefficient as real-time synchronization.

The interrupt, an asynchronous synchronization technique, allows the processor to execute other unrelated programs while waiting for a device to become DONE, but requires more hardware to request service from the processor. However, don't use them, just because they are available. Interrupts can occur anywhere in the execution of the program, and in some places they may cause bugs. Pinning down the cause of the bug is quite difficult. Such bugs often go undetected until, according to Murphy's law, their occurrence will cause the most embarrassment. By compar-ison, gadfly synchronization is much more predictable and easier to debug. If you have nothing else to do, so interrupt synchronization provides no advantage, you should use gadfly synchronization.

Interrupt polling usually requires a sequence of tests of each device's status ports. This method of polling is used by the Ariel operating system. However, the MMC2001 has a port, *NIPDR*, to read 32 normal interrupt requests at once, and the M·CORE processor has the instruction *FF1* (find first one) to determine the highest

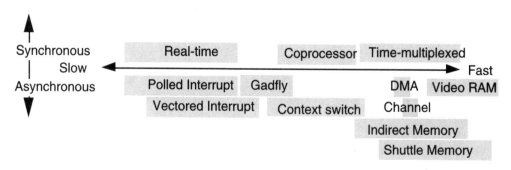

Figure 6.20. Synchronization Mechanisms Summarized

priority request in one instruction, which significantly reduces latency for lower priority devices.

The M·CORE processor provides for autovectored and for external vectored interrupts. However, the MMC2001 only implements autovectored normal and fast interrupts. Fast interrupts can be used for one high-performance device that needs service in a shorter time, while other devices use normal interrupts.

Note that the interrupt technique can be used together with the gadfly technique. With the gadfly technique, the interrupts are all disabled by clearing a bit in the interrupt module enable port (*NIER* or *FIER*) or by clearing a control bit in the device. Then the program can loop as it tests the device, without fear of being pulled out by an interrupt. When used together, gadfly is used for careful, individual stepping of an I/O system; and interrupt, for automatic, rapid feeding of data.

The DMA technique is useful for fast devices that require low latency. This technique can only store data in a buffer or read data from a buffer. DMA[2] (the I/O channel) can restart a DMA transfer a little bit faster than simple DMA. A variation of DMA, context switching, is almost as fast and flexible as the interrupt technique. A coprocessor uses a similar mechanism, and although it is generally used to execute data computation, it could be used for I/O. Shuttle, indirect, and time-multiplexed memories can be used for the fastest devices. What is somewhat surprising, DMA, which requires a fair amount of extra and expensive hardware, actually is most desirable for a rather limited range of synchronization timing. Indirect and shuttle memories can be used for much faster synchronization, and context switching for slightly slower synchronization.

The designer should carefully select the software technique for implementing the synchronization mechanism. He or she should consider assembler or C language programming to support the synchronization mechanism, or object-oriented or device driver techniques.

Assembly language real-time programming is often needed if the timing is very tight, such as in generating 100-KHz modulated X-10 frames. C language real-time programming is prone to changes of timing if a different compiler is used, or later version of the compiler is used, or if different optimization settings are selected, even for the same version of the same compiler. Nevertheless, C language real-time programming is of use in smaller microcontrollers such as the Motorola 6805.

Assembly language gadfly programming is often needed if the timing is very tight, such as in reading data from a floppy disk controller. C language gadfly programming is generally quite acceptable, and much easier to use, for most applications.

Object-oriented classes can be written for real-time or gadfly synchronized I/O devices. They are reasonably easy to write and test, and provide the basis for a library of prewritten, pretested, hardware-software packages that can be inserted into an application, even substituted for another object in an application late in the design.

Ariel device drivers should be used when several tasks share an I/O device. Ariel device drivers can be written for I/O devices that use real-time or gadfly synchronization, but they are significantly more complex than other I/O software, when we wish to permit other tasks to execute when we wait for the BUSY state to end, because the driver's procedures operate in the S/SE state that prohibits other tasks from being executed. The driver writer has to use some normal (T state) procedures that are coupled to the device driver S/SE state procedures, so that when the device is BUSY, the device driver procedure in S/SE state can exit to the normal (T state) procedure, where the task can be blocked, so that other tasks can be executed, until the device becomes DONE.

Interrupt synchronization can use assembly language programs, particularly in polling for the highest priority interrupt using the *FF1* instruction. However, the handler can be written almost exclusively in C or C++, with calls to subroutines written in assembler language to enable or disable interrupts. An object-oriented class can be written for an I/O device that uses interrupt synchronization, but the interrupt handler cannot be an object member function, because there is no way to pass the object pointer into it immediately after an interrupt occurs. But the interrupt handler can be an ordinary C procedure (with the key word interrupt before the procedure name) which calls on a member function. Then the member function can easily access the object's members.

Ariel device drivers can be written for I/O devices that implement interrupt synchronization. Ariel takes care of linking the interrupt handler to the hardware interrupt vector, so the programmer merely needs to put the handler's address into the appropriate structure to initialize the use of the interrupt. Handlers that fill or empty buffers are quite easy to implement, and can send a signal to a task to indicate when the buffer is DONE. However, interrupt handlers that use a queue are more difficult to implement. The driver writer has to use some normal (T state) procedures that are coupled to the device driver S/SE state procedures that push to or pull from the queue. This provides a way, when the queue is not available, for the device driver S/SE state procedure to exit to the normal (T state) procedure, where the task can be put in the blocked state, so that other tasks can be executed, until the device becomes DONE.

Fast I/O devices require extra hardware such as a DMA channel or a shuttle memory. The software coupled to such hardware again may use assembler or C language programs, object-oriented programming or device drivers. However, there being a large range of different types of hardware for fast I/O devices, we do not venture to offer any advice as to which is a better software approach for fast I/O devices.

6.5   Conclusions

We have discussed eight alternatives for solving the synchronization problem, and four software programming techniques. The programmer should select from these techniques for the best techniques that suit his or her application. This chapter has provided the basic theory, and may concrete examples, to help the programmer make that decision.

Do You Know These Terms?

See the End of Chapter 1 for Instructions.

synchronization
MPU is enabled
contact bounce
cached IO
start
MPU is masked
wait-and-see
cache
stop
MPU is disabled
unsolicited data
lazy management
complete
honour
  an interrupt

Direct memory
eager management
real-time
armed
  access (DMA)
I/O Channel
  synchronization
disarmed
context switch
I/O channel
  program
gadfly
  synchronization
enable
context

context switch
gadfly loop
disable
DMA channel
shuttle memory
handshake
critical section
DMA transfer
indirect memory
X-10
software disarm
steal a memory cycle
time-multiplexed
normal interrupt
poll

cycle steal model
  memory
fast interrupt
priority
burst mode
indirect memory
request an interrupt
round-robin
  priority
buffered I/O
transparent DMA

Problems

Problems 1, 10, 16, 19, and 27 are paragraph correction problems; for guidelines, refer to the problems at the end of Chapter 1. Guidelines for software problems are given at the end of Chapter 2, and guidelines for hardware problems, at the end of Chapter 3. Guidelines for Ariel Device Drivers are given at the end of Chapter 7.

1.* Synchronization is used to coordinate a computer to an input-output device. The device has busy, completion, and done states. The busy state is when data can be given to it or taken from it. The device puts itself into the done state when it has completed the action requested by the computer. A paper tape punch, by analogy to the paper tape reader, is in the idle state when it is not in use; in the busy state when it is punching a pattern that corresponds to the word that was output just before the done state was entered; and in the done state when the pattern has been punched. The busy and idle states are indistinguishable in an output device like this one, unless error conditions are to be recognized (in the idle state). An address trigger will generate a pulse whenever an address is generated. Its output should never be asserted if the CLK clock is high. Address triggers are often used to start a device or to indicate completion by the device.

2. Write a C program that punches paper tape using real-time synchronization. Analogous to the latter procedure `main()` in §6.1.1, data are output through *KCD* at a rate determined by the empirically evaluated constant *N*.

3. Write a hand-coded assembler language program that punches paper tape using real-time synchronization. Analogous to the last program in §6.1.1, data are output through *KCD* at a rate determined by the empirically evaluated constant *N*.

4. Write a `main()` procedure that punches paper tape using gadfly synchronization. Analogous to the procedure `main()` in §6.1.2, data are output through *KCD* when bit 7 of *EPDR* falls, until the next time it falls. Use a edge flag to detect the edge.

5. Write a hand-coded assembler language program PUNCH that punches paper tape using gadfly synchronization. Analogous to the last program in §6.1.2, data are output through *KCD* when bit 7 of *EPDR* falls, until the next time it falls. Use a edge flag to detect the edge.

6. The LED signal (Figure 6.5a) can be fully generated by the MMC2001 without using a 555 timer chip; when the output of *KCD* bit 0 is H (1) the LED is lit. To send a T (1) the LED should be pulsed at a rate of 38 KHz for 700 μsec. and be off for 350 μsec. To send a F (0) the LED should be pulsed at a rate of 38 KHz for 350 μsec. and be off for 350 μsec. Show a self-initializing procedure *void sendIr(int data)* that sends the least significant 11 bits of argument *data* through the infrared LED.

7. The MMC2001 and a 555 timer chip (Figure 6.5c) can generate a BSR X10 signal (Figure 6.7d); when the output of *EPDR* bit 0 is H (1) the 555 generates a 100-KHz pulse train on the 110-V. line; when this output is low, the 555 does not generate pulses. Rewrite the procedure `sendBsr` to send 16 bits through *EPDR* bit 0, synchronized to the 60-Hz waveform input on *EPDR* bit 1. A T (1) should be sent as a burst of 100-KHz pulses for 1/1080 s repeated each 1/360 s for 3 bursts after each edge of the 60-Hz waveform. An F (0) should be sent as no burst for 1/120 s. Show a self-initializing procedure `void sendBsr(int data)` to send the 16-bit `data`.

8. Rewrite §6.1.6's device driver so that it inputs or outputs until a carraige return is found, up to a maximum number of characters. Use §5.1.7.1's `main, put` and `get`.

9. Rewrite §6.1.6's device driver so that it inputs or outputs vectors having 16-bit `short` elements. Use §5.1.7.1's `main,` put and `get,` modified for `short` data.

10.* The real-time synchronization technique times the duration of external actions using the microcomputer CLK clock as a timing reference and the program counter as a kind of frequency divider. This technique uses the least amount of hardware, because the program itself contains segments that keep account of the busy-done state of the device. The program can be changed easily without upsetting the synchronization, because program segments execute in the same time regardless of the instructions in the segment. Computer scientists, for no good reason, abhor real-time synchronization, so it should never be used, even on a microcontroller dedicated to a single control function. Real-time synchronization cannot be used to synchronize error conditions, because we cannot predict the time of the next error. Gadfly synchronization uses hardware to track device state, and the program watches this

hardware's outputs. Therefore, feedback from the device controls I/O operation so it
can be completed as soon as possible. Nevertheless, real-time synchronization is
always faster than gadfly synchronization, because the former is always timed for the
minimum time to complete a device's action.

11. Write a C program that punches paper tape using interrupt synchronization.
Analogous to *main()* in §6.2.3.1, data are output through *KCD* from
buffer[0x80] each time an interrupt is generated by an edge on bit 7 of *EPDR*
causing *flag* to be set in the interrupt handler: *interrupt handler() { EPFR =
0x80; flag++ }*.

12. Write a C program that punches paper tape using interrupt synchronization.
Analogous to *main()* in §6.2.3.2, data are output through *KCD* from
buffer[0x80] each time an interrupt is generated by an edge on bit 7 of *EPDR*.

13. Design a hardware breakpoint device. When the address and RW signals are
determinate, this device compares them against a breakpoint address *ba* written in
KRDR, and a *brw* bit written in *EPDR* bit 0. The address is compared with *ba*, and the
RW signal with *brw*, by open collector quad 2-input exclusive NOR gates
(74HC266s: see Figure 6.21) whose output is low if its inputs differ. The 74HC266's
outputs are connected in a wire-AND bus with a 4.7 K pull-up resistor. If each pair
of inputs are equal, the outputs of the gates will be high, otherwise the outputs will
be low. They connect to *EPDR* bit 1 to generate an interrupt that stops the program
when the address matching the number in the output ports is generated.

a. Show a complete logic diagram of the system. Show all connections and pins
to 74HC gates. However, do not show pin numbers on the M·CORE itself, or
+ 5, or Gnd.

b. Show a self-initializing procedure *setBreakpoint(short a, char rw)*
to generate a edge *EPDR* bit 1 interrupt when the address *a* is written into if *rw* is
0, or read from if *rw* is 1.

14. Write a round-robin *interrupt handler()* (§6.2.4) that checks only EPDD's
edge interrupt bits. If *EPPAR* bits 0 and 1 are set so that *EPDR* bit 0 sees a rising or
falling edge, and *EPFR* bit 0 therefore becomes set, the service procedure *h0()* is
executed. The analogous actions occur for the other bits of *EPDR*. If *EPDR* bit 0 did

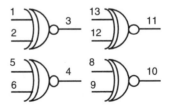

Figure 6.21. 74HC266

not see an edge, *EPDR* bit 1 is checked similarly, and so on. If *EPDR* bit n had an edge and procedure *hn()* was executed, 0 n < 8, then upon the next *EPDR* edge interrupt, bit $n+1$ is tested first, then bit $n+2$, and so on, where bit 0 is tested after bit 7. In this version of handling edge interrupts, the service procedures do not return a value and do not clear the interrupt, but `interrupt handler()` does clear the interrupt.

15. Compute the difference in (worst case) latency between polled interrupts and vectored interrupts for edge interrupts. Compare the first *interrupt handler()* in §6.2.4 to *interrupt handler1()* and *interrupt handler2()* in §6.2.5.

16.* Interrupts permit the computer to perform some useful function while waiting for a device to become busy or for an error condition to arise. Interrupts are always faster than a simple gadfly loop, because they save the state of the machine and restore it, while the gadfly technique has to loop a long time. When an edge device requests an interrupt, if the device is enabled by clearing a bit of the normal interrupt enable register of *NIER*, the flag flip-flop, which is in *EPFR* is set. When this flag flip-flop is set, the MPU immediately honors the interrupt, saving the values in all output registers on the stack and jumping directly to a handler routine. The handler may have just an RTS instruction to return to the program that was originally running. Vectored interrupts use external hardware to eliminate the polling routine in the device handler, so the interrupt handler can be executed immediately. Interrupts, and vectored interrupts in particular, should be used whenever the latency time requirement is critically small or something useful can be done while waiting for an interrupt; otherwise, real-time or gadfly synchronization should be used.

17. The coincident select keyboard shown in Figure 6.16c is modified to have sixteen columns. The left 8 columns of switches are connected to *KCD* as shown, and the right 8 columns are similarly connected to *EPDR*. Rewite the matrix, or coincident select handler, *interrupt handler()* in §6.2.7.1 to push a (low-order) 3-bit row concatenated with a (higher-order) 4-bit column number into the queue, where *EPDD's* bits correspond to columns 15 to 8 and *KCD's* bits correspond to columns 7 to 0.

18. The coincident select keyboard shown in Figure 6.16c is modified to have sixteen rows. The top 8 rows of switches are connected to *KRD* as shown, and the bottom 8 rows are similarly connected to *EPDR*. Rewite the matrix, or coincident select, handler *interrupt handler()* of §6.2.7.1 to push a (low-order) 4-bit row concatenated with a (higher-order) 3-bit column number into the queue, where *EPDD's* bits correspond to rows 15 to 8 and *KRD's* bits correspond to rows 7 to 0.

19. * Key bounce is a problem when the user bangs on a key repeatedly to get something done. It can be eliminated by a sample-and-hold circuit that takes a "snapshot" of the signal just once. Software debouncing is rarely done in microcomputers because it takes too much of the microcomputer's valuable time to monitor the key. Keyboards are often used on microcomputers because the user may want to enter different commands, and this is most easily done with a set of keys.

Linear selection is often used for a keyboard that has a lot of keys, like a typewriter. N-key rollover is a property of keyboards whereby the microcomputer can correctly determine what keys were pressed and in what order they were pressed, provided that no more than N keys are pressed at one time. Two-key rollover is commonly used in microcomputer systems but is rather inadequate for most adept users, who often hold several keys down at once.

20. Write a C program that follows §6.2.7.2's program, but outputs a five element vector where each element is 16-bits. Show only the $structs$ and procedures that are different from those in §6.2.7.2.

21. Write a C program that follows §6.2.7.2's program, but selects one of a group of five '589s and five '595s, there being eight such groups, using KCD. KCD's outputs are normally high, but one pin is asserted low to enable the '589s and '595s of a group; if bit 0 is asserted then group 0 of the '589s and five '595s is shifted in and out, and so on.

22. Write and Ariel device driver to store data input from KCD into $char$ $buffer[10]$ in the main program each time the interrupt handler is executed, in the manner of §6.2.9.1.

23. Write an Ariel interrupt device driver emptying a buffer, similar to §6.2.9.2's driver, enabling interrupts only when a buffer is set up for $\_pio\_cmd()$.

24. Write a program that uses an Ariel interrupt device driver for filling or emptying a buffer, similar to §6.2.9.2's program, in which the direction register can be changed at or after initialization, passing the port and direction address and initial value of the direction register as in §5.1.6.

25. Write an Ariel interrupt device driver using a queue for output from $\_pio\_cmd()$ to the interrupt handler, similar to §6.2.9.3's driver, enabling interrupts only when the queue has data in it is set up for $\_pio\_cmd()$. What advantage, if any, does this device driver have over that shown in §6.2.9.2?

26. Write a derived class of $Port$ that calls the Ariel device driver in §6.2.9.2 to output characters, ASCII character strings, or $char$ vectors. This class's input operations will not be used.

27. * Direct memory access is a synchronization technique that uses an extra processor that is able to move words from a device to memory, or vice versa. With an output device, when the device is able to output another word, it will assert a request to the DMA chip, which checks its busy-done state, and, if done, it requests that the microprocessor stop and release its control of R/W and address and data bus. The microprocessor will tell the DMA device when it has released control, the DMA device will output on the data bus and will send a signal to the I/O device to put a signal on the R/W line and an address on the address bus. A DMA chip itself is an

I/O device, whose busy state indicates that a buffer full of data has been moved. The busy state then is an interrupt request. Either gadfly or interrupt synchronization can be used to start a program when the buffer has been moved.

28. A pair of indirect memories using a MCM6264D-45 will be implemented, where one is shown in Figure 6.19, so that when one is being accessed by the MMC2001, the other is free to be used for I/O. The first memory is connected as shown in that figure. Show the logic diagram, including pin numbers, of the connections between the second memory and the MMC2001, which uses *KDDR* for the address and control signals, and *EPDR* for data. This memory's enable E1 is on *KCD* bit 7, its output enable G is on *KCD* bit 6, and its write control W is on *KCD* bit 5.

29. Complete the logical design of an indirect memory using a MCM6264D-45 (Figure 6.19) for input from a fast 8-bit data source *data*. Use 74HC163 counters (Figure 4.12) and 74HC244 tristate drivers (Figure 4.3) to supply addresses and a 74HC244 tristate driver to supply *data* to the MCM6264D-45 when the I/O device needs to write into the memory. When it does so, it pulses control signal WRITE low then high. The 74HC163 counters are written into by pulsing *KCD* bit 7, which is normally high, low and high when the address to be written into the counters is in *KCD* and *EPDR*.

a. Show the logic diagram for the complete circuit, but excluding connections already shown on Figure 6.19.

b. show a self-initializing procedure *void setAddress(int a)* that writes the address into the counter.

7

Timer Devices and Time Sharing

Chapter 6 covered synchronization techniques. This chapter continues this theme, especially where the external world is considerably slower than the micro-controller. In this case, hardware timer devices can be used to wait long periods of time, or to "sleep" and then "awaken" after a long interval of time, to control this external world.

The MMC2001 has a *Timer-Reset Module* and a *Pulsewidth Modulator* that can be used to precisely time events, from microseconds to days. This chapter shows how to program these devices, how to use interrupts with them, and how they can be used to time multiplex the use of the processor. The first section describes these hardware devices and their use for timing operations. The second section shows how these devices can be used to time multiplex the MPU. The third section shows how time sharing can be used with the Ariel operating system. This section shows how a prepackaged and debugged operating system can provide time-sharing capabilities with significantly less effort.

Upon completion of this chapter, the reader should be able to use each of the devices to generate interrupts or perform other functions the devices were designed for. He or she should be able to implement time sharing with these devices, and use the time sharing capabilities of the Ariel operating system.

7.1 Timer Devices

The MMC2001's *Pulsewidth Modulator* (PWM), discussed in the first subsection, can generate a pulse train signal on a pin which can be used to control external devices, and it can generate interrupts. The MMC2001's *Timer-Reset Module* (TRM) comprises a time-of-day module, a watchdog timer module, and a programmable interval timer. The first and last modules can be used to cause interrupts at precise times. The other module causes an interrupt if the program fails to access it at a regular rate. The PWM is discussed first, and the TRM is discussed in later subsections.

7.1.1   Pulsewidth Modulator

The MMC2001 has six pulsewidth modulators, each of which is like module 0 shown
in Figure 7.1, module i at an address 8 * *i* plus the addresses in Figure 7.1. The ports
for module 0 can be declared as:

```
volatile short PWMCR0@0x10005000, PWMPR0@0x10005002,
    PWMWR0@0x10005004, PWMCTR0@0x10005006;
```

The control register *PWMCR0's* bits 2 to 0 control a prescale counter, bit 3
enables the counter, bit 4 determines whether the output is a one-bit parallel port
with direction register or the output of the timer circuit, and bit 8 forces loading of
the module's period registers from period port *PWMPR0,* and width register from
width port *PWMWR0*. The counter port *PWMCTR0* is read-only; it permits the counter
value to be read. When the decrementing counter value becomes zero, it loads with
the period register. As long as the counter is greater than the width, the output is
asserted; then it becomes negated.

A pulse with modulator can be used to generate a repeating waveform with a
period and a pulsewidth that can be selected by numbers put in the control ports.
The procedure *pulseWidth* shown below will output a pulse train which is high for
(10 bit) *width* cycles and repeats after (10 bit) *period* cycles, where a cycle is
specified by the (3 bit) *preScale* value, as specified by Table 7.1.

```
enum { DOZE = 0x800, PWM_IRQ = 0x400, IRQ_EN = 0x200, LOAD = 0x100,
    DATA = 0x80, DIR = 0x40, POL = 0x20, MODE = 0x10, COUNT_EN = 8};
pulseWidth(short period, short width, short preScale){
    PWMPR0 = period; PWMWR0 = width;
    PWMCR0 = LOAD + MODE + COUNT_EN + preScale; /* start counter */
}
```

The width can be changed after each cycle. *putWidth* gadflies on the *PWMCTR0*
counter value. After it becomes zero, the port values are transferred to specify the
next period, and new values can be loaded. The counter decrements. Rather than
testing for zero, *putWidth* tests for when the value appears to increase, which
indicates that it passed zero. When it does, *putWidth* loads its argument into the
PWMWR0 port.

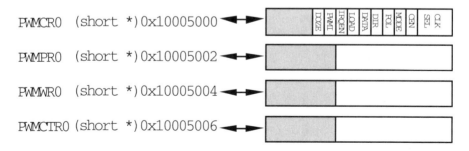

Figure 7.1. Pulsewidth Modulator

Table 7.1. PWM Prescale Values

preScale value	Divide CLK period by:	Period range (32 MHz CLK)	Resolution (32 MHz CLK)
0	4	0.25-127 µs	0.125 µs.
1	8	0.5-256 µs	0.25 µs
2	16	1 µs-0.5 ms	0.5 µs
3	64	4 µs-2 ms	2µs
4	256	16 µs-8.25 ms	8µs
5	2048	128 µs-32 ms	64 µs
6	16384	1 ms-0.5 s	0.5 ms
7	65536	4.1 ms-2 s	2.05 ms

```
putWidth(short width){ short lastValue;
    lastValue = PWMCTR0;
    while(PWMCTR0<lastValue) lastValue = PWMCTR0;
    PWMWR0 = width;
}
```

This procedure, *putWidth,* can modulate the periodic wave's duty cycle, which can be fed through a lowpass filter to generate an analog voltage proportional to the width. This can be used to generate audio signals, instead of using a D-to-A converter.

The pulsewidth modulator can generate interrupts to reload the port after the counter passes through zero. This interrupt can also be used to time-slice the MPU, as discussed in §7.2. The interrupt handler can be the normal interrupt handler of §6.2.2 or fast interrupt handler of §6.2.4. The program below illustrates how the PWM0 module can generate a normal interrupt each time the counter decrements to zero. We assume it is the only interrupt in this example. If this interrupt is polled among multiple interrupts as in §6.2.3, its address is put in *Jvector[10],* instead of location 28.

```
interrupt void handler(){short i=PWMCR0;/* rd control-status to clear int. */}

void main(){ char preScale = 0;
    disableInt(); PWMPR0 = period; PWMWR0 = width;
    PWMCR0 = IRQ_EN + LOAD + MODE + COUNT_EN + preScale; /* init ctrl */
    initInt(0x30000000, 0x30000200, handler, 10);
    NIER |= 1 << 10; /* enable normal interrupt on bit 10 */ enableInt();
}
```

We will use the traffic light controller example to illustrate how the PWM, and other timing devices in this section, can be used to synchronize to slow devices. We assume, in all these examples, that the traffic light table provides each light pattern's delay in seconds. The traffic light controller's *main* shown below is essentially the same as §5.2.3's interpreter, except that the delay is provided by the PWM, instead of a delay loop. This provides accurate timing, independent of compiler optimization options, and so on.

```
const unsigned char tbl[4][2]={{0x21,4},{0x22,1},{0x0c,6},{0x14,2}};
#define preScale 7
void main() { unsigned char i, j, preScale; short lastValue = 0;
     PWMCR0 = LOAD + MODE + COUNT_EN + preScale; /* start counter */
     KCDD = 0x3f; PWMPR0 = 488; PWMWR0 = 244; /* period to 1 s */
     do
         for(i=0;i<4;i++){
           KCD = tbl[i][0];
           for(j = 0; j < tbl[i][1]; j++)
                   {while(PWMCTR0 >= lastValue) lastValue = PWMCTR0;}
         };
       while(1);
}
```

7.1.2 Time-of-day Module

The time-of-day module (TOD), shown in Figure 7.2, maintains a running time of seconds, and fractions (1/256ths of a second). A port can hold a comparison alarm value of seconds and fractions. When the alarm seconds and fractions match the running time seconds and fractions, an interrupt can occur.

The time-of-day module's control-status port *TODCSR's* three least significant bits are, from least significant bit, a flag bit indicating an alarm match, an interrupt enable bit, and a counter enable bit. The seconds port *TODSR* is incremented each second, and port *TODFR* is incremented each 1/256th of a second. If *TODCSR's* three least significant bits are TTF, then if *TODSR = TODSAR*, and *TODFR = TODFAR*, a normal interrupt occurs if *NIER* bit 7 is set. The ports can be accessed as shown below:

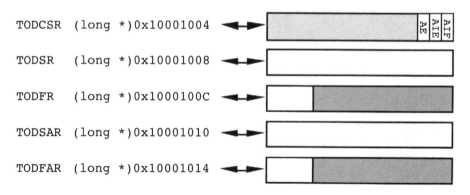

Figure 7.2. Time-of-Day Module

```
volatile long TODCSR@0x10001004, TODSR@0x10001008,
    TODFR@0x1000100C, TODSAR@0x10001010, TODFAR@0x10001014;
```

The procedure *main* below waits for *TODSR* = *TODSAR*, and *TODFR* = *TODFAR*. We initialize the seconds and fraction counters, and the alarm seconds and alarm fraction, and start the TOD module. It gadflies on the *AIF* bit of *TODCSR*.

```
void main(){
        TODSR = 0x12345678; TODFR = 0x12 << 24; /* set up current time */
        TODSAR = 0x89abcdef; TODFAR = 0x34 << 24; /* set up the alarm time */
        TODCSR = 4; /* start TOD */ while(TODCSR & 1) ; /* wait for timeout */
}
```

In the program below, the *handler* is entered when *TODSR* = *TODSAR*, and *TODFR* = *TODFAR*. The handler can initiate some operation when the alarm matches the counter.

```
interrupt void handler(){TODFAR = 0;/*write alarm fraction port to clear int.*/}
void main(){ disableInt();
        TODSAR = TODSR + 5; /* set up alarm s by adding time delay to s reg*/
        TODFAR = 1 << 24; /* set up fractional alarm time by just writing into it */
        TODCSR = 6; /* assertAE and AIE to turn on the module and enable interrupt */
        initInt(0x30000000, 0x30000200, handler, 10);
        NIER|= 1<<7; /* enable normal int. for TOD */ enableInt();
}
```

We again use the traffic light controller example to illustrate how the TOD can be used to synchronize to slow devices. This, too, provides accurate timing, independent of compiler optimization options, and so on. It illustrates the use of the TOD alarm to accurately synchronize events that are multiple seconds apart.

```
const unsigned char tbl[4][2]={{0x21,4},{0x22,1},{0x0c,6},{0x14,2}};

void main() { unsigned char i, j, preScale;
        TODCSR = 4; /* turn on TOD module */ KCDD = 0x3f;
        do
            for(i=0;i<4;i++){
                KCD = tbl[i][0]; TODSAR = TODSAR + tbl[i][1];
                TODFAR = 0; /*clr flag*/ while(TODCSR & 1) ; /*wait for timeout */
            };
        while(1);
}
```

7.1.3 Watchdog Timer

A *watchdog timer* (Figure 7.3) can be used to assure that a microcontroller is executing in a loop, so as to periodically service a watchdog timer. If is not serviced, an interrupt can pull the program execution back into the loop that periodically services

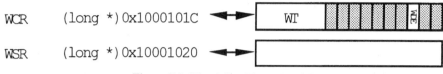

WCR   (long *) 0x1000101C

WSR   (long *) 0x10001020

Figure 7.3. Watchdog Timer Module

the watchdog timer. Obviously, such a device can be used to improve the reliability of a microcontroller, if the program running in it can service the watchdog timer within the watchdog period.

The watchdog timer is enabled when watchdog control register port *WCR* bit 2 (*WDE*) is asserted (*WCR* is a set port). The program must write 0x5555 followed by 0xaaaa into watchdog service register port *WSR* within each period. *WCR*'s bits 15 to 10 are the number of half-seconds in a period, minus 1. If the program fails to do this, then the MPU resets. *WCR* and *WSR* must be accessed as 32-bit ports, even though only the low 16 bits of each are used. *main* below initializes the watchdog timer for a 32 second period and, assuming that the code represented by the comment "some routine" therein takes less than 32 s, this program services the watchdog timer within each period.

```
volatile long WCR@0x1000101C, WSR@0x10001020;

void main (void) {
      WCR = 0xfc04; /* set WDE bit, and set WT to 0x3f */
      do { /* some routine */WSR = 0x5555; WSR = 0xaaaa; } while(1);
}
```

We again use the traffic light controller example to illustrate how a watchdog can be used to reset the MPU if the program falls out of its interpreter loop. We synchronize to the slow traffic light using §5.2.3's delay loop for simplicity, but the other delay mechanisms introduced in this chapter could be used in place of the delay loop. The watchdog timer will reset the MPU if we do not service the watchdog timer at least once each second and a half. Since the delay loop takes 1 second, the watchdog timer should be serviced each period. But if it is not serviced, the MPU is reset.

```
const unsigned char tbl [4] [2]={{0x21,4},{0x22,1},{0x0c,6},{0x14,2}};

void main() { unsigned char i, j; long k;
      KCDD = 0x3f; /* set port KCD to output on low-order 6 bits */
      WCR = 0xc04; /* set WT to 3 (period set to 1 1/2 s), start watchdog */
      do
          for(i=0;i<4;i++){
            KCD = tbl[i][0];
            for(j = 0; j < tbl[i][1]; j++) {
                  for(k = 0;k < 0xffffffff;k++) ; /* adjust for 1 s delay */
                  WSR = 0x5555; WSR = 0xaaaa; /* if not in loop, reset */
            }
          };
      while(1);
}
```

7.1.4   Programmable Interval Timer

Figure 7.4 shows the *programmable interval timer* device (PIT). It can be used, in a gadfly loop or generating an interrupt, at a predetermined rate or after a pre-determined delay. It often implements a time-sharing manager, as we show in the next section.

The PIT's control-status port *ITCSR's* five least significant bits are, from least significant bit, an enable bit *EN*, a reload bit *RLD*, an interrupt flag bit *ITIF*, an interrupt enable bit *ITIE*, and overwrite bit *OVW*. Setting the reload bit causes the counter to be loaded from the data port *ITDR*. Setting the overwrite bit causes the counter to be loaded with data written into data port *ITDR*. The other bits have obvious meanings. The counter decrements at a rate of 8192 Hz, having a period of about 122 μs, and can be read as alternate data register *ITADR*. The ports can be accessed using:

```
volatile long ITCSR@0x10001014, ITDR@0x10001028, ITADR@0x1000102C;
```

If *ITCSR's* five least significant bits are TFFFT, then writing *t* into the data port ITDR and gadflying on *ITCSR's* bit 2 waits *t* counter cycles:

```
void main(void){short t;ITCSR=0x11; ITDR=t;while(ITCSR&4); ITCSR|=4;}
```

The PIT can be used in a loop to make sure the loop is executed exactly once each *t* cycles (defined as in the previous example). If *ITCSR's* five least significant bits are FFFTT, then initializing *t* into the data port *ITDR* permits gadflying on *ITCSR's* bit 2 to wait t counter cycles each time the loop is executed:

```
void main(void) { short t;
      ITCSR=3; /* initialize control for free-running PIT */ ITDR=t; /* set loop time */
      do{
           while(ITCSR&4); ITCSR|=4; /* wait for PIT to count down, then clear flag */
           /* This program segment will execute exactly once each t cycles */
      } while(1);
}
```

In the program below, the *handler* is entered each time the PIT counter de-crements to zero. The normal interrupt module enables and recognizes PIT inter-rupts using bit 8 of the *NIER* port. If this is the only normal interrupt, the normal interrupt handler can clear the flag and then initiate some operation when the PIT times out each *t* cycles:

```
interrupt void handler(){ ITCSR |= 4; /* clear int. flag */ }
void main(){ short t;
      disableInt();ITCSR=0xB;/* initialize control for free-running PIT, enable int. */
      ITDR = t; /* set interrupt rate time */
      initInt(0x30000000, 0x30000200, handler, 10);
      NIER |= 1 << 8; /* enable normal interrupt on bit 8 */ enableInt();
      do ; while(1); /* wait here forever, permitting interrupts */
}
```

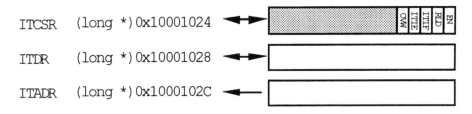

ITCSR   (long *)0x10001024

ITDR   (long *)0x10001028

ITADR   (long *)0x1000102C

Figure 7.4. Programmable Interval Timer

The second technique can be used to implement our traffic light controller, providing a delay using the PIT, instead of a delay loop. This also provides accurate timing, independent of compiler optimization options, direct memory access, and so on.

```
const unsigned char tbl[4][2]={{0x21,4},{0x22,1},{0x0c,6},{0x14,2}};
void main() { unsigned char i, j;
    KCDD = 0x3f; ITCSR = 3; ITDR = 8192; /* set PIT for 1 second rate */
    do {
        for(i=0;i<4;i++){
        KCD = tbl[i][0];
        for(j=0; j<tbl[i][1]; j++){ while(ITCSR & 4) ; ITCSR|=4; }
        };
    } while(1);
}
```

This example can be extended to many applications that require synchronization to very slow external devices. However, the MPU yet remains tied up in a gadfly loop. In the next section, we will permit the MPU to perform other functions while it is waiting to synchronize to a very slow I/O device.

7.2   Timesharing

A *multithread* scheduling technique, which is a primitive form of task scheduling done in a multitasking multiuser operating system, would permit some other work to be done while this routine waits for real-time interrupts. We generate real-time interrupts once every *tick*, where a tick is about 10 ms because faster times consume too much time switching between threads, while slower times make the threads run erratically in time, in the perception of human users. We maintain three different threads, where a *thread* is a part of the program that is independent of other threads and that can be executed to do useful work. A thread can be executed for one tick, and then another thread might be executed for a tick, and so on. We can put a thread to *sleep* for a number of ticks; the other threads will be able to execute without competition from a sleeping thread. Rather than gadflying when waiting while an I/O device is BUSY, after using part of a tick to initiate the I/O operation, we can sleep for the remaining number of ticks, until the I/O operation is completed and the DONE state is entered.

7.2.1   Multithread Scheduling

Multithread uses two simple ideas. First, the processor is tricked when returning from interrupts, to restore another thread's registers, in order to run it. Second, a simple, effective way chooses the next thread to be executed; essentially the thread waiting the longest to run is chosen to run. These two ideas are discussed now.

The information about a thread is maintained by a *struct*. Three threads can be maintained by a *struct* vector *threads[3]*. Members of each *struct* keep track of its *sleepTime, priority,* and *age* as well as the location of the thread's *stack*.

```
struct THREAD {
    volatile long sleepTime; unsigned char priority, age; char *stack;
} threads[3], *thisThread;
char *stackptr, nThreads, threadIndex;
```

We can use any timer device to generate periodic interrupts. The TOD alarm, the PWM device, or the PIT device can be used, but the PIT is best suited for this purpose, and is generally used for time slicing the MPU. The machine state needed for execution of a thread is frozen when the PIT interrupt occurs, by saving on the stack all its GPR registers, and the program counter and status register that are saved by a normal interrupt in control registers 2 (EPSR) and 4 (EPC). Several threads will have their registers saved this way. It should be noted that the normal GPR registers, rather than the alternate registers, need to be saved and restored, in this procedure. The PIT *handler* shown below, which calls the procedure *find-Next,* saving registers for one thread, may end by restoring these registers, but it may restore the registers for a different thread than the one that was just saved. That way a different thread can execute for the next tick.

```
#pragma NO_ENTRY
#pragma NO_EXIT
#pragma NO_RETURN
#pragma NO_FRAME
 interrupt void handler(){
    asm{
    SUBI  r0,68        ; allocate room to save all registers
    STM   r1           ; copy all but r0 onto the stack
    MFCR  r1,cr2        ; get saved status
    ST.L  r1,(r0,60)  ; put saved status on stack
    MFCR  r1,cr4        ; get return address
    ST.L  r1,(r0,64)  ; put return address on stack
    LRW   r1,L         ; get address of stackptr
    ST.L  r0,(r1,0)   ; save pointer
    }
```

```
    thisThread->stack=stackptr;  /* save stack pointer for the interrupted
        thread*/
    thisThread->age = thisThread->priority; /* update this thread's age */
    findNext(); /* determine which thread runs in the next tick time */
    stackptr = thisThread->stack; /* restore stack pointer for max age's thread */
    asm{
    LRW    r1,L          ; get address of stackptr
    LD.L   r0,(r1,0)     ; restore pointer
    LD.L   r1,(r0,60)    ; read status register
    MTCR   r1,cr2        ; put back status register
    LD.L   r1,(r0,64)    ; read return address
    MFCR   r1,cr4        ; restore saved return address
    LDM r1               ; restore all but r0 from the stack
    ADDI   r0,68         ; deallocate room used to save all registers
    RTE                  ; return to activated routine
L:  DC.W stackptr        ; address of global variable stackptr
    }
}
```

Several different procedures findNext can be used in the above handler, to implement different timesharing strategies. The simplest is to give each thread a time slice in strict round-robin order, so that thread 1 follows thread 0, and thread 2 follows thread 1. We show this procedure first. A more complex strategy is to maintain a thread's age, and assign priorities and sleep times to threads. This will be discussed later.

The procedure findNext below simply executes thread i + 1 after thread i, and thread 0 after thread n-1. It gives each thread a tick time every nThreads tick times.

```
void findNext(){
    thisThread=&threads[threadIndex=(threadIndex+1)%nThreads];
    }
```

The more powerful procedure findNext below searches all the threads, deciding which thread will be executed, by its sleepTime, priority, and age variables. If sleepTime is nonzero, sleepTime is decremented each tick, otherwise, a thread will be executed, among all the nonsleeping threads (sleepTime=0). A thread can be made to sleep N ticks by making sleepTime equal to N. A thread goes to sleep if sleepTime is made nonzero. Among nonsleeping threads, a thread with the oldest age will be executed. (sleepTime=0xffffffff will be used in the next section.) If no thread is "awake" so the "oldest age" j remains zero after all threads are examined, a gadfly loop waits for the next PIT timeout, and the search of the threads is repeated. The age of each non-executing thread is incremented (up to 0xff), but when a thread is executed, its age is reset to its priority.

The priority is essentially age's initial value. Nonsleeping threads share processing time, in some sense "proportional" to their priority values. If all threads have the same priority, they will share processing time equally, when they are not sleeping. If a thread has a much lower priority than all the other threads, it will execute when all the other threads are sleeping, or will execute very infrequently when they are not; such a thread is called a *background* thread. If a thread has a much higher *priority* than all the other threads, it will hog processing time; such a thread is called a *high-priority thread*.

The *sleep* procedure below, gadflies while *sleepTime* is nonzero, until a PIT interrupt selects another thread to run. When this thread runs again after waiting d tick times, its *sleepTime* becomes zero and it will exit the procedure.

```
void sleep(int d)
    { thisThread ->sleepTime = d; while(thisThread->sleepTime);}

void findNext(){ unsigned char i, j; THREAD *p;
    for (i = j = 0; i < 3; i++) { /* try each thread */

        p = &threads[i]; /* p is thread pointer */

    if ( p->sleepTime ) /* if sleeping */
        {if(p->sleepTime!=0xffffffff) -p->sleepTime;} /* sleep forever */
    else { /* if not sleeping */

        if(RTIFLG && (p->age!=0xff)) p->age++;/* inc. age ifn max. or PIT *
        if(p->age >= j) { j = p->age; thisThread = p;} /* find max age */
        }
            if(i == (nThreads - 1)) { /* if at end of search */
                RTIFLG = RTIF; /* remove interrupt */
                if(j == 0) { /* if no awake threads, restart search after next PIT */
                    while ( ! (RTIFLG & RTIF)) ; i = 0xff; /* search all the threads */
            }
        }
    }
}
```

A thread is started by the procedure *startThread*, which sets up the thread's execution entry point and its priority and age. The hardware stack is initialized to appear like the thread's stack just as the thread enters the PIT *handler*. This prepares the thread to run in the first tick time after being started, just as it will run in later tick times after it has been interrupted. However, just one thread, thread 0, continues to run as it starts other threads. Its stack is not initialized in *startThread*. Its stack is saved normally, when it is interrupted by the PIT device.

```
void startThread(long f,unsigned char p,long s){
    long *stack,t=nThreads++;
      threads[t].sleepTime=0;threads[t].age=threads[t].priority=p;

    if(t){ /* except for dummy thread 0, make stack space and initialize it */
      malloc(s);stack=(long *)(threads[t].stack =(char*)malloc(64));
      stack[15] = 0; stack[16] = f; /* initialize stack */
      }
}
```

The *restartThread* procedure below rewrites the program counter, and other thread variables, of an existing thread. It can be used after a thread is made to sleep "permanently" by setting its sleep time to the maximum value (0xffffffff) with no other way, in order to awaken it to execute other functions.

```
void restartThread(int f, unsigned char p, unsigned char t){long *stack;
    threads[t].sleepTime=0; threads[t].age=threads[t].priority=p;
    stack = (long *)threads[t].stack; stack[15] = 0; stack[16] = f;
}
```

Threads *main0*, *main1* and *main2*, listed below, "flash" an LED on *KCD* bit 4, at a rate of 1/2 Hz and an LED on *KCD* bit 3 at 1 Hz. These threads sleep, and awaken to flash the LEDs from time to time.

```
void main1(){ do { KCD ^= 0x10; sleep(100); } while(1); }

void main2(){ do { KCD ^= 8; sleep(50); } while(1); }
```

main initializes PIT with a 10 ms period, and calls *startThread*, to initialize its own thread as thread 0, and to initialize threads 1 and 2. The running thread, thread 0, calls *main0*. The other threads run *main1* and *main2* when they start executing, because the addresses of *main1* and *main2* are on these thread's stacks.

```
void main(){ short t;
        disableInt(); ITCSR = 0xB; /* enable PIT */
        ITDR = 82; /* 10 ms tick time */
        KCDD = 0xff; /* make KCD output, to evidence time slicing */
        initInt(0x30000000, 0x30000200, handler, 10);/* norm int. */
        NIER |= 1 << 8; enableInt();/* enable PIT interrupt */
        startThread(0, 0, 0); thisThread = &threads[0]; /* start threads */
        startThread((long)main1,50,64);startThread((long)-
        main2,50,64);
        main0(); /* begin executing one of the threads */
}
```

7.2.2   Threads for Timesharing

We now show threads that do some useful work while efficiently synchronizing to slow-speed devices. We will reexamine the keyboard and traffic light controller, and present a technique to support an alarm clock, and a Centronics parallel printer device.

§6.2.7.1's keyboard technique assumes that only one key is pressed at a time. If we allow two keys to be pressed simultaneously — as is often done by proficient keyboard users who press a new key before releasing the key being pressed — the program might keep picking up the first key, not seeing the new key while the first is in its scan. Any technique that can correctly recognize a new key even though n-1 keys are already pressed and are still down is said to exhibit *n-key roll-over*. Two-key roll-over is a common and useful feature that can be achieved with most keyboards, but for *n* greater than 2, one must guard against sneak paths through the keys. Sneak paths appear when three keys at three corners of a rectangle are pushed, and the fourth key seems to have been pushed because current can take a circuitous path through the pressed keys when the fourth key is being sensed. This can be prevented by putting diodes in series with each switch to block this sneak path current, but the solution is rather expensive and *n*-key rollover is not usually useful.

A 2-key roll-over that uses a queue to record keystrokes even while a previously key may have to be recorded or responded to, can be appended to the multithread scheduler of §7.2.1. Its *main* will include the initialization statement *KRDD = 0xff;* and the PIT *handler* calls the procedure *service* below, once each tick time.

```
char keys[8], previousKeys[8];
void service() { char r, c, i;
    for(i = 0; i < 8; i++){ /* scan 8 columns after a tick time is over */
        KRD = ~(1 << i); /* put a low on just one column */
        c = keys[i] & ~previousKeys[i] & ~KCD; /* analyze for new key */
        if(c) for(j = 0, r = 1; j < 8; j++, r <<= 1;) {/* scan the word */
            if ( c & r ) { push( i+8*j); thread[1].sleepTime = 0;
        }
        previousKeys[i] = keys[i]; /* save older copy of keys to detect an edge */
        keys[i] = ~KCD; /* save port data for next tick time */
}}
```

At the end of *service*, a first key pattern is copied to *keys*. After an 8-ms tick is executed, *handler* is again entered, which calls *service*. The first part of *service* is now executed, where the key pattern is compared to the key pattern 8 ms ago in *keys*, to determine if the key has been debounced. Moreover, a key pattern from 16 ms back is kept in *previousKeys*. This enables us to detect a falling edge of a debounced key. If a key input was high 16 ms ago, low 8 ms ago, and low now, the key has just been pressed, and push its code onto a queue. When a code is pushed onto the queue, a thread waiting for the data, thread 1 in this example, is awakened.

A *main1* program, which can substitute for §7.2.1's *main0, main1,* or *main2*, can pull key codes from the queue where the above procedure *service* pushes them. *main1* sleeps indefinitely until a key is detected, where *service* awakens it. As in §6.2.7.1's linear select example, each key commands a different procedure to be executed; key code 0 causes *p0()* to be executed, etc.

```
void main1() { do {
    sleep(0xffffffff);
    switch(Q.pull()) {
        case 0: p0(); break;
        case 1: p1(); break;
    }
} while(1); }
```

It should be clear that this keyboard example is an improvement over the previous example of keyboard software. This technique does not waste time in a wait loop to get a couple of samples of the key signals to debounce the switch and establish a leading edge. Instead, it lets other threads run the computer, using up a tick time to wait and see if the key is still pressed. It further can handle n-key rollover if diodes are put in series with the switches. The main1 program simply awaits key inputs, sleeping forever *(thread[1].sleepTime=0xffffffff)*. It is awakened when *service* gets something for it to do *(thread[1].sleepTime = 0)*. This "wait until" scheduling mechanism is used in the Macintosh; it waits on a next event queue to tell it what to do next.

We next look at our familiar traffic light controller's use of a multithread scheduler. Compared to §7.1.4, this example puts the thread running the traffic light to sleep a number of tick times indicated by the time a light is to be left on.

```
const unsigned char tbl[4][2]={{0x21,4},{0x22,1},{0x0c,6},{0x14,2}};
void main1() { unsigned char i, j, k;
    KCDD = 0x3f; /* set port KCD to output on low-order 6 bits */
    WCR = 0xc04; /* set WT to 3 (period set to 1 1/2 s), start watchdog */
        do{
        for(i=0;i<4;i++){
        KCD = tbl[i][0];
        sleep(tbl[i][1] * 122); /* sleep tbl[i][1] s */
        } while(1);
}
```

Note that sleeping for a specific number of ticks is a "wait for" elapsed time scheduling mechanism. You can use it to make a thread wait for a rather long time until it is next able to do something. While it is waiting, other threads can use the computer without competition from this thread. However, the scheduler will not be able to guarantee that the thread will execute when it awakens. When it awakens, it only competes for time slices, along with all other nonsleeping threads. The one with the largest age will be given the use of the time slice.

An alarm clock can be implemented using a thread. Numbers corresponding to events to be executed are stored in a vector *procedure*, in the order they are to be executed, and a vector *times* stores in element *i* the number of ticks from *procedure[i-1]* to *procedure[i]*. The procedure main1 will execute *procedure[k]* after *times[0]* + *times[1]* + ... *times[k]* ticks have occurred.

```
int times[10], procedure[10];
void main2() { char i; do {
    sleep(times[i]);
    switch(procedures[i++]) {
      case 0: p0(); break;
      case 1: p1(); break;
   }
} while(1);
}
```

We illustrate a thread using gadfly and real-time interrupt synchronization for the *Centronics printer* example (Figure 7.5). Figure 7.5a shows the "Centronics" parallel printer connector. A character is printed by putting its ASCII code on the data lines, and asserting Stb low for at least 1 μs. When the printer accepts the character, it pulses Ack low (Figure 7.5b). We connect the printer data lines to *KCD*, connect negative logic Centronics signals Ack to *EPDR* bit 0, and Stb to *EPDR* bit 1.

A personal computer mechanical printer usually has a buffer in it, and can quickly put the character in the buffer, but if the buffer is full of data, the printer

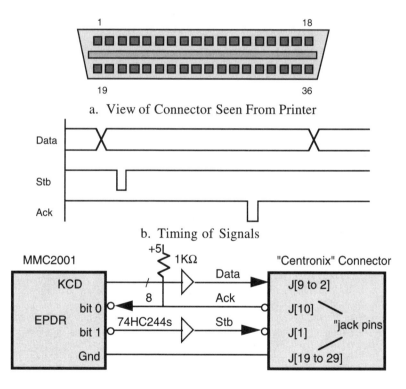

a. View of Connector Seen From Printer

b. Timing of Signals

c. Connections

Figure 7.5. "Centronics" Parallel Printer Port

must wait ms for a character to be mechanically printed before it has more room and
can store the incoming character in the buffer. Thus, the time from Stb to Ack will be
a few μs if the buffer is not full, or a few ms if full.

Different response times indicate using different synchronization mechanisms.
Fast response, when the printer's buffer is not full, indicates gadfly synchronization,
but the slow response, when it is full, indicates real-time interrupt.

We can use both gadfly and real-time interrupt synchronization. After writing
data to *KCD*, we produce a negative pulse on *EPDR* bit 1, which pulses Stb. This
should cause the printer to assert Ack low, to put a falling edge on *EPDR* bit 0.
Recalling that the printer might respond quickly or slowly, we check key wakeup bit
0 right after Stb is pulsed, as in a gadfly loop. Interrupts are disabled while this bit is
checked, for if they were enabled, the interrupt handler would be entered before the
key wakeup bit would be checked. If the printer responded quickly and this bit is set,
then we re-enable interrupts and return. Otherwise, because we anticipate a long
wait, in the procedure *sleep*, the current thread's *sleepTime* is set to make the
thread sleep "forever" when the next tick occurs and the current tick is wasted using
a gadfly loop on *sleepTime*. When the key wakeup interrupt occurs, its handler
clears the thread's *sleepTime*, thus waking up the thread. Sleeping forever avoids
the possibility that decrementing *sleepTime* will restart the thread when the
printer has not responded. Of course this will hang up the thread if the printer is not
on and does not respond. The user is supposed to recognize and fix this.

```
void put (char data) {
    KCD = data; disableInt ();
    EPDR &= ~2; EPDR |= 2;
    if (EPFR & 1) { /* if Ack returns very quickly */
        EPFR=1; enableInt ();
        return; /* permit interrupts. */
    }
    enableInt ();
    sleep (0xffffffff);
}
interrupt void handler1 (void) {EPFR=1; threads [1].sleepTime=0;}
void main2 () { char i, string [10]; disableInt ();
    KCD=0xff; FIER|=1<<21; EPFR=1; EPPAR=1; EPDDR=2;
    initInt (0x30000000, 0x30000200, handler1, 10);
    enableInt (); for (i = 0; i < 10; i++) put (string [i]);
    sleep (0xffffffff); /* sleep kills the thread; it will not be awakened. */
}
```

7.2.3   Object-oriented Classes for Timesharing

Object-oriented programming provides both protection for each thread, and sleep
capability in place of delay or gadfly loops that permits the microcontroller to
perform useful work while waiting for an I/O operation. These concepts are covered
in this section.

To use objects for programs like §7.2.1's *main1* and *main2*, we define a class *thread* which contains data members *sleepTime, age, priority* and *stack* used for time-slicing as well as functions and data members common to two or more threads, and we generally define a derived class such as *thread1* for each thread, unless it is completely identical to another thread. The *thread* class will have a function main that will be overridden in each derived class. The function *main* of class *thread1* will contain the starting procedure for thread 1. The *main* procedure that is started after reset, as *main* of §7.2.1 was started, will call class *thread1*'s constructor which allocates its stack and initializes its data members. The real-time interrupt will start each thread's *main* function in turn, and when the thread's *age* is largest, run if for a time tick.

Object-oriented threads provide protection and polymorphism. Each thread has its own scope of names for function and data members; these can be declared private to protect them from other threads, and names can be reused. True global variables can be used to share information among threads and interrupt handlers.

In a multi-thread, object-oriented classes can use *sleepTime* in lieu of real-time delay loops or gadfly loops. In the earlier class *IQFPort* (§6.2.8.1), if the input queue is empty, *get* gadflies on the queue size until an interrupt pushes some data, then *get* pulls this data from the queue. By modifying the routines, *get* instead can sleep indefinitely, until an interrupt pushes some data and wakes it up. The interrupt service routine is modified to awaken the thread by clearing its *sleepTime*. Then, when the thread is selected to run (having the highest age), *get* can pull the data from the queue in the interrupt handler. These remarks also apply to the class *OQFPort* (§6.2.8.2) modifying the function *put*. When *put* is executed but the queue is filled and this function member should put more data into it, the member function will be rewritten so the thread will sleep indefinately. An interrupt pulls some data to make room in the queue for more data, so it should wake up the thread by clearing its *sleepTime*. When *put* resumes execution, it can push some data and return. In both cases, the interrupt handler must know which thread to wake up. That can be done if each I/O device, and therefore each interrupt handler, is associated with exactly one thread, or if the thread number or thread pointer is kept by the device when the thread is put to sleep indefinately

A class *Pipe* is useful in linking a thread with another thread in a *pipeline*. The following templated class is a simple but effective pipe.

```
template <class T> class Pipe : public Port<T> {
    Queue <T>*Q; public : THREAD *thread1, *thread2;
    Pipe(unsigned char size) : Port(0) { Q = new Queue<T>(size); }

    virtual void put(T data) {
        while(Q->size>=Q->maxSize){thread1=thisThread;
        sleep(0x);}
        Q->push(data); if(thread2) thread1->sleepTime=0; thread2=0;
    }
```

```
virtual T get (void) { T v;
    while (Q->size  ==  0)  { thread2 = thisThread;
    sleep (0xffffffff); }
    v = Q->pull (); if (thread1) thread1->sleepTime = 0; thread1 = 0;
}
virtual T operator = (T data) {put (data); return data; } // assignment
operator T () { return get (); }; // cast
};
```

The procedure *main1*, executing in *threads[1]*, "outputs" data to the procedure *main2*, which "inputs" it:

```
Pipe<char> pipe (10);
void main1 () { char c; do { pipe = c} while (1); }
void main2 () { char c; = pipe; }
```

The procedure *main1*, executing in *threads[1]*, "outputs" data to the procedure *main2*, executing in *threads[2]*, where it appears as an "input". Since *pipe* is declared as a global object, its constructor is called before *main1* or *main2* are called. *main1*'s use of the overloaded assignment operator calls *Pipe*'s *put* function member, and *main2*'s use of the overloaded cast operator calls *Pipe*'s *get* function member. The queue holds *main1*'s output data until *main2* is ready to use it. One thread "outputs" to the pipe, while the other "inputs" from the pipe, as if the pipe is an I/O device. However, the pipe is merely a queue that holds "output" data until it is "input" to the other thread.

Practically all the techniques illustrated in this section can also be implemented using the Ariel operating system, as we will see in the next section.

7.3 Ariel Time Management

The Ariel kernel provides two time-management mechanisms, *time wait* and *time-of-day*. See Table 7.2. In addition, timer coordination mechanisms are built into most service calls that can be delayed if their action might not be completed immediately.

Timer services are provided by a real-time interrupt that occurs at a fixed time interval (called a *tick*). Actual wait time is (with an error of less than one tick) the product of the number of ticks by the tick interval. However, when a task stops waiting, becoming unblocked, it enters the ready state, and competes with other ready state tasks for the use of the MPU. A higher priority task can prevent a task, that becomes ready, from running.

Table 7.2. Time and Time-of-Day Services

_tim_wait ()	Wait for given time interval.
_tim_cancel ()	Continue given task if it is paused for time interval.
_tim_get ()	Get number of ms since system was started.
_tod_get ()	Get Time-Of-Day clock/calendar string.
_tod_set ()	Set Time-Of-Day clock/calendar.
_tod_wait ()	Wait for given Time-Of-Day.

A task can wait for a given interval of time, during which the task is blocked from executing. A wait can be canceled by another task, providing a kind of signal. A task can wait forever (usually until the wait is canceled).

Any task can submit an ASCII clock/calendar string to the time-of-day mechanism. Thereafter, the string will be updated every second. The current value of such a string can be read by any task. A task can also be blocked until a given definite time-of-day, such as 12 noon, or a relative time-of-day, such as 30 min after the current hour.

Various services have time-dependent options. For example, a given event flag can be set, or a given signal can be sent to a task, after a specified time interval. In most service calls requiring coordination, a maximum delay can be specified, after which the task becomes ready (the service call returning an error message). Finally, to execute a task periodically, it can be terminated, and automatically restarted after a given delay.

A task may wait for a number (<256) of units, which may be any of: _TIME_MS, _TIME_MS_TENS, _TIME_MS_HUNDREDS, _TIME_SECONDS, _TIME_MINUTES, _TIME_HOURS, or _TIME_DAYS. For instance,

$$\_tim\_wait(50 \mid \_TIME\_MINUTES);$$

puts the task in the blocked state for 50 min. Special values such as in the service call

$$\_tim\_wait(1 \mid \_TIME\_MS);$$

puts the task in the blocked state for one tick. The constant, _TIME_NEXT_TICK, can be used instead of 1 | _TIME_MS, and should be used to improve clarity. The service call

$$\_tim\_wait(0 \mid \_TIME\_MS);$$

puts the task in blocked state indefinitely. The constant, _TIME_UNLIMITED, can be used to improve clarity, instead of 0 | _TIME_MS. A task, with an id of id, blocked by the timer, can be made ready by _tim_cancel(id).

If s is a 21-byte char string, _tod_get(s); returns the time, such as in "07 DEC 2001 13:50:19." Similarly, using this string, time can be set using _tod_-set(s). Using a six-character string s, _tod_wait(s); waits until the time in the character string. In this service call, "??" may be substituted for any of the substrings, as in "??1400," which means 14 min after the hour.

We again repeat our familiar traffic light controller's use of a multithread scheduler to illustrate the use of Ariel's timer mechanisms. Compared to §7.2.2, this example uses an Ariel service call _tim_wait() to cause a one-s delay.

As in §7.2.2, waiting for a specific number of seconds is a "wait for" elapsed time scheduling mechanism. You can use it to make a task wait for a rather long time until it is next able to do something. While it is waiting, other tasks can use the computer without competition from this task. However, the scheduler will not be able to guarantee that the task will execute when it awakens. When it awakens, it only competes for time slices, along with all other tasks that are ready. The one with the highest priority will be given the use of the time slice.

```
const unsigned char tbl[4][2]={{0x21,4},{0x22,1},{0x0c,6},{0x14,2}};

void main1() { unsigned char i;
    KCDD = 0x3f; /* set port KCD to output on low-order 6 bits */
    do{
        for(i=0;i<4;i++){
        KCD = tbl[i][0];
        _tim_wait(tbl[i][1]|_TIME_SECONDS); /* sleep tbl[i][1] s */
        }
    } while(1);
}
```

Most service calls that can experience unpredictable delays, or that cause an action after a delay, have a parameter like _tim_wait() shown above. Table 7.3 lists these service calls. As an example of such a call, a keyboard scan can be done in a task using a message buffer, the task to be restarted each time tick using _tsk_exitrestart().

```
char keys[8], previousKeys[8], msg[5]; uxid_t mbx; rv_t result;
void kbdScan() { char fallingEdge, r, c, i;
    for(i = 0; i < 8; i++) { /* scan 8 columns after a tick time is over */
        KRD = ~(1 << i); /* put a low on just one column */
        c = keys[i] & ~previousKeys[i] & ~KCD; /* analyze for new key */
        if(c) for(j = 0, r = 1; j < 8; j++, r <<= 1;) {/* scan the word */
            if ( c & r ) {
            msg[0]=msg[1]=msg[2]= 0; msg[3] = 1; msg[4] = (i<<3) + j;
            _mbx_send(mbx, (uptr_t)&msg, &result,_CONT_NOCOOR,100L);
            );
        }
        previousKeys[i] = keys[i]; /* save older copy of keys to detect an edge */
        keys[i] = fallingEdge & ~KCD; /* save port data for next tick time */
    }
    _tsk_exitrestart(SUCCESS, _TIME_NEXT_TICK);
}
```

This procedure follows the same strategy as §7.2.2's service(). However, rather than a user-defined queue, Ariel's mailbox queue is used; _mbx_send() pushes a key code. We assume the mailbox has been created: mbx = _mbx_create{}; has been executed in another task to put its id into mbx. When a key is pressed and recognized, a message is composed by putting its length in msg[0 to 3] and the key code is put into msg[4]. A consumer task executes _mbx_recv(); to get the key code.

Service calls _drv_waitstd(), _drv_cancelstd(), _drv_waitaux(), and _drv_cancelaux() use a standard and auxiliary timer that are automatically associated with each I/O device driver. A typical use of these timers is to retract the head of a floppy disk drive after a delay. However, device driver procedures cannot block the task that called the driver, because these procedures execute in S/SE states that must be serially executed. This causes some complications in writing device

Table 7.3 Service Calls with a Wait Limit

_cmp_alloc()	Allocate from a common memory pool
_csv_wait()	Wait for controlled shared variables to be true
_efg_timesend()	Wait until event flags are set
_efg_wait()	Set event flags after a time interval
_fbp_alloc()	Allocate from a fixed block memory pool
_mbx_recv()	Receive a message from a mailbox
_mbx_send()	Send a message to a mailbox
_pio_cmd()	Perform I/O
_sem_wait()	Wait for a semaphore to be free
_sig_timesend()	Send a signal after a given time interval
_sig_wait()	Wait until a signal arrives
_tim_wait()	Wait for given time interval
_tsk_exitrestart()	Terminate the requesting task with automatic restart
_tsk_start()	Start a task
_drv_waitstd()	Start (standard) timer pre-assigned to the unit
_drv_cancelstd()	Cancel (standard) timer
_drv_waitaux()	Start (auxiliary) timer pre-assigned to the unit
_drv_cancelaux()	Cancel (auxiliary) timer

drivers that should block their tasks when I/O cannot be completed. A calling routine in T state must block the task.

```
const drv_t function0={initFun,svcFun,0,0,0,0,0};
ucd_t ucd0 = {_EXTNAME('P','T','R','O'),&function0,0,0,0,0,{0,0}};

rv_t status; uxid_t i_o_Id;

void put(char *data, short n){ cdev_t charDev;
    charDev.aub = data; charDev.lub = n; charDev.axb = charDev.spl = 0;
    while(charDev.lub != 0){
        _pio_cmd(i_o_Id,_PIO_WRITE,0L,&charDev,&status,_CONT_NOCOOR);
        if(charDev.spl == 1) time_wait(_TIME_MS | 5);
        else time_wait(_TIME_MS | 10);
    };
}
int32 initFun( ucd_t *ucp ) {KCDD = 0xff; KRDD = 1; return SUCCESS;}

void svcFun(ucb_t *ucb, int fun, cdev_t *dparams) {
    if(params->spl == 0) {/*executed on even numbered entries to _pio_cmd()*/
        KCD = *params->aub++; /* output data from buffer */
        KRD = 1; params->spl = 1; return; /* pulse, return for 5 ms delay */
    } /* below is executed on odd numbered entries to _pio_cmd() */
    KRD = 0; params->spl = 0; /* remove pulse, return for 10 ms delay */
    if((-params->lub) == 0) /* count down number of bytes to output */
        _drv_pio_done(ucb,SUCCESS,0); return; /* if all gone, indicate end */
}
main(){char s[12] = ''Hello World'';
    if(!ISSVCERR(i_o_Id = _pio_create(&ucd0))) put(s, 12);
}
```

Timing services can be accessed in a T state procedure that calls a device driver. The previous example illustrates a paper tape punch, which pulses the output bit that drives a solenoid on *KRD* bit 0 for 5 ms. to punch the data pattern and advance the paper tape, and removes the pulse for 10 ms to retract the solenoids.

7.4   Conclusions

The timer and time-of-day hardware, time-sliced software, as well as Ariel service calls, provide the MMC2001 programmer easy access to time synchronization. The time-of-day module (TOD) illustrates one of MMC2001's main design goals, which is to make the hardware as close as possible to the application, making it easy for an application to use standard time units like seconds. If timesharing is implemented using the real-time interrupt, a procedure can sleep a number of tick times. Finally, if Ariel is used, its service calls can be utilized to wait for a time, or until a time.

The programmer has three good options to choose from. The Ariel-based service calls to wait for a delay seem to be warranted whenever there is another reason to use the operating system, such as to support multitasking. However, you do not need to use any operating system to be able to utilize time-sliced sharing of the MPU. The object-oriented technique, especially using classes derived from *Port*, are capable of supporting a rich selection of capabilities, without the user having to program these routines. But these routines are easy to reprogram, since the source code is shown in this and earlier chapters. Finally, the user can directly access the TOD or PWM ports in assembly or C language programs, to provide for time delay operations. But we recommend this only when tight synchronization is needed. Object-oriented and device driver interfacing provides the application programmer a measure of detachment from the hardware, reducing the cost of writing large applications.

Do You Know These Terms?

See the End of Chapter 1 for Instructions.

Timer-Reset Module (TRM)	tick thread	priority sleepTime	Centronics printer pipe
pulsewidth modulator (PWM)	sleep age	*n*-key roll-over 2-key roll-over	time wait time-of-day
multithread scheduling			

Problems

Problem 9 is a paragraph correction problem; for guidelines, refer to the problems at the end of Chapter 1. Guidelines for software problems are given at the end of Chapter 2, and guidelines for hardware problems, at the end of Chapter 3. In the problems in this chapter assume the MMC2001 CLK frequency is 32 MHz.

1. Write a self-initializing procedure `square(short i)`. You should generate a square wave having the nearest possible frequency, given the PWM's available combinations of prescale, width, and period values.

2. Write a self-initializing procedure `train(short i, short j)`, to generate a waveform having the nearest possible period, being high for i μs and low for j μs, given the PWM's available combinations of prescale, width, and period values.

3. Write a self-initializing procedure `pulse(short w)` to generate a positive pulse of width w μs, having the nearest possible pulsewidth, given the PWM's available combinations of prescale, width, and period values.

    a. Write `pulse()` using gadfly synchronization on PWM device 4.
    b. Write `pulse()` using interrupt synchronization on PWM device 4.

4. A conventional "step-and-repeat" telephone will be dialed using the PWM. Write a `main()` procedure and an interrupt handler for the PWM device 1 that will cause an interrupt every 5 ms. The most significant bit of *KCD* is given a value 1 to close the relay in series with the dial contacts. A digit '0' is represented by a sequence of 10 closures. Use global variables to keep track of what part of the sequence of numbers, what part of the number, and what part of the pulse has been output.

    a. Write the handler to output just one digit number, which is in global variable `char number`.
    b. Write the handler to output the seven numbers in the vector `char numbers[7]`.

5. To generate sounds, a sequence of pulses will be produced at PWM device 2's pin, which will be passed through a lowpass filter to an audio amplifier and speaker. The audio level is proportional to the pulse train's duty cycle, so to generate a waveform, the data in `short sequence[8000]` is to be output through the PWM. A procedure `main()` will output the pulse train which will have a period of approximately 8 KHz and a pulsewidth of `sequence[i]` * 4 CLK periods.

    a. Write `main()` using gadfly synchronization on PWM device 5.
    b. Write `main()` using interrupt synchronization on PWM device 5.

6. An "alarm clock" can start a procedure at a specified time, using PWM device 0. Write a `main()` procedure and a handler that interprets a table of times that the

"alarm" is supposed to "go off," so that when this happens a program corresponding to the "alarm" will be called from *main()*. If no procedure is to be
executed, *main()* calls a procedure *dummy(){}*. Suppose that a table of 10
"alarms" is stored:

> struct{ long *T*, void (*GO)(); } alarm[10];

where *T* is a time in 1/4 μs cycles, when it will go off for row *i*, and *GO* is the address
of a subroutine to be started when that interval is over. Each such subroutine ends
with an RTS.

7. Repeat problem 6 using the TOD module, where *T* is a time seconds.

8. Repeat problem 6 using the PIT module, where *T* is a time in 1/4 s.

9. * Key bounce is a problem when the user bangs on a key repeatedly to get
something done. It can be eliminated by a sample-and-hold circuit that takes a
"snapshot" of the signal just once. Software debouncing is rather rarely done in
microcomputers because it takes too much of the microcomputer's valuable time to
monitor the key. Keyboards are often used on microcomputers because the user may
want to enter different commands, and this is most easily done with a set of keys.
Linear selection is often used for a keyboard that has a lot of keys, like a typewriter.
N-key roll-over is a property of keyboards whereby the microcomputer can correctly
determine what keys were pressed and in what order they were pressed, provided that
no more than *N* keys are pressed at one time. Two-key rollover is commonly used in
microcomputer systems but is rather inadequate for most adept users, who often
hold several keys down at once.

10. Assume that three threads *thread[1]*, *thread[2]*, and *thread[3]*, have
priorities 5, 3 and 3. Identify which will be executed during each tick time, from time
t0 when they are all forked by *thread[0]*, for the next 30 ticks. Give a general rule
for assigning priorities to threads (e.g. give the same priority to thus and such
threads, higher priority to thus and such threads) and give a rough estimate of the
amount of CPU time a thread gets as a function of its priority for two cases: (1) when
n threads all have the same priority p, and (2) when there are just two threads having
priorities p1 and p2.

11. Replace procedures *main1()* and *main2()* of §7.2.2 with a single procedure
main1() so that each thread control a different traffic light, but both threads
execute only one procedure *main1()*. The first thread's north and south lights are
red, and the east and west lights are green, for 10 s, then north and south lights are
red, and the east and west lights are yellow, for 2 s, then north and south lights are
green, and the east and west lights are red, for 16 s, then north and south lights are
red, and the east and west lights are yellow, for 2 s. This pattern is stored in global
vector *char tbl1[4][2]*. The second thread's light pattern is the same as the first,
except north and south are exchanged with east and west. This pattern is stored in

global vector *char tbl2[4][2]*. (Hint: *main()* should check if *thisThread* is
&threads[1] or *&threads[2]*, to set a local variable pointer *tbl* to global
vector *tbl1* or *tbl2*.

12. Replace procedures *main1()* and *main2()* of §7.2.2 so that *main1()* reads
tape and *main2()* punches tape. Initially *main1()* goes to sleep; when *EPDR* bit
0's signal falls *main1()* wakes up and reads data from *KCD*, pushing it into a queue.
Then *main1()* goes to sleep again until another byte arrives. If data are available in
the queue, *main2()* pulls a byte into *KRD*, asserts *EPDR* bit 1 for one tick time, and
negates *EPDR* bit 1 for two tick times. These threads continue working this way
indefinitely. Use §2.2.2's procedures *pstop* and *plbot*, and assume the queue does not
overflow.

13. Replace procedures *main1()* and *main2()* of §7.2.2 so that *main1()* punches
tape and procedure *main2()* sends Morse code for the global null-terminated
ASCII string:

> *char message[] = "Now is the time for all good men to come to the aid of their country.";*

The *main()* procedure sets up 1/256 s ticks (programmable timer interrupts) and
then forks *main1()* and *main2()*. Either *main1()* or *main2()* may execute
before the other or might be used without the other, so each should initialize its I/O
as if the other does not run, and should not interfere with the other thread's I/O.
main1() causes the characters in *message* to be punched on paper tape. The paper
tape punch will energize solenoids to punch holes if a T (true, 1) bit is in a corresponding *KCD* register, and to punch a sprocket hole if a T is in *KRD* bit 7, otherwise
if an F (false, 0) is in *KRD* bit 7 the paper tape is pulled forward by a motor (unless
tape runs out). The holes are to be punched for about 16.4 ms and paper tape is to be
advanced for about 50 ms After all characters are punched, *main1()* will "kill"
itself by sleeping indefinitely as tape runs out. *main2()* causes the characters in
message to be sent on *KRD* bit 6 in Morse code. The Morse code for characters are
stored in the global vector *int Morse[2][128]* where *Morse[0][]* is the
number n of dots and dashes, and *Morse[1][]* is the (right-justified) pattern; a 0 is
a dot, a 1 is a dash, and unused bits are 0. Do not write the vector *Morse*. A dot is
sent by making *KRD* bit 6 T (true, 1) for about 1 s, and a dash sent by making *KRD* bit
6 T (true, 1) for about 3 s; there is a 1-s F (false, 0) between dots and dashes in a
letter, and a 3-s F (false, 0) between letters. After all characters in *message* are sent,
main2() will "kill" itself by sleeping indefinitely and outputting an F.

14. Rewrite *main()* in §7.2.1 to use the MMC2001 STOP instruction when all
threads other than thread 0 sleep. This reduces the MMC2001's power consumption
when no threads need to be executed.

15. Repeat problem 12 using a class *pipe* object to hold the data moved from
main1() to *main2()*. *main1()* reads paper tape, and *main(2)* punches it. Both
sleep when they await data.

16. Write an Ariel task to scan the keyboard each tick time, putting the encoded key into a *struct code(char row, col; };* sent in a message buffer rather than a mailbox.

17. Write Ariel service calls to:

    a. Get the time-of-day into global *char s[21]*.
    b. Wait for 600 ms.
    c. Cancel the wait for the task whose id is in global variable *tskId*.
    d. Send a signal 0 to the task that sent it, after 112 ms.
    e. Exit this task with return code SUCCESS, restarting it after 5 min.

18. Write a pair of Ariel tasks to read paper tape, and punch paper tape, as described in problem 12. Use *_msb_put()* and *_msb_get()* service calls to pass the characters, one at a time, between the producer, which reads the tape, and the consumer, which punches tape.

19. Write an Ariel device driver for the Centronix Printer (Figure 7.5). This driver should have the same characteristics as the program in §7.2.2, but follow the style of §7.3.

20. Write an Ariel device driver to send Morse code as described in problem 13. Use *_drv_time_wait()* service calls to wait between changes in the output. The code pattern is output on KRD bit 6.

8

Embedded I/O Device Design

The MMC2001 has some uncommitted chip area in which the application may put additional special-purpose hardware. The application designer now has the opportunity to design functional hardware, such as a character string search engine, or an I/O device such as an interface to a local area network, into the same chip that contains the M·CORE processor, SRAM and ROM.

This capability expands the design alternatives that the applications engineer should consider. He or she now should consider designing an I/O device using various combinations of external and internal hardware, as well as combinations of software such as assembly language and C statements, C++ objects, and Ariel device drivers.

In order to make intelligent design decisions, the designer now needs to understand the capabilities of on-chip hardware, as well as the techniques for designing such hardware, which affect the cost and time to design it. This chapter will show how to design hardware implemented in a programmable logic array that is to be put on the MMC2001 chip.

Hardware can be designed using hardware description languages, or using graphical design programs. In the first section of this chapter, we provide an overview of a hardware description language, verilog, which is used by many hardware design teams. It provides an adequate means of describing the hardware that can be put on the MMC2001. From your understanding of this particular language, you should be able to adapt to other hardware description languages or graphical design programs, should you use another means to describe your hardware.

After introducing verilog, the next section of this chapter describes the interface to the MMC2001. We list the control signals and bounds on the gate capacity of this microcontroller.

Then we present an architecture for special-purpose processing in user-designed hardware. This architecture uses your understanding of the various types of ports and synchronization introduced in Chapters 5 and 6.

We conclude with some design examples, and with discussion of the hardware-software tradeoffs that the applications designer needs to address in deciding how to implement I/O devices and other special-purpose hardware that can be put on the MMC2001.

8.1   Verilog

A *hardware description language* is a text-oriented description of hardware. One such language, the very high speed integrated circuits hardware description language (*VHDL*) was developed about 1983 by companies dealing with the U.S. Department of Defense. An IEEE standard (1076–1993), it is based on the programming language ADA, and combines syntactic features of sequential languages, concurrent languages for parallel computation, net-list languages for specifying interconnections, languages to express timing, and waveform-generating languages. Another, *verilog*, was developed about the same time as a proprietary language for a commercial design package. An IEEE standard (1364–1995), it uses some C syntax as well as ADA syntax. We will use verilog in this chapter because it is easier to learn, especially if the reader is familiar with C. Also, it is widely used by applications designers, especially within Motorola. From verilog, the reader can learn to use VHDL, or other languages, to describe his or her design.

In this section, we present enough verilog to design devices in the remainder of this chapter. We are mainly interested in how hardware can be described to be implemented in the Altera Flex EPF10K100ABC356-1 FPGA (*field-programmable gate array*) on Motorola's FPGA01 board. This hardware description can be delivered to Motorola to implement the same logic, that was put on the FPGA, into the MMC2001 chip.

Verilog describes a hardware *module* either structurally, by showing what is connected to what within it, or behaviorally, by showing how the module's outputs can be generated from its inputs, using a procedural language somewhat like C. Consider a two-bit decoder (Figure 8.1). The structural description for this decoder is shown below:

```
module decoder(out, a, b);   // comments may appear after the double "slash"
    output [0:3]  out;       // operands declared, this first arg. is output vector
    input  a, b;             // second and third args are scalar inputs
    wire   abar, bbar;       // internal "local variable" has no storage of data
    not #1                   // lists all the inverters; they have a delay of 1
        C0(abar, a), C1(bbar, b);  // gate names are like function names,
    nand #1                  // lists all the NAND gates; they have a delay of 1
        N0(out[0], abar, bbar); N1(out[1], abar, b ); // operands are
        N2(out[2], a    , bbar); N3(out[3], a , b ); // connections
endmodule
```

The *structural description* shows the components and interconnections between them. The **module** declaration specified the name and a list of the input/output ports between parentheses, and the end of the description is indicated by the **endmodule** statement. In between these statement, we can declare variables and list gates, as well as other things which will be discussed later.

The output ports, input ports, and internal connections (like local variables) are typically declared first. Each declaration consists of a key word, an optional range for vectors, and a list of variable names. An output is declared with a key word **output**, an input is declared with a key word **input**, and an input/output is

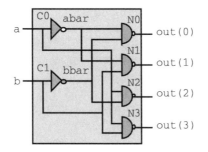

Figure 8.1. A Two-Bit Decoder

declared with a key word **inout**. Internal logical variables can be declared to have memory, using the key word **reg** or to be links that have no memory, having the key word **wire**.

The gates are listed in groups as if they are function calls. Each group of gates begins with the gate type, followed by properties, such as delay which is specified by a pound sign "#" followed by the number of unit delays. Each gate in that group has the name of the particular gate, followed by a list of connections, where the output connection(s) are listed first, then the input connection(s), and the tristate enable(s). The names of the gates are represented by key words such as **and**, **nand**, **or**, **nor**, **xor**, **xnor**, **buf**, and **not**.

The *behavioral description* shows a high-level "program" to get the outputs from the inputs. One needs to think of this description in terms of a simulator, as it has a number of commands needed to properly and efficiently run the simulator. A behavioral description of the two-input decoder is shown below.

```
module decoder(out, a, b);   // the first few lines are identical to the structural desc.
    output   [0:3] out;
    input    a, b;
    reg  [0:3]  out;   // outputs are declared to be storage elements for simulator
    integer i;   // procedural local variable used in the for loop
    always @ ( a or b ) begin    // proceed simulation if either a or b changes
            for(i = 0; i < 3; i++)
            out[i] = ((2*a + b) == i)?0:1;   // make ab output low, others high
    end
endmodule
```

The behavioral description consists of the module declaration, its name, inputs and outputs, exactly as in the structural description. Most statements are written using C/C++ syntax, but additional ADA-like statements are inserted to run the simulator. Also, local variables (e.g. loop counter i) can be declared in order to write the description, and key word **reg** can be used to hold binary data in the simulator, even though the variable will become a wire in the structural description. A purely behavioral description body can have one or more statements beginning with key word **always** followed by a statement that is simulated as an infinite loop. Key word **@** indicates an *event control* with a variable; it means "wait here until the

indicated variable changes". Key words **or** permit listing several variables that can
allow the simulation to proceed past the event control, and **posedge** and **negedge**
to permit the simulator to proceed if a positive edge or a negative edge appears. In
verilog, ADA key words **begin** and **end** (instead of C/C++'s curly brackets)
enclose a compound statement used in place of a single statement.

A minor difference between verilog and C/C++ is the syntax of loop expressions. You can use C/C++'s *for* loops because they are recognized by the Altera
verilog compiler, but it does not use *while* or *do while* loops, and it has a **repeat**
loop and a **forever** loop. A more significant difference between their syntaxes shows
up in case expressions, and bit and other base constants. In addition to the usual
digit values, x or X represents a don't care and, z or Z represents a high-impedance
output. As an example, Figure 8.1's decoder can also be described, in place of the for
loop, as follows:

```
case ({a,b})                    // concatenation of variables is indicated by a comma
    2'b 00: out = 4'b 0111  ;   // 2'b or 2'B represents 2-bit binary
    2'b 01: out = 4'b 1011      // 4'b or 4'B represents 4-bit binary
    2'b 10: out = 4'b 1101      // d or D represents decimal, o or O represents octal
    2'b 11: out = 4'b 1110  ;   // h or H represents hexadecimal
endcase                         // note the absence of break in verilog case stmnt.
```

To write data into registers and memories, declare their variable names using key
word **reg** and write conventional C/C++ assignment statements. Nonstorage
variables can be declared as **wire** as we discussed earlier, or as **tri**, **wand**, or **wor**,
as well as a few others; these having the meaning of tristate, wire-and, or wire-or
connections. When assigning a value to any non-storage variable, the key word
assign precedes the assignment statement. For instance, we illustrate below the
74HC374 (Figure 4.3d).

```
module C74HC374(q, d, clk, en);     // list all of the names of this chip's pins
    output [7:0]      q;            // eight-bit output
    input  [7:0]  d;                // eight-bit input
    input  clk, en;                 // scalar inputs
    reg    [7:0]  r;                // internal register
    always @ ( posedge clk ) r = d; // write data inputs
    assign q = en ? (8'b z) : r;    // if en is high, output is tristate-open, else r
endmodule
```

As shown in Figure 8.2, two copies of a previously defined module, such as the
C74HC374, can be connected. In a structural description, the larger module,
reg16, is described by a list of its component modules, C74HC374s, and connections to a component module are shown in its list of operands. Parameters can be
put in fixed positions, the leftmost actual argument (name used in the higher-level
module description, reg16) being connected to the leftmost formal argument (name
used in the lower-level module description, C74HC374), or can be matched to formal
arguments. In the latter technique, for each argument, after a period (.) a formal
parameter name such as q is followed by the actual parameter name, such as q [7:0],
in parenthesis.

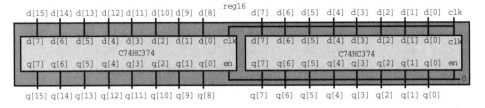

Figure 8.2. Module `reg16` Built From Module `C74HC374`

The verilog description, corresponding to Figure 8.2, is listed below.

```
module reg16 (q, d, clk);          // list all of the names of this chip's pins
    output [15:0]    q;            // eight-bit output
    input  [15:0]    d;            // eight-bit input
    input  clk;                    // scalar inputs
    C74HC374         r1(.q (q[ 7:0]), .d (d[ 7:0]), .clk (clk), .en (0));
    C74HC374         r2(.q(q[15:8]),.d(d[15:8]),.clk(clk),.en(0));
endmodule
```

Modules can be defined in a mixture of structural and behavioral descriptions. Also, modules can be *parameterized* to avoid the writing of nearly identical descriptions. A parameterized behavioral description of four exclusive-or gates is shown in module `xor_chain` (Figure 8.3) The module block structural description defines `xor_chains` parameter width, written after key word **parameter**, to be 8, shown within parenthesis after the pound sign (#). block uses a version of the `xor_chain` module that has 8 gates.

```
module xor_chain(out, a, b);
    parameter width = 4;           // parameter defaulting to 4
    output     [1:width] out;      // (one-dimensional) array output
    input [1:width] a, b           // (one-dimensional) array inputs
    assign out = a ^ b;
endmodule

module block;
    wire [1:8] a, b, c             // some signals
    xor_chain #(8) x1(a, b, c);// parameter 8 specifies width
endmodule
```

Figure 8.3. Parameterized `xor_chain` Module

Modules can also be defined by an *array of instances*. An index range is appended to the module name to designate a (one-dimensional) array of modules. Then a (one-dimensional) array of arguments can be connected to it. Parameterized modules and array of instances simplify the description of modules used in I/O devices, that are generally connected by busses, which are arrays of wires. We illustrate how neatly these techniques work by describing common modules using tristate driver and a D edge-triggered flip-flop shown below. The tristate driver TRI drives its output out with the signal a when the enable en is true; this is behaviorally defined below. The flip-flop will described soon.

```
module TRI(out, a, en);
    output out; input a, en; assign out = en ? a : 1'b z; endmodule
```

In the module driver, an array of instances of a scalar tristate driver TRI is used (see Figure 8.4). In driver, array wire element w2[1] is connected to the output, wire w1[1] is connected to the data input, and scalar signal cnt is connected to enable en, of array instance TRI[1], and so on.

```
module driver;
    wire [1:8] w1, w2; wire ctl;         // some bus and single wire signals
    TRI[1:8]   t(w2, w1, ctl);           // array of tristate drivers (ctl is replicated)
endmodule
```

The flip-flop DFF stores its d input value into the register q whenever the clk signal rises. The output qbar is always the complement of the stored variable q.

```
module DFF(q, qbar, d, clk); output q, qbar; input d, clk; reg ,q;
    wire qbar; assign qbar = ~ q; always @ (posedge clk) q = d;
endmodule
```

A serial-in-serial-out shift register (Figure 8.5) can be defined as an array of DFF modules, as shown in module shift_register below. In it, the expression { a,b,c } means the concatenation of a, b, and c. The left DFF output is connected to the shift_register output out, the right DFF input is connected to the shift_register input in, and the remaining DFF modules are connected through wire d such that the output of each shift register is connected to the input of the shift register to its left.

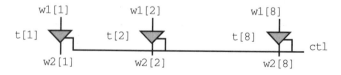
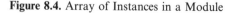

Figure 8.4. Array of Instances in a Module

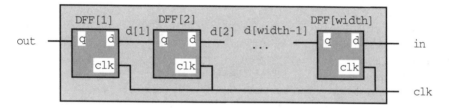

Figure 8.5. Shift Register

```
module shift_register(out, in, clk);
    parameter width = 4; output out; input in, clk; wire [1:width-1] d;
    DFF[1:width] ({out, d[1:width - 1] }, { d[1:width - 1], in }, clk );
endmodule
```

A binary ripple counter (Figure 8.6) is described similarly. Its rightmost flip-flop is clocked with counter's clk input; each flip-flop's d input connects to its qbar output, and each successive flip-flop is clocked by the qbar output of the flip-flop to its right.

```
module counter(out, clk); parameter width = 4;
    output [1:width] out; input clk; wire [1:width] w;
    DFF [1:width] (out, w, { w[2:width], clk } );
endmodule
```

Neither ROMs nor RAMs should be broken down into primitive gates; rather, each ASIC foundry such as Motorola has a library of memory modules, which the applications engineer can select from. However, for simulation and descriptive purposes, we often instantiate a memory with a behavioral description. An h-word, dw-bit-per-word ROM, with aw address bits (h $\leq 2^{aw}$) is described below. Note the array of **reg** "instances" forms the memory. The *initial directive*, which is executed only when the module is first used, initializes the ROM's contents (... means more assignment statements). This directive, **initial**, can also be used to clear, or otherwise initialize, registers.

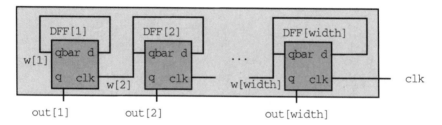

Figure 8.6. Counter

```
module rom(out, a, en);
    parameter dw=4, aw=5, h=32; // data width, address width, and mem height
    output [dw-1,0] out; input [aw-1:0] a; reg [dw-1,0] m [0:h-1];
    initial begin m[0] = dw'b 01100; m[1] = dw'b 10100; ....end
    always @ (en) out = en ? m[a] : dw'b z;
endmodule
```

An h-word, dw-bit-per-word RAM, with an aw-bit address, is described in similar manner, except that it need not be initialized, and there would be means to write an input word into it.

Finally, files can be included, and symbolic names can be defined, to facilitate the writing of large descriptions. Note that this is similar to C/C++, but in place of the pound sign (#) verilog uses the single quote ('). Also in the body of the description, all the defined symbolic names are prefixed by the single quote (') wherever they are used.

```
'include file1     // copy of file1 is inserted here
'define SIZE 2     // symbolic name SIZE is defined
module block;
    wire [1:'SIZE] a, b, c;           // SIZE is used to specify dimension
    xor_chain #('SIZE) x1(a, b, c);   // SIZE is used to specify parameter
endmodule
```

8.2   MMC2001 Environment for Additional Hardware

In order to use the FPGA, you need to know the names and purposes of signals that are available for your design. The verilog module cmb1200_wrapper listed below gives the signals we use in examples in this chapter (almost all MMC2001 pins can be used as inputs or outputs, but we will constrain our examples to using the External Interface Module pins shown here). You should write your description in place of the second to last line of this description. Table 8.1 shows a part of an .acf file, which is an Altera file that is input to the verilog compiler and that defines the pins for module cmb1200_wrapper.

```
module cmb1200_wrapper ( fpga_out_b, m_data_l_z, m_address, m_cs,
    m_clkout, m_rw, reset_b, m_rstin_b, m_eb_A_b, m_eb_B_b, m_oe_b,
    fpga_io0, fpga_io5, fpga_io6, fpga_io10, fpga_io18, fpga_io19,
    fpga_io20, fpga_io21, fpga_io25, fpga_io29, fpga_io30, fpga_io31,
    fpga_io32, fpga_io33, fpga_io36, fpga_io40, fpga_io41, fpga_io42,
    fpga_io43, fpga_io44, fpga_io46, fpga_io47, fpga_io52, fpga_io53,
    fpga_io54, fpga_io55, fpga_io56, fpga_io57, fpga_io58, fpga_io59,
    fpga_io60, fpga_io61, fpga_io62, fpga_io63, fpga_io64, fpga_io65,
    fpga_io66, fpga_io67, fpga_io68, fpga_io69, fpga_io70, fpga_io71
);
```

Table 8.1. PLA Pin Definitions (read top-to-bottom in each of two columns)

```
CHIP cmb1200_wrapper
BEGIN
  |fpga_out_b        : OUTPUT_PIN = G5;      |fpga_io0    : INPUT_PIN = AF13;
  |m_data_l_z15      : BIDIR_PIN = A3;       |fpga_io5    : INPUT_PIN = AA25;
  |m_data_l_z14      : BIDIR_PIN = B5;       |fpga_io6    : INPUT_PIN = AA26;
  |m_data_l_z13      : BIDIR_PIN = C6;       |fpga_io10   : INPUT_PIN = Y25;
  |m_data_l_z12      : BIDIR_PIN = B3;       |fpga_io18   : INPUT_PIN = U24;
  |m_data_l_z11      : BIDIR_PIN = B2;       |fpga_io19   : INPUT_PIN = T22;
  |m_data_l_z10      : BIDIR_PIN = C5;       |fpga_io20   : INPUT_PIN = T2;
  |m_data_l_z9       : BIDIR_PIN = C4;       |fpga_io21   : INPUT_PIN = R23;
  |m_data_l_z8       : BIDIR_PIN = C3;       |fpga_io25   : INPUT_PIN = N23;
  |m_data_l_z7       : BIDIR_PIN = AF4;      |fpga_io29   : INPUT_PIN = L25;
  |m_data_l_z6       : BIDIR_PIN = AD8;      |fpga_io30   : INPUT_PIN = K23;
  |m_data_l_z5       : BIDIR_PIN = AE5;      |fpga_io31   : INPUT_PIN = J24;
  |m_data_l_z4       : BIDIR_PIN = AD6;      |fpga_io32   : INPUT_PIN = J2;
  |m_data_l_z3       : BIDIR_PIN = AF2;      |fpga_io33   : INPUT_PIN = J22;
  |m_data_l_z2       : BIDIR_PIN = AD5;      |fpga_io36   : INPUT_PIN = G22;
  |m_data_l_z1       : BIDIR_PIN = AD4;      |fpga_io40   : INPUT_PIN = E24;
  |m_data_l_z0       : BIDIR_PIN = C18;      |fpga_io41   : INPUT_PIN = D26;
  |m_address19       : INPUT_PIN = B20;      |fpga_io42   : INPUT_PIN = C11;
  |m_address18       : INPUT_PIN = B19;      |fpga_io43   : INPUT_PIN = AF5;
  |m_address17       : INPUT_PIN = AF22;     |fpga_io44   : INPUT_PIN = B10;
  |m_address16       : INPUT_PIN = AF21;     |fpga_io46   : INPUT_PIN = A8;
  |m_address15       : INPUT_PIN = A21;      |fpga_io47   : INPUT_PIN = AF6;
  |m_address14       : INPUT_PIN = A22;      |fpga_io52   : INPUT_PIN = C12;
  |m_address13       : INPUT_PIN = C17;      |fpga_io53   : INPUT_PIN = AF9;
  |m_address12       : INPUT_PIN = AD17;     |fpga_io54   : INPUT_PIN = A11;
  |m_address11       : INPUT_PIN = AE18;     |fpga_io55   : INPUT_PIN = AF10;
  |m_address10       : INPUT_PIN = B18;      |fpga_io56   : INPUT_PIN = B12;
  |m_address9        : INPUT_PIN = AF20;     |fpga_io57   : INPUT_PIN = AD12;
  |m_address8        : INPUT_PIN = AD16;     |fpga_io58   : INPUT_PIN = A12;
  |m_address7        : INPUT_PIN = AE17;     |fpga_io59   : INPUT_PIN = AF12;
  |m_address6        : INPUT_PIN = A19;      |fpga_io60   : INPUT_PIN = A15;
  |m_address5        : INPUT_PIN = C16;      |fpga_io61   : INPUT_PIN = B15;
  |m_address4        : INPUT_PIN = AF18;     |fpga_io62   : INPUT_PIN = AF15;
  |m_address3        : INPUT_PIN = B17;      |fpga_io63   : INPUT_PIN = A16;
  |m_address2        : INPUT_PIN = AE16;     |fpga_io64   : INPUT_PIN = AD15;
  |m_address1        : INPUT_PIN = A18;      |fpga_io65   : INPUT_PIN = A17;
  |m_address0        : INPUT_PIN = B16;      |fpga_io66   : INPUT_PIN = AF17;
  |m_cs3             : INPUT_PIN = N4;       |fpga_io67   : INPUT_PIN = C21;
  |m_cs2             : INPUT_PIN = N5;       |fpga_io68   : INPUT_PIN = AE22;
  |m_cs1             : INPUT_PIN = N3;       |fpga_io69   : INPUT_PIN = B24;
  |m_cs0             : INPUT_PIN = N2;       |fpga_io70   : INPUT_PIN = AD21;
  |m_clkout          : INPUT_PIN = A14;      |fpga_io71   : INPUT_PIN = C22;
  |m_rw              : INPUT_PIN = P24;      |m_eb_A_b    : INPUT_PIN = C;
  |reset_b           : INPUT_PIN = AB4;      |m_eb_B_b    : INPUT_PIN = A9;
  |m_rstin_b         : INPUT_PIN = E2;       |m_oe_b      : INPUT_PIN = E1;
  DEVICE = epf10k100abc356-1;
END;
```

```
    output fpga_out_b;                // Buffer control for FPGA access
    inout [15:0] m_data_l_z;          // m_data_l_z[15:0] - MCU, D[15:0]
    input [19:0] m_address;           // m_address[19:0] - MCU, A[19:0]
    input [3:0] m_cs;                 // CS3 pos logic, CS2 to CS0 neg logic
    input m_clkout;                   // System clock (MCU's CLKOUT)
    input m_rw;                       // MCU's R/W (1 = READ, 0 = WRITE)
    input reset_b;                    // MCU's RSTOUT
    input m_rstin_b;                  // MCU's RSTIN
    input m_eb_A_b;                   // MCU's Enable byte A - EB[1]
    input m_eb_B_b;                   // MCU's Enable byte B - EB[0]
    input m_oe_b;                     // MCU's OE (output enable)
    input fpga_io0, fpga_io5, fpga_io6, fpga_io10, fpga_io18,
fpga_io19, fpga_io20, fpga_io21, fpga_io25, fpga_io29, fpga_io30,
fpga_io31, fpga_io32, fpga_io33, fpga_io36, fpga_io40, fpga_io41,
fpga_io42, fpga_io43, fpga_io44, fpga_io46, fpga_io47, fpga_io52,
fpga_io53, fpga_io54, fpga_io55, fpga_io56, fpga_io57, fpga_io58,
fpga_io59, fpga_io60, fpga_io61, fpga_io62, fpga_io63, fpga_io64,
fpga_io65, fpga_io66, fpga_io67, fpga_io68, fpga_io69, fpga_io70,
fpga_io71;                            // Each pin can be declared output or inout
// fpga_out_b must be driven low during a read of the FPGA by the MCU.
    assign fpga_out_b = (m_cs[3] && m_rw) ? 0          1;

    example_top example_top ( ); // Instantiate your module here!

endmodule
```

Modules that you can use to build up your design are listed below and in Figure 8.7.

```
module NAND2(out, a, b);
    output out; input a, b; assign out = ~ a & b; endmodule

module NAND3(out, a, b, c);
    output out; input a, b, c; assign out = ~ a & b & c; endmodule

module NAND4(out, a, b, c, d);
    output out; input a, b, c, d; assign out = ~ a & b & c & d;
endmodule

module NOR2(out, a, b);
    output out; output a, b; assign out = ~ a | b; endmodule
module NOR3(out, a, b, c);
    output out; input a, b, c; assign out = ~ a | b | c; endmodule

module NOR4(out, a, b, c, d);
    output out; input a, b, c, d; assign out = ~ a | b | c | d;
endmodule

module BUFFER(out, a);>output out; input a; assign out = a; endmodule
```

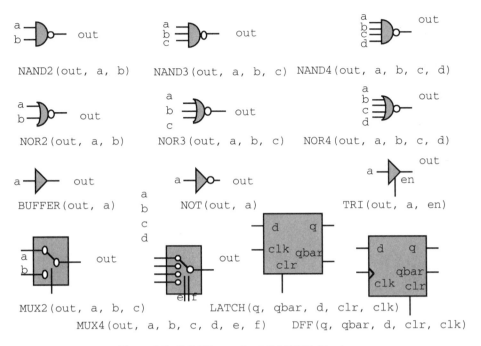

Figure 8.7. Cell Library for MMC2001 Hardware

```
module NOT(out, a);
    output out; input a; assign out = ~ a; endmodule

module TRI(out, a, en);
    output out; input a, en; assign out = en ? a : 1'b z; endmodule

module MUX2(out, a, b, c);
    output out; reg out; input a,b,c; assign out=c?a:b; endmodule

module MUX4(out, a, b, c, d, ,e, f);
    output out; output a, b, c, d, ,e, f; reg out;

    always case ({ e, f })
        2'b 00: out=a; 2'b 01: out=b; 2'b 10: out=c; 2'b 11: out=d;
    endcase

endmodule
module LATCH(q, d, clr, clk);
    output q; input d, clr, clk; reg q; wire qbar; assign qbar = ~q;
    always @ (clk or clr) begin
        if(clr) q = 0; else if(clk) q = d;
    end
endmodule
```

```
module DFF(q, d, clr, clk);
output q; input d, clr, clk; reg q; wire qbar; assign qbar = ~q;
    always @ (posedge clk or clr)
        begin if(clr) q = 0; else if(posedge clk) q = d; end
endmodule
```

The general synthesis design problem starts with a high-level description, and breaks it down into primitive modules that can be laid out on the MMC2001 chip. The designer who has studied previous chapters in this book can often start with an object-oriented device's member functions description, or their equivalent procedures, then implement them in a verilog behavioral description. Also, most hardware description language compilers have modules that are functionally equivalent to the standard 74HC series of integrated circuits. The design of hardware for the MMC2001 can begin with a logic diagram of these integrated circuits. Ultimately, its components can be implemented with cells in the *cell library* shown in Figure 8.7.

This list of cells is representative of the kind of primitives that Motorola would want in order to lay out the design on the MMC2001. However, the modules in a typical cell library include a collection of cells, for each one shown in Figure 8.7, with different power consumption and delay characteristics. For instance, the two input NAND gate might have 15 alternative cell designs. The library for the Altera Flex EPF10K100ABC356-1 FPGA has parameterized gates that permit up to 12 inputs to a NAND gate. However, the MMC2001 can only directly implement four-input gates. The design process leads to the primitives shown in Figure 8.7. But, there is still considerable layout work to be done, to put these primitives on the surface of the MMC2001 chip with minimal waste of surface area, and acceptable delays. The more of this work that is done by the applications designer, the lower Motorola's cost of having the design put on the MMC2001 chip.

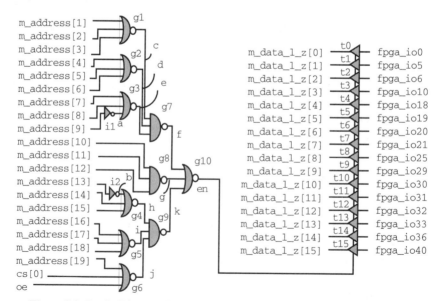

Figure 8.8. Logic Diagram for a Completely Decoded Input Device (revised)

The basic input port (Figure 5.1), now connected as in Figure 8.8, is described thus:

```
module cmb1200_wrapper (
    assign     fpga_out_b = (m_cs[3] && m_rw) ? 0 : 1;
    ... // as shown earlier in this section
  wire     a, b, c, d, e, f, g, h, i, j, k;
  NOT    i1(a, m_address[9]), i2(b, m_address[13]);
NOR3     g1(c, m_address[1], m_address[2], m_address[3]),
         g2(d, m_address[4], m_address[5], m_address[6]),
         g3(e, m_address[7], m_address[8], a),
         g4(h, b, m_address[14], m_address[15]),
         g5(d, m_address[16], m_address[17], m_address[18]),
         g6(d, m_address[19], cs[0], oe), g10(en, f, g, k),
NAND3    g7(f, c, d, e), g9(k, i, h, j),
         g8(d, m_address[10], m_address[11], m_address[12]);
TRI  t0(m_data_1_z[0], fpga_io0, en),t1(m_data_1_z[1],fpga_io5, en),
    t2(m_data_1_z[2], fpga_io6, en), t3(m_data_1_z[3], fpga_io10, en),
    t4(m_data_1_z[4], fpga_io18, en), t5(m_data_1_z[5], fpga_io19, en),
    t6(m_data_1_z[6], fpga_io20, en), t7(m_data_1_z[7], fpga_io21, en),
    t8(m_data_1_z[8], fpga_io25, en), t9(m_data_1_z[9], fpga_io29, en),
    10(m_data_1_z[10], fpga_io30, en),t11(m_data_1_z[11], fpga_io31,en),
    12(m_data_1_z[12], fpga_io32, en), t13(m_data_1_z[13], fpga_io33,en),
    t14(m_data_1_z[14], fpga_io36,en),t15(m_data_1_z[15], fpga_io40,en),
endmodule
```

Currently, the MMC2001 can support as many as 40,000 additional gates. However, some circuits that do not lay out efficiently may have somewhat less capacity. Also, the MMC2001 has 144 pins. At the time of writing of this text, there is a discontinuity in integrated circuit chip package pricing, around 176 pins. Packages below 176 pins are significantly cheaper than those having more than this number. Therefore, a limit of 32 pins for logic added to the chip is reasonable, and will be assumed in examples and problems in this chapter. It is possible to exceed these gate and pin limitations, but implementation cost increases substantially when either limit is exceeded.

8.3   A MOVE Architecture for I/O Devices

The additional hardware that is inserted in the MMC2001 chip may be more than just a simple port that has been described in earlier chapters. It may be a processor, or a parallel array of processors. The reader, having understood the various ports in Chapter 5 and the various synchronization primitives of Chapter 6, can use these concepts to design a processor to be inserted in the MMC2001 chip. In this section, we present an architecture for a MOVE processor that can take advantage of these ports and primitives.

As illustrated in Figure 8.9, there is a program memory, generally implemented in ROM, and a data memory, which contains dual-ported SRAM (which can be read from and written into at different addresses in the same memory cycle), ROM, and memory-mapped I/O ports, including one or more program counters. In each memory cycle, a program counter supplies an address to the program memory, which reads out two addresses, the data source address and the data destination address. These are sent to the data memory, to transfer a word from the source to the destination. Unless the destination is the program counter port, the program counter automatically increments, so that it reads the next location in program memory during the next memory cycle. This architecture just MOVEs data, repetitively. We call it a *MOVE processor*.

A MOVE processor is really just a *vertical microprogrammed controller*, but the applications designer can "grow his or her own instruction set and execution environment." A macro assembler can be used to write programs that use symbolic names for MOVE processor addresses, and for sequences of MOVEs to perform common operations. By fully exploiting the techniques described in earlier chapters, when they offer an advantage, he or she can get some very simple but powerful processing techniques, with which he or she can build a single I/O processor, or parallel processors for I/O.

For purposes of illustration, we define a generic MOVE processor controller using a program counter pc, saved program counter retAdd, condition selector condition, condition sense (one dimensional) array wire c[], stack pointer sp, and data memory m. See Figure 8.10. The following Verilog case and if statements indicate how the data being transferred, on dataBus, is obtained and used, and how pc is modified, as a function of the instruction's source address sa and the instruction's destination address da provided by the control memory when it fetches a (move) instruction expressed as a pair (sa, da).

```
case (sa) // partial listing          case (da) // partial listing
    0: dataBus = retAdd;
    1: dataBus = condition;               1: condition = dataBus;
    2: dataBus = sp;                      2: sp = dataBus;
```

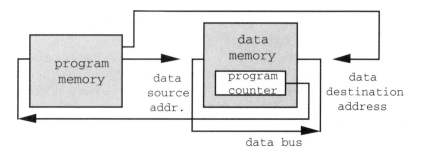

Figure 8.9. Architecture for a MOVE Processor

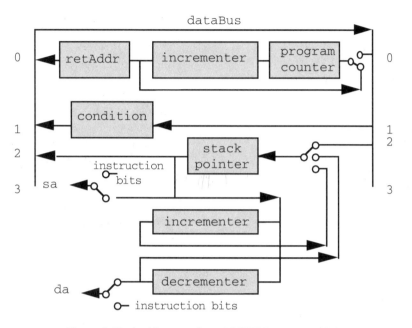

Figure 8.10. Architecture for a MOVE Processor ALU

```
3: dataBus = m[sp++];                    3: m[--sp] = dataBus;
default: dataBus = m[sa];                default: m[da] = dataBus;

endcase                                  endcase
if((da == 0) && c[condition]) { retAdd = pc + 1; pc = dataBus; }
else pc++;
```

The array wire c[], like one's sensor nerves, has individual wires connected to various flag flip-flops, counters, and comparators scattered through the processor, which have to be tested in conditional jump instructions. Input condition c[0] is always 1, and condition register condition can be explicitly or automatically cleared before unconditional jumps. The memory having constants stored in ROM, a jump to location $2'h$ 34 in program memory can be implemented by putting the number $2'h$ 34 into a ROM location such as $3'h$ 103 in data memory (assuming condition is 0). Then a transfer from $3'h$ 103 to location 0, designated by ($3'h$ 103, 0), puts m[$3'h$ 103], which is $2'h$ 34, into pc so the program jumps to $2'h$ 34. A jump to subroutine is executed the same way, the return address being automatically put in retAdd whenever a jump is made. If the subroutine has no jump (or jump to subroutine) instructions, then to return from the subroutine, execute the move: (0, 0).

A *stack pointer* can be read and written at location 2. A word moved on dataBus can be pushed if *da* is 3, and a word can be pulled into dataBus, if sa is 3. In a subroutine, this stack can therefore be used to save and restore the pc, if the subroutine needs to call other subroutines or even to execute jumps. A move (0, 3) can push the saved program counter, and a move (3, 0) can return from a subroutine. Note that the program counter should be pushed before any jump or jump to subroutines is executed; it can be done as the first move instruction executed inside the subroutine.

MOVE processor input and output ports can be, respectively, outputs and inputs of a module, like an adder which has registers (a and b) on its inputs. The verilog descriptions below put this adder at locations 2′h 20 and 2′h 21 in the data memory map, although it could be put almost anywhere else. See Figure 8.11.

```
case (sa) // partial listing          case (da) // partial listing

   . . .                                 . . .
   case 2'h 20: dataBus = a + b;         case 2'h 20: a = dataBus;
   case 2'h 21: dataBus = m[a + b];      case 2'h 21: b = dataBus;

   . . .                                 . . .
endcase                               endcase
```

To add two numbers, putting their sum in a destination, transfer one number to adder input register a, transfer the second to the other register b, and transfer the adder's output to the destination. An input from location 2′h 21 to dataBus will use the sum as an address, to read m[a + b] to dataBus. This is especially useful when looking up a table in ROM. A similar technique can be used with da to write data into a vector using an index. In place of the adder, a general ALU can be used; to get different functions, such as OR, AND, and SUB, the data can be read in different locations. For instance,

```
case (sa)  // partial listing         case (da) // partial listing

   . . .                                 . . .
   2'h 20: dataBus = a + b;              2'h 20: a = dataBus;
   2'h 21: dataBus = a + b + 1;          2'h 21: a = ~ dataBus;
   2'h 22: dataBus = a | b;              2'h 22: b = dataBus;
   2'h 23: dataBus = ~(a | b);           2'h 23: b = ~ dataBus;

   . . .                                 . . .
endcase                               endcase
```

provides most useful arithmetic functions. Going further, though, a specialized module could be built to efficiently compute a communication frame's CRC (cyclic redundancy check), MPEG's DCT (discrete cosine transform), or Huffman coder or decoder translation.

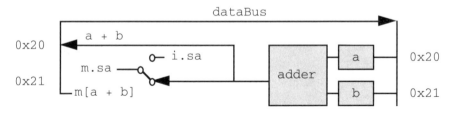

Figure 8.11. Adder Module

Additional modules, and registers at different memory map addresses, can input data to, or output data from, the MOVE processor, outputs from and inputs to the MMC2001 address, data, and control busses, or the outside world. From earlier chapters, note that memory-mapped I/O ports, including those in this architecture, can be input ports, or basic, readable, and shadowed output, set and clear ports, address triggers and address register outputs. Set and clear ports can be implemented with fewer gates than other ports.

We often move I/O data to or from a buffer or queue. The description of a module, which provides an address to move data to or form a buffer, is similar to that of a stack. A queue can be implemented with two such pointers. In both cases, a counter register can keep track of the number of elements in the buffer or queue. A condition sense such as c[1], can be wired to a buffer's counter, and the corresponding index 1 can be put in *condition* so that a jump can be conditionally made on whether the count is zero.

Address triggers are particularly useful; reading data from, or writing data to, a location, or fetching an instruction from program memory, can automatically trigger some other activity, such as loading condition from the low-order bits of sa, da, or pc. The following example (See Figure 8.12) loads a buffer from an input port using gadfly synchronization. Data input from port at location 3'h 031 is stored using an autoincrementing index register x which is at location 4, as a down-counter count at location 6, counts the number of words to be moved. Gadfly synchronization uses a falling edge-triggered flip-flop busy, which is cleared when the external world has data, and is set when port is read, if count is not zero. The program below transfers five bytes from port to the buffer. Its first two lines initialize address x, count c, and busy. Then the number 1 is explicitly moved into condition. This makes jumps, which load the program counter, conditionally dependent on wire s[1], which is connected to busy. The fourth move continually loads the constant 3'h 2b4 into pc as long as busy is 1, gadflying until new data arrive. When data arrive, they are moved from port to destination 5, being written at the address given by x, after which x is incremented (i.e., the data are written into m[x++]). In addition, as a result of an address trigger, count is decremented and the low-order two bits of this destination address, which are 2'b 01, are loaded into condition. This makes jumps, which load the program counter, dependent on wire s[2]. This wire, s[2], is asserted (and busy is set) exactly as long as count is not zero. This jump is taken until the counter reaches zero, after which the jump is not taken, whereupon the MMC2001 exits this program segment. However, when pc is 3'h 2b6, then condition is loaded with 1, selecting condition s[1], This program segment is listed below, showing the location and contents of each move in program memory.

Note how efficiently vertical microprogramming can execute a gadfly loop and move data to a buffer, using techniques developed to access I/O ports and synchronize to events. This gadfly loop has a latency of one memory cycle, because hardware tied one of the processor's condition wires c[i] to the I/O device signals that had to be tested.

Program Memory

3′h 2b1 (3′h 121, 4)   // initialize x to point to start of buffer in SRAM (3′h 84)
3′h 2b2 (3′h 120, 6)   // initialize count to number of bytes to move (5), set busy
3′h 2b3 (3′h 122, 1)   // write 1 into condition to choose the bit for the condition
3′h 2b4 (3′h 123, 0)   // jump to this same location as long as busy is 1
3′h 2b5 (3′h 031, 5)   // port to m[x], inc. x, dec. count, 01 to condition 0 to busy
3′h 2b6 (3′h 123, 0)   // if x is not zero, jump to 4th instruction, at
                        location 3′h 2b3

```
case (sa)  // partial listing          case (da)     // partial listing
                                         0:  begin
                                               pc = dataBus; // make jumps
                                             end
                                         4:  x = dataBus;
   3′h 031: dataBus = port;              5:  begin
                                               m[x++] = dataBus;
                                               if (count != 0) busy = 1;
   3′h 120: dataBus = 3′h 005; (ROM)            condition = da; count;
   3′h 121: dataBus = 3′h 084; (ROM)          end
   3′h 122: dataBus = 3′h 001; (ROM)    6:  begin
   3′h 123: dataBus = 3′h 2b4; (ROM)
count = da; busy = 1;                        end
                                         3′h 084: SRAM = dataBus; ...

endcase                                 endcase
```

   The design of the condition wire system is fairly simple. If all procedures executed in the processor sense n signals with wires c[n], then make condition's
dimension m just big enough so each input is on a separate wire ($n \leq 2^m$). For those
transfers that should reload condition, put their locations at addresses such that
the low-order m bits are distinct from each other. Then the low-order m bits of the
address can be transferred to condition whenever certain locations are addressed,
by address triggers. The designer should also look for groups of transfers that always
occur together; these can be implemented as one transfer on *dataBus* and address
triggers that cause other activities to occur. Additionally, a *shadow port* in the *da*
decoder, can simultaneously transfer data to two or more ports in one memory cycle.
For instance, if one output port register is clocked by an incomplete decoder that
matches address 2′b 10101x, and another output port register is clocked by an
incomplete decoder that matches address 2′b 10100x, then if the destination address
is 2,b 101011, both ports are written in one move instruction.
   Clearly, all the synchronization techniques can be implemented in the MOVE
processor. A delay loop can be used to implement real-time synchronization, and a
gadfly loop can be implemented by attaching one of the condition sense lines, such as
c[9], and putting 9 into condition. *Interrupts* can be implemented by an initial
three-memory-cycle operation, which first pushes retAd and condition onto the
stack, and then transfers a constant to pc. Obviously, an address trigger can clear
the source of the interrupt. *Direct memory access* can also be implemented by forcing
a hardware-generated constant onto *sa* and *da*, which steals a cycle to move data.

Using two program counters, we can implement context switching; one program counter, having higher priority, can take care of input from the outside world, and the other can communicate with the MMC2001 whenever the higher priority program counter is not used. We can also utilize shuttle memory; a memory can be switched into and out of the MOVE processor's address map, from or to the MMC2001 address map, and a pair of these memories can be used to avoid waiting for a connection to a shuttle memory.

8.4   Examples

The previous sections provided a brief tutorial on verilog, a description of the environment of the Altera Flex EPF10K100ABC356-1 FPGA, and the architecture of a MOVE processor. We now put these pieces together.

In this section, we will give an example of a hardware design to be added to an MMC2001, from a behavioral description to the specification of modules in the cell library (Figure 8.7), and the loading of this design into the FPGA. We will consider the design of a character string search engine. It will compare characters fed to it from a hard disk, or fast communication network, to a preloaded character string that is stored in an internal 32-byte memory.

The first step in this design is to develop an accurate behavioral description of the design. This can start with a word description, a design in another medium such as the 74HC family of integrated circuits, or a C/C++ program. We assume that the word we search for is a normal 7-bit ASCII text word having no more than 31 characters and ending in a space, and for simplicity, that upper and lower case letters are considered different (it is *case sensitive*), a word begins after a space and ends just before the next space, and a space is not followed immediately by another space. Furthermore, we do not consider how the comparison string is written into. This example exhibits all the key elements of the design; a realistic design, which removes the simplifying assumptions, can be derived from it. Here is the C/C++ procedure that specifies the behavioral description:

```
char search(char *p){char i, m = 1, f = 0; static char s[32];
    do{
        if(*p != s[i]) m = 0;
        if(*p == 0x20) { f |= m; m = 1; i = 0; }
        else i++;
    } while(*p++);
    return f;
}
```

We convert this procedure into a verilog behavioral description. The C/C++ description is passed a pointer to the null-terminated input string. The verilog description is passed a character at a time, and when a character is input to this module, a module clock clk causes the module to execute the comparison of one character c. The verilog description follows this strategy:

```
module search(f, c, clk);
    output f; input [6:0] c; input clk; integer i;
    reg [6:0] s[0:31]; /* search string memory */ reg m, f; /* status flags */
    always @ ( posedge clk ) begin
        if(c != s[i]) m = 0;
        if(c == 2'h 20) { f = m; m = 1; i = 0; } else i++;
    end
endmodule
```

In *top-down design*, the original module is broken down into component modules. The previous verilog behavioral description is translated into a verilog description with high-level structural components that have behavioral descriptions within them. In top-down design, you see what components are needed to fit the higher-level module's operations, rather than designing ill-fitting components and trying to add extra hardware to the higher-level module to use them. This description is shown in Figure 8.12.

```
module search(f, a, clk, reset);
    output f; input [6:0] c; input clk, reset;
    reg m, f; wire [6:0] b; wire [4:0] address; wire c, d;
    compare m1(c, a, b); space m2(d, a);
    counter m3(address, d, clk); memory m4(b, address);
    DFF   ff1(.q (m), .d (h), .clr (reset), .clk (clk));
          ff2(.q (f), .d (k), .clr (reset), .clk (clk));
    NAND2 g1(e, m, c), g3(h, e, g), g4(i, d, m), g5(k, i, j);
    NOT   g2(g, d), g6(j, f);
endmodule
module compare(result, a, b); // returns a 1 if the characters a and b match
    output result; input [6:0] a, b; assign result = a == b;
endmodule
module space(result, a); // returns a 1 if character a is a space character
    output result; input [6:0] a; assign result = a == 2'h 20;
endmodule
```

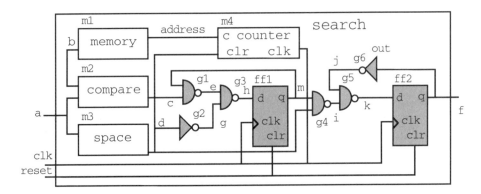

Figure 8.12. Search Module

```
module memory(result, a); // returns the word selected by the address a
    output result; input [4:0] a; reg [6:0] s[0:31];
    assign result = s[a];
endmodule
module counter(count, clr, clk); // returns count, cleared if clr is asserted
    output [4:0] count; input clr, clk; reg [4:0] count;
    always @ ( posedge clk ) if(clr) count = 0; else count++;
endmodule
```

Continuing our top-down design, the component modules are broken down into library cells (see Figure 8.13), which are described below.

```
module space(result, a);
    output result; input [6:0] a; wire c, d, e;
    nor3 g1(e, a[0], a[1], a[2]), g2(d, a[3], a[4], a[6]);
    nand3 g2(c, d, a[5], e); not g4(result, c);
endmodule
```

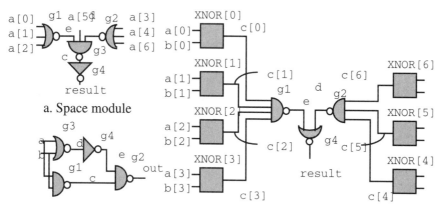

a. Space module

b. XNOR module

c. Compare module

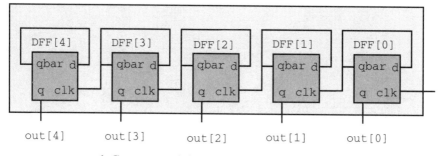

out[4]   out[3]   out[2]   out[1]   out[0]

d. Counter module (essentially Figure 8.6)

Figure 8.13. Component Modules

```
module xnor(out, a, b);
    output out; input a, b; wire c, d, e;
    nand2 g1(c, a, b), g2(out, c, e); nor2 g3(d, a, b); not g4(e, d);
endmodule
module compare(result, a, b);
    output result; input [6:0] a, b; wire [6:0] c; wire d, e;
    xnor[6:0] m1(c, a, b); nor2 g0(result, d, e);
    nand4 g1(e, c[0], c[1], c[2], c[3]); nand3 g2(d, c[4], c[5], c[6]);
endmodule
```

The space module (Figure 8.13a) is merely a decoder. The compare module (Figure 8.13c) is composed of exclusive-NOR gates which are not in the cell library. We use a more complex edge-triggered flip-flop that But these are easily built from cells, as shown in Figure 8.13b. To build the counter, we use an edge-triggered flip-flop that also has the complemented output available (problem 9 here). The memory module is not further broken down into logical cells; it should be implemented with a foundry's library of static RAM layouts.

We now build a parallel array of four search modules, to search for up to four character strings simultaneously. (This extension of the design can obviously be expanded to more than four modules.) Figure 8.14 shows search module connections. However, realizing that a character can be compared each clock cycle, but the MMC2001 can move a byte each five cycles (using ld.b r1,(0,r2) std.b r1,(0,r3), loopt r4), we design a faster processor that can input a character each clock cycle. We will use the MOVE processor, and use a DMA to transfer data arriving on the input port port at 2'h a3 as long as DmaEn is set. The search modules all obtain their a input from a register a[6:0], which is written into at location 2'h 1b, which also generates an address trigger attached to clk. Other parts and program segments of the processor set and clear this DmaEn control bit, but once it is set, the instruction (2'h a3, 2'h 1b) is executed repreatedly. Outputs of search are NANDed; the output f of this NAND gate indicates, if it is 0, that all four character strings have been found somewhere in the input string. The MOVE processor can later use this signal to conditionally jump to a program segment that responds to a successful search.

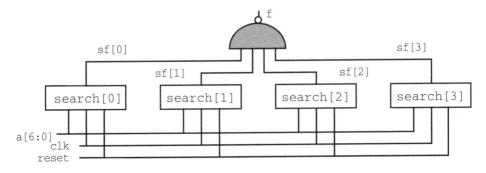

Figure 8.14 MOVE Processor Using Search Modules

```
module parallel_search(f);
   output f; reg [6:0] a; reg DmaEn; wire [7:0] sa, da;
   wire clk, reset; wire [1:4] sf; wire [6:0] dataBus;
   search [1:4] srch (sf,a,clk,reset); NAND4(f, sf[1], sf[2], sf[1],
     sf[1]);
   if(DmaEn) {sa = 2'h a3; da = 2'h 1b;}// else read instruction, increment pc
   case (sa) // partial listing              case (da) // partial listing
   ...                                        ...
     case 2'h a3: dataBus = port;              case 2'h 1b: begin a = dataBus;
                                                 assign clk = 1; end
   ...                                        ...
   endcase                                    endcase
endmodule
```

We deliberately made a mistake in the design above to illustrate the advantage of top-down design. The problem to be solved, is to search for four character strings simultaneously. We first designed the hardware to search one string, and now we simply replicated it. However, this lead to the duplication of the module space in each module search, which all do exactly the same thing. Had we used top-down design, breaking up the module parallel_search, we would realize that we need one copy of the module space and four copies of the module search without the module space in it. We should use top-down design, and pay attention to shared hardware, so that we do not duplicate it in the component modules.

A final few remarks about implementing either search or parallel_-search: the implementation of verilog modules in the FPGA is described in the Altera manual *Max+Plus II Verilog HDL* and in the Motorola manual, *MMCFPGA1200UM/D*. However, the following additional requirements need to be conveyed. In order to access the FPGA, the MMC2001 CLKOUT, which disabled at reset, must be enabled for the FPGA to function properly, and chip selects must be initialized properly in order to access the FPGA. The initialization sequence below should be put in your source code.

```
*((long *) 0x10001000) = 0x00000100;   // Reset source/chip configuration reg.
*((long *) 0x10004018) = 0x00000018;   // External interface module control reg.
*((long *) 0x10004000) = 0x0000F861;   // External interface module CS0 ctl reg.
*((long *) 0x10004004) = 0x00003021;   // External interface module CS1 ctl reg.
*((long *) 0x10004008) = 0x00000021;   // External interface module CS2 ctl reg.
*((long *) 0x1000400C) = 0x000015A1;   // External interface module CS3 ctl reg..
```

Also, the FPGA is able to drive tristate outputs from the chip. However, it does not have an internal tristate bus for a MOVE processor. Inputs to the tristate bus can be implemented using multiplexers, so that the FPGA will emulate the tristate bus for your design. However, when conveying the design to Motorola, you can substitute the multiplexer module with tristate drivers on the outputs of modules that feed data to the MOVE processor bus.

8.5   Conclusions

The designer needs to understand the capabilities and the techniques for the design of external hardware that can be inserted in the MMC2001 chip. External hardware can be designed, using just the right logic for the application, to handle tasks in parallel, or perform sequences of computation in a single clock cycle, so that it can be considerably faster than can be done using the MMC2001 or equivalent general purpose computer. However, if the latter is unable to do some useful work while external hardware unloads some work from it, and if the latter is fast enough, external hardware is unnecessarily expensive and takes longer to design than software running in the general purpose computer.

If you do see the advantage of conveying all or part of your design from software running in the general purpose computer to hardware in a special-purpose module, then this chapter should give you some feel for the design effort to implement special-purpose hardware. The good news is that hardware descriptions languages, like verilog, make the transition from software to hardware fairly smooth. But, as knowing a few words in a foreign language does not make one conversant in that language, just knowing the syntax of verilog does not make one a good hardware designer. We have shown an architecture that is very flexible, and that can take advantage of the techniques you used to interface to the outside world in a conventional processor. This architecture provides a good way to design the hardware that you may add to the MMC2002, thus giving you some advantage over your competitors.

If you wish to further develop your understanding of verilog, we recommend *The Verilog Hardware dsecription Language*, by Thomas and Moorby, Kluwer Academic Publishers, Boston, 1996. This book includes a verilog simulator that can run on various platforms. If you wish to use VHDL, an excellent text is *Digital Systems Design Using VHDL*, by Roth, PWS Publishing, Boston, 1998.

Do You Know These Terms?

See the end of Chapter 1 for instructions.

hardware	module	initial	top-down design
description	structural	directive	stack pointer
language	description	.acf file	shadow port
VHDL	behavioral	cell library	interrupt
verilog	description	MOVE processor	Direct
field-	parameterized	vertical	memory
programmable	description	micro-	access
gate	array of	porgrammed	event control
array (FPGA)	instance	controller	
	bus	case sensitive	

Problems

1. Redesign the module decoder so that it has an enable input; all outputs are high if the enable is high, otherwise one output is low, as selected by inputs a and b in the original module.

    a. as a logic diagram similar to Figure 8.1
    b. as a verilog structural description using NAND gates and inverters
    c. as a verilog behavioral description using a for loop
    d. as a verilog behavioral description using a case statement,

2. Redesign the module decoder so that it has three outputs and eight outputs.

    a. as a logic diagram similar to Figure 8.1
    b. as a verilog structural description using NAND gates and inverters
    c. as a verilog behavioral description using a for loop
    d. as a verilog behavioral description using a case statement

3. Write a parameterized verilog description of a rising edge triggered *width*-bit wide register using the module DFF in §8.1. The module contains width flip-flops.

4. Write a parameterized verilog structural description of a rising edge triggered, synchronously cleared serial-in-serial-out shift register using the module DFF in §8.1. The module contains width flip-flops. When input clk rises, if input clr is asserted, the flip-flop is cleared, otherwise it shifts as in shift_register.

5. Write a parameterized verilog structural description of a rising edge triggered, parallel-in-serial-out shift register using the module DFF in §8.1. The module contains width flip-flops. When input clk rises, if input ld is asserted, the flip-flop is loaded with parallel inputs in[1:width], otherwise it shifts as in shift_register.

6. Write a parameterized verilog description of a rising edge triggered, serial-in-parallel-out shift register using the module DFF in §8.1. The module contains width flip-flops. When input clk rises data shifts as in shift_register. The data is available on outputs out[1:width].

7. Write a parameterized verilog structural description of a parallel load down-counter using the module DFF in §8.1. The module contains width flip-flops. When input clk rises, if load is asserted, data in[1:width] is loaded, otherwise the binary number is decremented. The counter value is available on outputs out[1:width].

8. Write a parameterized verilog structural description of an up-counter using the module DFF in §8.1. The module contains width flip-flops. When input clk rises,

the binary number is incremented. The counter value is available on outputs
out[1:width].

9. Write a verilog behavioral description of a latch that can be loaded from input d,
if clk is asserted, set if set is asserted, or cleared is clear is asserted, and outputs
the complement of the stored data as well as the uncomplemented stored data.

10. Write a verilog behavioral description of an edge-triggered flip-flop that can be
loaded from input d, if clk rises, set if set is asserted, or cleared is clear is
asserted, and outputs the complement of the stored data as well as the uncomplemented stored data.

11. Write a verilog behavioral description of an edge-triggered flip-flop that can be
loaded from input d, if clk rises, set if set is asserted, or cleared is clear is
asserted, and outputs the complement of the stored data as well as the uncomplemented stored data.

12. Write a parameterized verilog structural description of a parallel load down-
counter using the module in problem 11. The module contains width flip-flops.
When input clk rises, if load is asserted, data in[1:width] is loaded, otherwise
the binary number is decremented.

13. Show the logic diagram similar to Figure 8.8, and verilog structural description,
which implements the basic output port of Figure 5.2's decoder, using only the cells
in Figure 8.7. Write the verilog structural description of this design, as a part of
module cmb1200_wrapper.

14. Show the logic diagram similar to Figure 8.8, and verilog structural description,
which implements the readable output port of Figure 5.3's decoder, using only the
cells in Figure 8.7. Write the verilog structural description of this design, as a part of
module cmb1200_wrapper.

15. Show the logic diagram similar to Figure 8.8, and verilog structural description,
which implements the set port of Figure 5.5, using only the cells in Figure 8.7. Write
the verilog structural description of this design, as a part of module
cmb1200_wrapper.

16. Show the logic diagram similar to Figure 8.8, and verilog structural description,
which implements the address trigger of Figure 5.6, using only the cells in Figure 8.7.
Write the verilog structural description of this design, as a part module
cmb1200_wrapper.

17. Show the logic diagram that implements the program counter and condition
register of Figure 8.9, and verilog structural description, using only the cells in
Figure 8.7. The incrementer uses a chain of exclusive-OR gates. There are 513 words
in the program memory, and 11 conditions to be tested at one time or another. Show

the multiplexer with unspecified inputs. Write the verilog structural description of this design, as a part `module cmb1200_wrapper`.

18. Show the logic diagram that implements the stack pointer of Figure 8.9, and verilog structural description, using only the cells in Figure 8.7. The incrementer and decrementer use chains of exclusive-OR gates. There are 753 words in the data memory. Write the verilog description of this design, as a part `module cmb1200_wrapper`.

19. Show the logic diagram that implements the last ALU verilog module behavioral description given in §8.3. Give the verilog structural description, using only the cells in Figure 8.7. The adder use ripple carry, and the ALU 12-bits long Write the verilog structural description of this design, as a part `module cmb1200_wrapper`.

20. Show the logic diagram that implements the last ALU of §8.3 which has OR as well as ADD primitives, and verilog structural description, using only the cells in Figure 8.7. The adder use ripple carry, and the ALU 12-bits long Write the verilog structural description of this design, as a part `module cmb1200_wrapper`.

21. Show the logic diagram that implements the last ALU of §8.3 which has OR as well as ADD primitives, and verilog structural description, using only the cells in Figure 8.7. The adder use ripple carry, and the ALU 12-bits long Write the verilog structural description of this design, as a part `module cmb1200_wrapper`.

22. Show the verilog behavioral description, and a program segment, that outputs data from a buffer, similar to the buffer input example at the end of §8.3.

23. Show the verilog behavioral description, and a program segment, that inputs data to a queue, similar to the buffer input example at the end of §8.3. The queue contains 2^n elements, and begins at $k * 2^n$ so that wrap-around is handed by index register increment/decrement truncation.

24. Show the verilog behavioral description, and a program segment, that outputs data from a queue, similar to the buffer input example at the end of §8.3. The queue contains 2^n elements, and begins at $k * 2^n$ so that wrap-around is handed by index register increment/decrement truncation.

25. Show the logic diagram and verilog structural description of a module that compares characters to a character string in an internal memory, in which one or more spaces, commas, periods, single or double quotes, or tabs, begins and ends a string. Show only the parts and modules that are different from those shown in Figures 8.12 and 8.13.

26. Show the logic diagram and verilog structural description of a module that compares characters to a character string in an internal memory, in which upper and lower case characters are considered the same (comparison is not case sensitive). Show only the parts and modules that are different from those shown in Figures 8.12 and 8.13.

9

Communication Systems

The microcomputer has many uses in communication systems, and a communication system is often a significant part of a microcomputer. This chapter examines techniques for digital communications of computer data.

Attention is focused on a microcomputer's communication subsystem — the part that interfaces slower I/O devices like typewriters and printers to the microcomputer. This is often a universal asynchronous receiver transmitter (UART). Because of their popularity in this application, UARTs have been used for a variety of communications functions, including remote control and multiple computer intercommunications. However, their use is limited to communicating short (1-byte) messages at slow rates (less than 1000 bytes per second). The synchronous data link control (SDLC) is suitable for sending longer messages (about 1000 bytes) at faster rates (about 1,000,000 bits per second) — for sending data between computers or between computers and fast I/O devices. The IEEE-488 bus, for microcomputer control of instruments like digital voltmeters and frequency generators, and the SCSI bus, for communication to and from intelligent peripherals, send a byte at a time rather than a bit at a time.

The overall principles of communication systems, including the ideas of levels and protocols, are introduced in the first section. The signal transmission medium is discussed next, covering some typical problems and techniques communications engineers encounter in moving data.The UART and related devices that use the same communications mechanisms are fundamental to I/O interface design. Therefore, we spend quite a bit of time on these devices, imparting basic information about their hardware and software. They will probably find use in most of your designs for communicating with teletypes or teletype-like terminals, keyboards, and CRTs, as well as for simple remote control. We look at the more complex communications interfaces used between large mainframe computers to control test and measurement equipment in the laboratory and to connect intelligent I/O. Finally, we give a detailed example of a SCSI-2 interface.

Communications terminology is rather involved with roots in the (comparatively ancient) telephone industry and in the computer industry, with some terminology stemming uniquely from digital communications. Communications design is almost a completely different discipline from microcomputer design. Moreover, one kind of system, such as one using UARTs, uses quite different terminology than that used to describe another, similar system, such as one using SDLC links. While it is important to be able to talk to communication system designers and learn their terminology, we are limited in what we can do in one short chapter. We will as much as possible use terminology associated with the so-called X.25 protocol, even for discussing UARTs, because we want to economize on the number of terms that we must introduce, and the X.25 protocol appears to be the most promising protocol likely to be used with minicomputers and microprocessors. However, you should be prepared to do some translating when you converse with a communications engineer.

On completing this chapter, you should have a working knowledge of UART communications links. You should be able to use the MMC2001 UART module, connect a UART to a microcomputer, and connect a UART to a remote control station so it can be controlled through the MMC2001 UART0 module or UART. You should understand the basic general strategies of communication systems, and the UART, SDLC, IEEE-488, and SCSI bus protocols in particular, knowing when and where they should be used.

9.1   Communications Principles

In looking at the overall picture, we will first consider the ideas of peer-to-peer interfaces, progressing from the lowest level to the higher level interfaces, examining the kinds of problems faced at each level.

Data movement is *coordinated* in different senses *at different levels of abstraction*, and by different kinds of mechanisms. At each level, the communication appears to take place between *peers* which are identifiable entities at that level; even though the communication is defined between these *peers* as if they did indeed communicate to each other, they actually communicate indirectly through peers at the next lower level. (See Figure 9.1.)

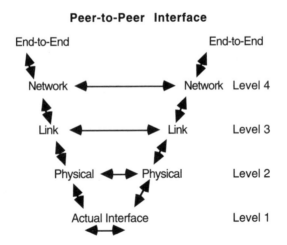

Figure 9.1. Peer-to-Peer Communication in Different Levels

Consider this analogy. The president of company X wants to talk to the president of company Y. This is called *end-to-end* communication. The job is delegated to the president's secretary, who calls up the secretary of the other president. This is referred to as *network control*. The secretary does not try to holler to the other secretary but dials the other secretary on the telephone. The telephone is analogous to the *link control* level. But even the telephone "delegates" the communication process to the electronics and the electrical circuits that make the connection at the telephone exchange. This is the *physical control* level. End-to-end communication is done between user (high-level) programs. User programs send information to each other at the level of the presidents in the analogy. Network control is done at the operating system level. As with the secretary, this software must know where the communications object is and how to reach this object. Link control is done by I/O interface software and is responsible for setting up and disconnecting the link so the message can be sent. Physical control actually moves the data. In the design of I/O systems, we are concerned primarily with link control and secondarily with physical control.

The peer-to-peer interfaces are defined without specifying the interface to the next lower level. This is done for the same reasons that computer architecture is separated from computer organization and realization, as we explained in §1.1.1. It permits the next lower level to be replaced by another version at that level without affecting the higher level. This is like one of the presidents getting a new secretary: the presidents can still talk to each other in the same way, even though communication at the next lower level may be substantially changed.

We now discuss some of the issues at each of the levels. At the lowest level, the main issue is the medium, and a secondary issue is the multiplexing of several channels on one link. The technique used to synchronize the transmission of bits may be partly in the physical interface level and partly in the link control level.

The *medium* that carries a bit of information is of great concern to the communications engineer. Most systems would probably use voltage level to distinguish between true and false signals. In other systems, mechanical motion carries information, or radio or light beams in free space or in optical fibers carry information. Even when the carrier is electric, the signal can be carried by current rather than voltage, or by presence or absence of a particular frequency component. The signal can be conveyed on two frequencies: a true is sent as one frequency while a false is sent as another frequency (*frequency shift keying*). More than one signal can be sent over the same medium. In *frequency multiplexing, n* messages are sent, each by the presence or absence of one of n different frequency components (or keying between n different pairs of frequencies). In *time multiplexing, n* messages can be sent, each one in a time slot every n*th* time slot. A frequency band or a time slot that carries a complete signal, enabling communication between two entities at the link control level, is called a *channel*. Each channel, considered by itself, may be *simplex* if data can move in one direction only, *half-duplex* if data can move in either direction but only one direction at a time, or *full duplex* if data can move in both directions simultaneously.

Usually, a bit of information is sent on a channel over a time period, the *bit time period*, and this is the same time for each bit. The *baud rate* is the inverse of this bit time period (or the shortest period if different periods are used). The *bit rate*, in contrast, is the rate of transfer of information, as defined in information theory. For simplicity in this discussion, the bit rate is the number of user information bits sent per time unit, while the baud rate is the total number of bits — including user information, synchronization, and error-checking bits — per time unit.

In general, a clock is a regular occurrence of a pulse, or even of a code word, used to control the movement of bits. If such a (regular) clock appears in the channel in some direct way, the system is *synchronous*, otherwise it is *asynchronous*. In a synchronous system, the clock can be sent on a separate line. The clock can also be sent on the same wire as the data — every other bit being a clock pulse and the other bits being data — in the so-called *Manchester code*. Circuitry such as a phase-locked loop detects the clock and further circuitry uses this reconstructed clock to extract the data. Finally, in an asynchronous link, the clock can be generated by the receiver, in hopes that it matches the clock used by the sender.

Link control is concerned with how data is moved as bits, as groups of bits, and as complete messages that are sent by the next higher level peer-to-peer interface. Link control is usually implemented by I/O device chips.

At the *bit level*, individual bits are transmitted; at the *frame level*, a group of bits called a frame or packet is transmitted; and at the *message level*, sequences of frames, called messages, are exchanged. Generally, at the frame level, means are provided for detection and correction of errors in the data being sent, because the communication channel is often noisy. Because the frame is sent as a single entity, it can have means for synchronization. A frame, then, is some data packaged or framed and sent as a unit under control of a communications hardware/software system. The end-to-end user often wishes to send more data — a sequence of frames — as a single unit of data. The user's unit of data is known as the message.

At each level, a coordination mechanism is used, which is called a *protocol*. A protocol is a set of conventions that coordinate the transmission and receipt of bits, frames, and messages. Its primary functions in the link control level are the synchronization of messages and the detection or correction of errors. This term protocol suggests a strict code of etiquette and precedence countries agree to follow in diplomatic exchange, so the term aptly describes a communication mechanism whereby sender and receiver operate under some mutually acceptable assumptions but do not need to be managed by some greater authority like a central program. Extra bits are needed to maintain the protocol. Since these bits must be sent for a given data rate in bits per second, the baud rate must increase as more extra bits are sent. The protocol should keep efficiency high by using as few as possible of these extra bits. Note that a clock is a particularly simple protocol: a regularly occurring pulse or code word. An important special case, the *handshake protocol*, is an agreement whereby when information is sent to the receiver, it sends back an acknowledgment that the data is received either in good condition or has some error. Note, however, that a clock or a protocol applies to a level, so a given system can have a bit clock and two different protocols — a frame protocol and a message protocol.

A collection of individual protocols, each at a different level, is called a *stack*. Don't confuse this stack with the stack data structure described in §1.2.2, §2.1, and §2.2. This stack defines the overall protocol at all levels of interest to the discussion.

The third level of peer-to-peer interface is the network level. It is concerned about relationships between a large community of computers and the requirements necessary so that they can communicate to each other without getting into trouble.

The *structure* of a communication system includes the physical interconnections among stations, as well as the flow of information. Usually modeled as a graph whose nodes are stations and whose links are communications paths, the structure may be a loop, tree graph, or a rectangular grid (or sophisticated graph like a banyan network).

A path taken by some data through several nodes is called *store and forward* if each node stores the data for a brief time, then transmits it to the next node as new data may be coming into that node; otherwise if data pass through intermediate nodes instantaneously (being delayed only by gate and line propagation), the path is called a *circuit* from telephone terminology. If such a path is half duplex, it is sometimes called a bus because it looks like a bus in a computer system.

Finally, the communication system is *governed* by different techniques. This aspect relates to the operating system of the system of computers, which indirectly controls the generation and transmission of data much as a government establishes policies that regulate trade between countries. A simple aspect of governance is whether the decision to transmit data is centralized or distributed. A system is *centralized* if a special station makes all decisions about which stations may transmit data; it is decentralized or *distributed* if each station determines whether to send data, based on information in its locale. A centralized system is often called a *master slave* system, with the special station the master and the other stations its "slaves." Other aspects of governance concern the degree to which one station knows what another station is doing, or whether and how one station can share the computational load of another.

9.2 Signal Transmission

The signal is transmitted through wires or light pipes at the physical level. This section discusses the characteristics of three of the most important linkages. Voltage or current amplitude logical signals, discussed first, are used to interconnect terminals and computers that are close to each other. The digital signal can be sent by transmitting it at different frequencies for a true and for a false signal (frequency shift keying). This is discussed in the next subsection.

9.2.1 Voltage and Current Linkages

In this section, we discuss the line driver and line receiver pair, the 20-mA current loop, and the RS232 standard.

Standard high current TTL or LSTTL drivers can be used over relatively short distances, as the IEEE-488 standard uses them for a bus to instruments located in a laboratory. However, slight changes in the ground voltage reference or a volt or so of noise on the link can cause a lot of noise in such links. A *differential line* is a pair of wires, in which the variable in positive logic is on one wire and in negative logic on the other wire. If one is high, the other is low. The receiver uses an analog comparator to determine which of the two wires has the higher voltage, and outputs a standard TTL signal appropriately. If a noise voltage is induced, both wires should pick up the same noise so the differential is not affected and the receiver gets the correct signal. Similarly, imperfect grounding and signal ringing affect the signal on both wires and their effect is cancelled by the voltage comparator. A number of driver and receiver integrated circuits are designed for differential lines, but some require voltages other than $+5$, which may not be used elsewhere in the system. An integrated circuit suitable for driving and receiving signals on a half duplex line, using a single 5-V supply, is the SN75119, shown in Figure 9.2a. If driver enable DE (pin 7) is high, then the signal on IN (pin 1) is put on line LA (pin 3) and its complement is put on line LB (pin 2); otherwise the pins LA and LB appear to be (essentially) open circuits. If receiver enable RE (pin 5) is high, then the output OUT (pin 6) is low if the voltage on LA is less than that on LB, or high if the voltage on LA is greater than that on LB; if RE is low, OUT is (essentially) an open circuit. The *RS442 standard* (RS means recommended standard) uses basically this differential line, but a driver such as the Am26LS30 has means to control the slew rate of the output signal.

The 20-mA current loop is often used to interface to teletypes or teletype-like terminals. A pair of wires connect driver and receiver so as to implement an electrical loop through both. A true corresponds to about 20 mA flowing through the loop, and a false corresponds to no current or to -20 mA in the loop (for "neutral working" or "polar working" loops, respectively). A current, rather than a voltage, is used because it can be interrupted by a switch in a keyboard and can be sensed anywhere in the loop. A current is also used in older equipment because the 20-mA current loop was used to drive a solenoid, and a solenoid is better controlled by a current than a voltage to get faster rise times. The current is set at 20 mA because the arc caused by this current will keep the switch contacts clean.

A 20-mA current loop has some problems. A loop consists of a current source in series with a switch to break the circuit, which in turn is in series with a sensor to sense the current. Whereas the switch and sensor are obviously in two different stations in the circuit, the current source can be in either station. A station with a current source is called *active*, while one without is *passive*. If two passive stations, one with a switch and the other with a sensor, are connected, nothing will be communicated. If two active stations are connected, the current sources might cancel each other or destroy each other. Therefore, one station must be active while the other is passive, and one must be a switch and the other must be a sensor. While this is all very straightforward, it is an invitation to trouble. Note also that the voltage levels are undefined. Most 20-mA current loops work with voltages like $+5$ or -12 or both, which are available in most communication systems; but some, designed for

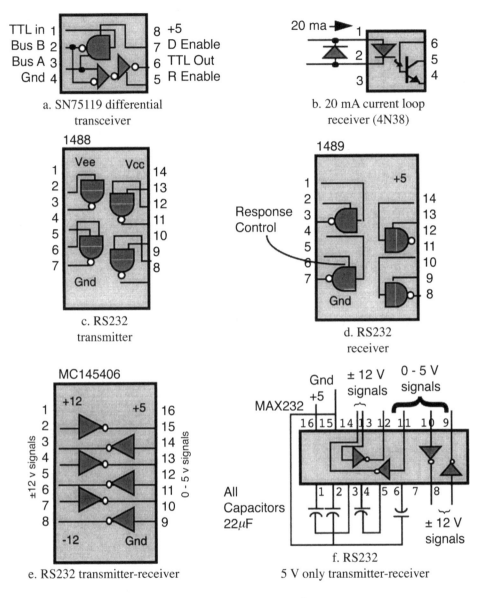

Figure 9.2. Drivers and Receivers

long distance communication, utilize "telegraph hardware" with voltages upwards of 80 V. Therefore, one does not connect two 20-mA current loop stations together without checking the voltage levels and capabilities. Finally, these circuits generate a fair amount of electrical noise, which gets into other signals, especially lower level signals, and the switch in such a circuit generates noise that is often filtered by the sensor. This noise is at frequencies used by 1200-baud lines, so this filter can't be used in other places in a communication subsystem. The circuitry for a 20-mA current loop can be built with an opto-isolator, as shown in Figure 9.2b. If the current through the LED is about 20 mA, the phototransistor appears to be a short circuit; if the current is about 0 mA, it is an open circuit and the output is high. The diode across the LED is there to prevent an incorrect current from destroying the LED.

An interface standard developed by the Electronic Industries Association (EIA) and other interested parties has evolved into the RS232-C (recommended standard 232 version C). A similar standard is available in Europe, developed by the Comite Consultatif Internationale de Telegraphie et Telephonie (CCITT), and is called the CCITT V.24 standard. These standards are supposed to be simple and effective, so that any driver conforming to it can be connected to any receiver conforming to it, covering the voltage levels used for the signals as well as the pin assignments and dimensions of the plugs. Basically, a false variable is represented by any voltage from $+15$ to $+5$ V, and a true by any voltage from -5 to -15 V (negative logic is used.) A number of specifications concerning driver and receiver currents and impedances can be met by simply using integrated circuit drivers and receivers that are designed for this interface — RS232 drivers and RS232 receivers. The MC1488 is a popular quad RS232 line driver, and the MC1489 is a popular receiver. (See Figure 9.2c.) The driver requires $+12$ V on pin 14 and -12 V on pin 1. Otherwise, it looks like a standard quad TTL NAND gate whose outputs are RS232 levels. The four receiver gates have a pin called response control (pins 2, 5, 9, and 12). Consider one of the gates, where pin 1 is the input and pin 3 is the output. Pin 2 can be left unconnected. It can be connected through a (33K) resistor to the negative supply voltage (pin 1) to raise the threshold voltage a bit. Or it can be connected through a capacitor to ground, thus filtering the incoming signal. This controls the behavior of that gate. The other gates can be similarly controlled. The MC145406 is a chip that combines three transmitter and three receiver gates in one chip (Figure 9.2e); and the MAX232 (Figure 9.2f), made by MAXIM, has two transmitters and two receivers, and a charge pump circuit that generates ±10 V needed for the transmitter, from the 5-V supply used by the microcomputer. (This marvelous circuit is just what is needed in many applications, but the currently available chips have a small problem: if the 5-V supply turns on too fast, the charge pump fails to start; put a small (10 Ω) resistor in series with the 5-V pin and put a large (100 μF) capacitor from that pin to ground.)

The RS232 interface standard also specifies the sockets and pin assignments. The DB25P is a 25-pin subminiature plug, and the DB25S is the corresponding socket — both of which conform to the standard. The pin assignments are shown in Table 9.1. For simple applications, only pins 2 (transmit data), 3 (receive data), and 7 (signal ground) need be connected; but a remote station may need to make pins 5 (clear to send), 6 (data set ready), and 8 (data carrier detect) 12 V to indicate that the link is in working order, if these signals are tested by the microcomputer. These can be wired to -12 V in a terminal when they are not carrying status signals back to the microcomputer.

Table 9.1. RS232 Pin Connections for D25P and D25S Connectors

Pin	Name	Function
1	Protective Ground	Connects machine or equipment frames together and to "earth"
2	Transmitted Data	Data sent from microcomputer to terminal
3	Receive Data	Data sent from terminal to microcomputer
4	Request to Send	(Full Duplex) enables transmission circuits (Half Duplex) puts link in transmit mode and disables receive circuitry
5	Clear to Send	Responds to Request to Send; when high, it indicates the transmission circuitry is working
6	Data Set Ready	(telephone links) The circuitry is not in test, talk, or dial modes of operation so it can be used to transmit and receive
7	Signal Ground	Common reference potential for all lines. Should be connected to "earth" at just one point, to be disconnected for testing
8	Data Carrier Detect	A good signal is being received
9	+P	+12 volts (for testing only)
10	−P	−12 volts (for testing only)
11 25		Used for more elaborate options

9.2.2   Frequency Shift-Keyed Links Using Modems

To send data over the telephone, a *modem* converts the signals to frequencies that can be transmitted in the audio frequency range. The most common modem, the Bell 103, permits full duplex transmission at 300 baud. Transmission is originated by one of the modems, referred to as the *originate modem*, and is sent to the other modem, referred to as the *answer modem*. The originate modem sends true (mark) signals as a 1270-Hz sine wave and false (space) signals as a 1070-Hz sine wave. Of course, the *answer modem* receives a true as a 1270-Hz sine wave and a false as a 1070-Hz sine wave. The answer modem sends a true (mark) as a 2225-Hz sine wave and a false (space) as a 2025-Hz sine wave. Note that the true signal is higher in frequency than the false signal, and the answer modem sends the higher pair of frequencies.

Some modems are originate only. They can only originate a call and can only send 1070- or 1270-Hz signals and receive only 2025- or 2225-Hz signals. Most inexpensive modems intended for use in terminals are originate only. The computer may have an answer-only modem, having the opposite characteristics. If you want to be able to send data between two computers, one of them has to be an originate modem. So an answer/originate modem might be used on a computer if it is expected to receive and also send calls. Whether the modem is originate-only, answer-only, or answer/originate, it is fully capable of sending and receiving data simultaneously in full duplex mode. The originate and answer modes determine only which pair of frequencies can be sent and received, and therefore whether the modem is capable of actually initiating the call.

Modems have filters to reject the signal they are sending and pass the signals they are receiving. Usually, Bessel filters are used because the phase shift must be kept uniform for all components or the wave will become distorted. Sixth order and higher filters are common to pass the received and reject the transmitted signal and the noise, because the transmitted signal is usually quite a bit stronger than the received signal, and because reliability of the channel is greatly enhanced by filtering out most of the noise. The need for two filters substantially increases the cost of answer/originate modems.

The module that connects the telephone line to the computer is called a *data coupler*, and there is one that connects to the originator of a call and another that connects to the answerer. The data coupler isolates the modem from the telephone line to prevent lightning from going to the modem, and to control the signal level, using an automatic gain control; but the data coupler does not convert the signal or filter it. The data coupler has three control/status signals. *Answer Phone* ANS is a control command that has the same effect on the telephone line as when a person picks up the handset to start a call or answer the phone. *Switch hook* SH is a status signal that indicates that the telephone handset is on a hook, if you will, so it will receive and transmit signals to the modem. Switch hook may also be controlled by the microcomputer. Finally, *ring indicator* RI is a status signal that indicates the phone is ringing.

Aside from the fact that data are sent using frequency analog signals over a telephone, there is not much to say about the channel. However, the way an originate modem establishes a channel to an answer modem and the way the call is terminated is interesting. We now discuss how the Motorola M6860 modem originates a call and answers a call. Calling a modem from another, maintaining the connection, and terminating the connection involve handshaking signals *data terminal ready* DTR and *clear to send* CTS in both originate and answer modems. (See Figure 9.3a for a diagram showing these handshaking signals.) If a modem is connected to an RS232C line, as it often is, data terminal ready can be connected to request to send (pin 4) and clear to send can be connected to the clear to send (pin 5) or the data set ready (pin 6), whichever is used by the computer. Figure 9.3b shows the sequence of operations in the modems and on the telephone line, showing how a call is originated and answered by the Motorola M6860 modem chip.

The top line of Figure 9.3b shows the handshaking signals seen by the originator, the next line shows signals seen by the originator modem, the center line shows the telephone line signals, the next line shows signals seen by the answer modem, and the bottom line shows the handshaking signals seen by the answerer. As indicated, the originator asserts the switch hook signal. This might be asserted by putting the telephone handset on the modem hook or by an output device that asserts this signal. This causes the command ANS (answer phone) to become asserted, which normally enables the data coupler electronics to transmit signals. The telephone is now used to dial up the answerer. (17 s is allowed for dialing up the answerer.) The answering modem receives a command RI (ring indicator) from the telephone, indicating the phone is ringing. It then asserts the ANS signal to answer

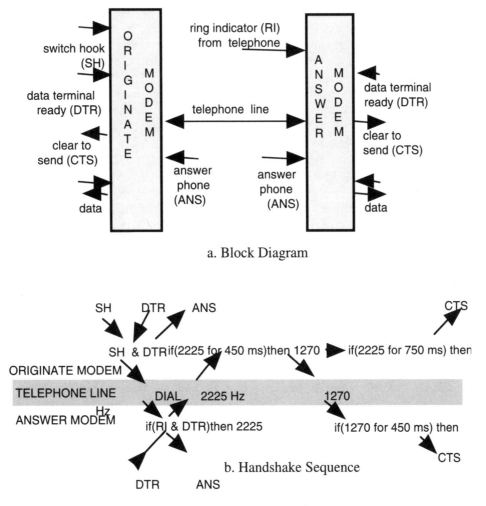

a. Block Diagram

b. Handshake Sequence

Figure 9.3. Originating a Call on a Modem

the phone, enabling the data coupler to amplify the signal. The answerer puts a true signal, 2225 Hz, on the line. The originator watches for that signal. When it is present for 450 ms, the originator will send its true signal, a 1270-Hz sine wave. The answerer is watching for this signal. When it is present for 450 ms, the answerer asserts the CTS command and is able to begin sending data. The originator meanwhile asserts CTS after the 2225-Hz signal has been present for 750 ms. When both modems have asserted CTS, full duplex communication can be carried out.

Some answer modems will automatically terminate the call. To terminate the call, send more than 300 ms of false (space) 1070 Hz. This is called a *break* and is done by your terminal when you press the "break" key. The answer modem will then hang up the phone (negate ANS) and wait for another call. Other modems do not have this automatic space disconnect; they terminate the call whenever neither a high nor a low frequency is received in 17 s. This occurs when the telephone line goes dead or the other modem stops sending any signal. In such systems, the "break" key and low frequency sent when it is pressed can be used as an "attention" signal rather than a disconnect signal.

9.2.3   Infrared Links

Another scheme to transmit serial data is to send it on an infrared carrier, according to the IRDA Serial Infrared Physical Layer Specification. A logic 0 (F) is sent as emission of infrared light for 3/16 of a bit time, followed by 13/16 of a bit time of darkness. A logic 1 (T) is sent as a bit time of darkness. Data are sent by driving current through an infrared diode, and data are received by having the infrared light shine on a phototransistor, much as in an opto-isolator. The main advantage of this scheme is that the two communicating devices need not have any electical connections between them, so there is significantly less danger of electrical shock and noise.

9.3   UART Link Protocol

By far the most common technique for transmitting data is that used by the *Universal Asynchronous Receiver Transmitter* (UART). This simple protocol is discussed in this section. Software generation of UART signals, discussed first, is quite simple and helps to show how they are sent. A special remote control chip that uses the UART protocol is discussed in the next subsection. The UART chip is then discussed. A system inside the MMC2001 that is capable of UART signal generation and reception, the MMC2001 UART, is discussed last, because it has several useful but nonstandard extensions to the UART protocol.

9.3.1   UART Transmission and Reception by Software

As noted earlier, the Universal Asynchronous Receiver-Transmitter (UART) is a module (integrated circuit) that supports a frame protocol to send up to eight bit frames (characters). We call this the *UART protocol*. However, the UART protocol can be supported entirely under software control, without the use of a UART chip or its equivalent. A study of this software is not only a good exercise in hardware-software tradeoffs, but is also an easy way to teach the protocol; the software approach also is a practical way to implement communication in a minimum cost microcomputer. However, we do warn the reader that most communication is done with UART chips or their equivalent, and low-cost microprocessors such as the MMC2001 already have a built-in "UART" on the microprocessor chip itself.

The UART frame format is shown in Figure 9.4. (The UART protocol is contained within the UART frame format.) When a frame is not being sent, the signal is high. When a signal is to be sent, a *start bit*, a low, is sent for one bit time.

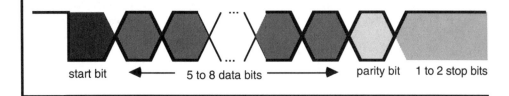

start bit   ◄──────── 5 to 8 data bits ────────►   parity bit   1 to 2 stop bits

Figure 9.4 Frame Format for UART Signals

The frame, from 5 to 8 bits long, is then sent 1 bit per bit time, least significant bit first. A parity bit may then be sent and may be generated so that the parity of the whole frame is always even (or always odd). To generate even parity, if the frame itself had an even number of ones already, a low parity bit is sent, otherwise a high bit is sent. Finally, one or more *stop bits* are sent. A stop bit is high, and is indistinguishable from the high signal that is sent when no frame is being transmitted. In other words, if the frame has n stop bits (n = 1, 1 1/2 or 2) this means the next frame must wait that long after the last frame bit or parity bit of the previous message has been sent before it can begin sending its start bit. However, it can wait longer than that.

In addition to the format above, the protocol has some rules for sampling data and for error correction. A clock, used in the receiver, is 16 times the bit rate, and a counter, incremented each clock time, is used to sample the incoming data. (The same clock is used in the transmitter to generate the outgoing data.) The counter is started when the input signal falls, at the beginning of a frame. After 8 clock periods, presumably in the middle of the start bit, the input is sampled. It should be low. If it is high, the falling edge that started the counter must be due to some noise pulse, so the receiver returns to examine the input for the leading edge of a start bit. If this test passes, the input is sampled after every 16 clock periods, presumably in the middle of each bit time. The data bits sampled are reassembled in parallel. The parity bit, if one is used, is then sampled and checked. Then the stop bit(s) are checked.

The following are definitions of error conditions. If the parity bit is supposed to be even, but a frame with odd parity is received, a *parity error* is indicated. This indicates that one of the frame bits or the parity bit was changed due to noise. Note that two errors will make the parity appear correct — but two wrongs do not make a right. Parity detection can not detect all errors. Even so, most errors are single-bit errors, so most errors are detected. If a stop bit is expected, but a low signal is received, the frame has a *framing error*. This usually indicates that the receiver is using the wrong clock rate, either because the user selected the wrong rate or because the receiver oscillator is out of calibration. However, this condition can arise if the transmitter is faulty, sending frames before the stop bits have been timed out, or if more than one transmitter is on a link and one sends before the other's stop bits are completely sent. Finally, most UART devices use a buffer to store the incoming word, so the computer can pick up this word at leisure rather than at the precise time that it has been shifted in. This technique is called *double buffering*. But if the buffer is not read before another frame arrives needing to fill the same buffer, the first frame is destroyed. This error condition is called an *overrun error*. It usually indicates that the computer did not empty the receive buffer before a subsequent message arrived.

The UART communication technique is based on the following principle. If the frame is short enough, a receiver clock can be quite a bit out of synchronization with the transmitter clock and still sample the data somewhere within the bit time when the data are correct. For example, if a frame has 10 bits and the counter is reset at the leading edge of the frame's start bit, the receiver clock could be 5% faster or 5% slower than the transmitter clock and still, without error, pick up all the bits up to the last bit of the frame. It will sample the first bit 5% early or 5% late, the second 10%, the third 15%, and the last 50%. This means the clock does not have to be sent with the data. The receiver can generate a clock to within five percent of the transmitter clock without much difficulty. However, this technique would not work for long frames, because the accumulated error due to incorrectly matching the clocks of the transmitter and receiver would eventually cause a bit to be mis-sampled. To prevent this, the clocks would have to be matched too precisely. Other techniques become more economical for longer frames.

A C procedure *SUart* to generate a signal compatible with the UART protocol is quite simple. The procedure is shown below and its description follows it. It uses the same pins (*UODR* bits 1 and 0) that will be used later with MMC2001's UART0 device examples.

```
void SUart(char c) { unsigned char i, parity;
    UODDR = 2; parity = UODR = 0 ; delay(N);
    for(i = 8; i > 0; i-- ) {
        if(c & 1) { UODR = 2; parity++; }
        else UODR = 0; c >>= 1; delay(N);
    }
    if(parity & 1) UODR = 2; else UODR = 0;
    delay(N); UODR = 2; delay(N); delay(N);
}
void main() { char c = RUart(); SUart(c); }
```

In the above procedure we use the following delay procedure, whose argument is the time delay. Let N be the parameter that delays for the time to send one bit.

```
                    void delay(int t) { while(--); }
```

The start bit is output from the least significant bit of c, and a delay subroutine is called to delay one bit time. Then the bits are written to the output port so the least significant bit is sent out the serial channel, and parity is updated with the exclusive-OR of parity and data so the least significant bit is the parity of the data sent. This is repeated for 8 data bits. Then the parity bit is output, and the stop bit is output. Appropriate delays are inserted between each bit that is sent serially.

A C procedure *RUart* to receive a UART frame is also quite simple. Again, the subroutine is shown below and its description follows it.

The while loop waits for the input to go low, and the do while loop confirms that it is still low after a half a bit time (using the procedure delay to delay a half bit time). Then, after a delay of a bit time, the least significant bit is picked up, and is exclusive-

ORed with the computed parity bit. For eight steps, another bit is picked up, the parity is updated, and a bit delay is wasted. Then the transmitted parity bit is combined with the computed parity bit to determine if a parity error occurred, and the stop bits are checked.

```
char RUart() { unsigned char i, parity, c;
    do {while(UODR & 1) ; delay(N/2); } while (UODR & 1);
    parity = c = 0; delay(N);
    for(i = 8; i > 0; i-- )
      { if(UODR & 1) c |= 0x80; parity++; c >>= 1; delay(N); }
    if(UODR & 1) parity++; if (parity & 1) { /* report parity error */;}
    delay(N); if(!(UODR & 1)) { /* report framing error */;}
    delay(N); if(!(UODR & 1)) { /* report framing error */;}
    return c;
}
```

Both preceding C procedures are simple enough to follow. They can be done in software without much penalty, because the microprocessor is usually doing nothing while frames are being input or output. In an equivalent hardware alternative, essentially the same algorithms are executed inside the UART chip. The hardware alternative is especially valuable where the microcomputer can do something else as the hardware tends to transmitting and receiving the frames, or when it might be sending a frame at the same time it might be receiving another frame (in a full duplex link or in a ring of simplex links). In other cases, the advantages of the hardware and software approaches are about equal: the availability of cheap, simple UART chips favors the hardware approach, while the simplicity of the program favors the software approach. The best design must be picked with care and depends very much on the application's characteristics.

9.3.2   The UART

The UART chip is designed to transmit and/or receive signals that comply with the UART protocol (by definition). This protocol allows several variations (in baud rate, parity bit, and stop bit selection). The particular variation is selected by strapping different pins on the chip to high or to low. The UART can be used inside a microcomputer to communicate with a teletype or a typewriter, which was its original use, or with the typewriter's electronic equivalent, such as a CRT display. It can also be used in other remote stations in security systems, stereo systems controlled from a microcomputer, and so on. Several integrated circuit companies make UARTs, which are all very similar. We will study one that has a single-supply voltage and a self-contained oscillator to generate the clock for the UART, the Intersil IM6403.

The UART contains a transmitter and a receiver that run independently, for the most part, but shares a common control that selects the baud rate and other variations for both transmitter and receiver. We discuss the common control first, then the transmitter, and then the receiver. The baud rate is selected by the crystal con-

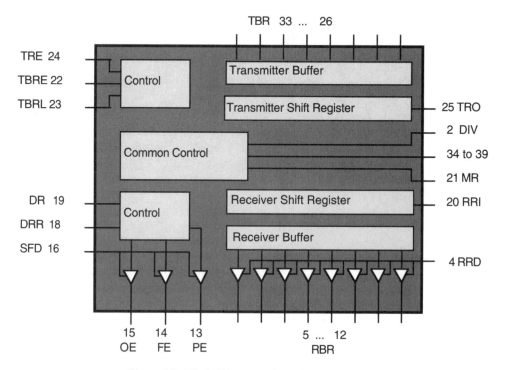

Figure 9.5. Block Diagram of a UART (IM6403)

nected to pins 17 and 40 and by the divide control DIV on pin 2. If DIV is high, the oscillator frequency is divided by 16; if low, by 2^{11}. If the crystal is a cheap TV crystal (3.5795 MHz) and DIV is low, the baud rate is close to 110, which is commonly used for teletypes. When master reset MR, on pin 21, is high, it resets the chip; it is normally grounded. The other control bits are input on pins 39 to 35 and are stored in a latch inside the chip. The latch stores the inputs when pin 34 is high. This pin can be held high to defeat the storage mechanism, so the pin levels control the chip directly. Pin 36 selects the number of stop bits: low selects 1 stop bit, high selects 2 (except for an anomaly of little interest). If pin 35 is high, no parity bit is generated or checked, otherwise pin 39 selects even parity if high, odd if low. Pins 37 and 38 select the number of data bits per frame; the number is five plus the binary number on these pins. The user generally determines the values needed on these pins from the protocol he or she is using, and connects them to high or low. However, these inputs can be tied to the data bus of a computer, and pin 34 can be asserted to load the control latch to effect an output register. When the computer executes the reset handler, it can then set the control values under software control.

The operation of the transmitter and receiver is compactly and simply explained in the data sheets of the 6403 and are paraphrased here. The transmitter has a buffer register, which is loaded from the signals on pins 33 (msb) to 26 (lsb) when transmitter buffer register load TBRL (pin 23) rises. If n < 8 bits are sent, the rightmost n bits on these pins are sent. Normally, these pins are tied to the data bus to make the buffer look like an output register, and TBRL is asserted when the register is to be loaded. When this buffer is empty and can be loaded, transmitter buffer register empty TBRE (pin 22) is high; when full it is low. (SFD, pin 16, must be low to read out TBRE.) The computer may check this pin to determine if it is safe to load the buffer register. It behaves as a BUSY bit in the classical I/O mechanism. The data in the buffer are automatically loaded into the transmitter shift register to be sent out as transmitter register output TRO (pin 25) with associated start, parity, and stop bits as selected by the control inputs. As long as the shift register is shifting out part of a frame, transmitter register empty TRE (pin 24) is low. Figure 9.6 shows a typical transmission, in which two frames are sent out. The second word is put into the buffer even as the first frame is being shifted out in this double buffered system. It is automatically loaded into the shift register as soon as the first frame has been sent.

The receiver shifts data into a receiver shift register. When a frame has been shifted in, the data are put in the receiver buffer. If fewer than 8 bits are transmitted in a frame, the data are right justified. This data can be read from pins 5 to 12, when receive register disable RRD (pin 4) is asserted low. Normally these pins are attached to a data bus, and RRD is used to enable the tristate drivers when the read buffer register is to be read as an input register. If RRD is strapped low, then the data in the read buffer are continuously available on pins 5 to 12. When the read buffer contains valid data, the data ready DR signal (pin 19) is high, and the error indicators are set. (DR can only be read when SFD on pin 16 is high.) The DR signal is an indication that the receiver is DONE, in the classical I/O mechanism, and requests the program to read the data from the receive buffer and read the error indicators if appropriate. The error indicators are reloaded after each frame is received, so they always indicate the status of the last frame that was received. The error indicators, TBRE, and DR can be read from pins 15 to 13 and 22 and 19 when SFD (pin 16) is asserted low, and indicate an overrun error, a framing error, and a parity error, and that the transmit buffer is empty and that the receive buffer is full respectively, if high. The error and buffer status indicators can be read as another input register by connecting pins 22 and 19, and 15 to 13 to the data bus, and asserting SFD when this register is selected; or, if SFD is strapped low, the error and buffer status indicators can be read directly from those pins. When the data are read, the user should reset the DR indicator by asserting data ready reset DRR (pin 18) high. If this is not done, when the next frame arrives and is loaded into the buffer register, an overrun error is indicated.

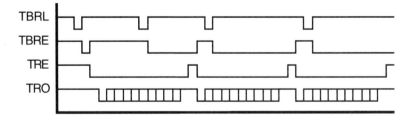

Figure 9.6. Transmitter Signals

The UART can be used in a microcomputer system. The control bits (pins 35 to 39) and the transmit buffer inputs (pins 26 to 33) can be inputs, and the buffer status and error indicators (pins 22 and 19, and 15 to 13) and receive data buffer outputs (pins 5 to 12) can be outputs. All the inputs and outputs can be attached to the data bus. TBRL, SBS, SFD, and RRD (pins 23, 36, 16, and 4) are connected to an address decoder so that the program can write in the control register or transmit buffer register, or read from the error indicators or the read buffer register. The TBRE signal (pin 22) is used as a BUSY bit for the transmitter; and the DR signal (pin 19) is used as a DONE bit for the receiver. When the UART is used in a gadfly technique,which can be extended to interrupt or even DMA techniques, the program initializes the UART by writing the appropriate control bits into the control register. To send data using the gadfly approach, the program checks to see if TBRE is high and waits for it to go high if it is not. When it is high, the program can load data into the transmitter buffer. Loading data into the buffer will automatically cause the data to be sent out. If the program is expecting data from the receiver in the gadfly technique, it waits for DR to become high. When it is, the program reads data from the receive buffer register and asserts DRR to tell the UART that the buffer is now empty. This makes DR low until the next frame arrives.

The UART can be used without a computer in a remote station that is implemented with hardware. Control bits can be strapped high or low, and CRL (pin 34) can be strapped high to constantly load these values into the control register. Data to be collected can be put on pins 33 to 27. Whenever the hardware wants to send the data, it asserts TBRL (pin 23) low for a short time, and the data get sent. The hardware can examine TBRE (pin 22) to be sure that the transmitter buffer is empty before loading it, but if the timing works out so that the buffer will always be empty there is no need to check this value. It is pretty easy to send data in that case. Data, input serially, are made available and are stable on pins 5 to 12. Each time a new frame is completely shifted in, the data are transferred in parallel into the buffer. RRD (pin 4) would be strapped low to constantly output this data in a hardware system. When DR becomes high, new data have arrived, which might signal the hardware to do something with the data. The hardware should then assert DRR high to clear DR. (DR can feed a delay into DRR to reset itself.) The buffer status and error indicators can be constantly output if SFD (pin 16) is strapped low, and the outputs can feed LEDs, for instance, to indicate an error. However, in a simple system when the hardware does not have to do anything special with the data except output them, it can ignore DR and ignore resetting it via asserting DRR. In this case, the receiver is very simple to use in a remote station.

9.3.3   The UART Device in the MMC2001

The MMC2001 has UART-like systems, UART0 and UART1. We describe the UART0 device shown in Figure 9.7. We describe its baud rate, data, control and status ports. Then we will show how the UART can be used for gadfly and interrupt synchronization.

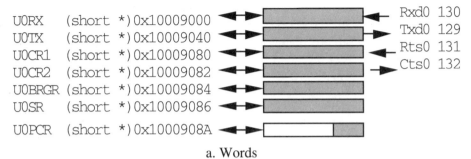

UORX   (short *)0x10009000   Rxd0 130
U0TX   (short *)0x10009040   Txd0 129
U0CR1   (short *)0x10009080   Rts0 131
U0CR2   (short *)0x10009082   Cts0 132
U0BRGR (short *)0x10009084
U0SR   (short *)0x10009086
U0PCR   (short *)0x1000908A

a. Words

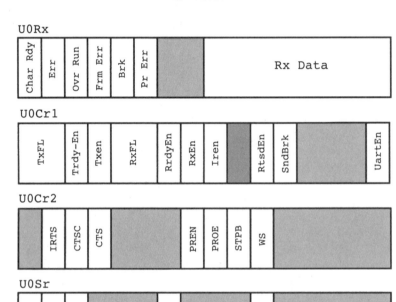

b. Ports in Words

Figure 9.7. MMC2001 UART0.

The clock rate is established by the 12-bit *U0BRGR* port (0x10009086). The number put in this port is the CLK clock divided by 16 times the desired baud rate. For example, if the CLK clock is 32 MHz, to get 9600 baud, put 208 into *U0BRGR*.

UART0 has a pair of data registers that are connected to shift registers. Eight bits of the data written into the least significant byte of *U0TX* at 0x10009040 are put into the shift register and shifted out, and 8 bits of the data shifted into the receive shift register can be read as the least significant byte of *U0RX* at 0x10009000. For both input and output, a hardware queue of 16 characters holds characters not yet transferred. Additionally the high bits of *U0RX*, from most significant bit, are

CHAR_RDY, which is asserted when data has been received, *ERR* when an error occurs, *OVR_RUN*, which is asserted when an overrun caused the error, *FRM_ERR*, which is asserted when an framing error occurred, *BRK*, which is asserted when a break caused the error (when a low signal lasts longer than a frame), *OVR_RUN*, which is asserted when an overrun caused the error, and *PR_ERR*, which is asserted when a parity error occurred.

Status port *UOSR* at 0x10009086 has, from most significant bit *TX_MPTY*, which is 1 when the transmitter queue and shift register are empty, *RTSS*, which is 1 if the RTS pin is asserted low, and *TRDY*, which is 1 if the transmit buffer has room for more data. Bit 9, *RRDY*, is 1 if the receive buffer has room in it, and bit 5, *RTSD*, is 1 if the RTS input changes state. *RTSD* is cleared by writing a 1 into it (a clear port).

The 16-bit control port, *UOCR1*, at 0x10009080, has two most significant bits *TxFL*, which determine when *UOSR* bit *TRDY* is set. If these bits are 00, that bit is set if the output queue has room for at least one character; if 01, for at least four characters; if 10, for at least eight characters; if 11, for at least fourteen characters. If bit 13, *TRDY_EN*, is set, then if *UOSR* bit *TRDY* is set, an interrupt is requested, asserting *NIPND* or *FIPND* interrupt bit 16. Bits 12 to 10, *RxFL* and *RRDY_EN*, determine when *UOSR* bit *RRDY* is set in the same manner as *TxFL*, and *TRDY_EN* control the transmitter, asserting *NIPND* or *FIPND* interrupt bit 18. Port *UOCR1* bit 5, *RTSD_EN*, enables an interrupt when RTS changes. Bit 4, *SND_BRK*, sends a break (a low signal longer than a frame). Finally, bit 0, *UART_EN*, must be set to apply power to the UART, before using it.

The 16-bit control port, *UOCR2*, at 0x10009082, has bits 14 to 12, *IRTS*, *CTSC*, and *CTS*, which control output pins, and bits 8 to 5, *PREN*, *PROE*, *STPB*, and *WS*, that determine the transmit and receive frame structures. The rightmost four bits of port control register *UOPCR* control (left-to-right) the *CTS*, *RTS*, *TXD*, and *RXD* pins; set the bits to let the UART use them, otherwise they are parallel port bits (Figure 5.7).

We illustrate the use of simple gadfly synchronization with UART0. `main` initializes UART0 for 9600 baud, 8 data bits, no parity bit and one stop bit. Automatic RTS handshaking is disabled (we discuss this feature and enable it in a later example). The `put` procedure gadflies on transmitter ready (*TRDY*); when it is ready, `put` outputs its argument to *UOTX*. The `get` procedure gadflies on receive ready (*RRDY*), when the receiver has data, `get` returns the data from *UORX*.

```
volatile short UORX@0x10009000, UOTX@0x10009040, UOCR1@0x10009080,
    UOCR2@0x10009082, UOBRGR@0x10009084, UOSR@0x10009086,
    UOPCR@0x1000908A; }

enum{IRTS = 0x4000, WS = 0x20, TRDY = 0x2000, RRDY=0x200, ERR=0x4000,
    TXEN = 0x1000, RXEN = 0x100, UART_EN = 1};

void put(char d) { while( ( UOSR & TRDY ) == 0 ) ; UOTX = d; }

int get() { while( ( UOSR & RRDY ) == 0 ) ; return UORX; }
```

```
void main() { char i, j;
    UOBRGR=208; /* 9600 baud */ UOPCR = 3; // Rxd, Txd pins used by UART0
    UOCR1 = UART_EN + TXEN + RXEN; // enable UART, transmitter, receiver
    UOCR2 = IRTS + WS; // disable hardware RTS control, have 8-bit data frame
    put(0x55); i = get(); // output, and then input, data
}
```

The simple gadfly synchronization mechanism above is supported by 16-element hardware queues for both input and output. In many applications, this facilitates asynchronous input and output, allowing the program to *put* as many as 16 characters before finishing outputting the first one on the Txd0 pin, or to be *getting* a character that is 16 characters behind the character currently being input on the Rxd0 pin. However, if queues having more than 16 bytes are needed, interrupts using queues (§2.2.2) can be used. These software queues become extensions to the hardware queues built into the UART0 device. We use 8-element queues that do not have critical sections (§6.2.3.5).

UART0 initialization enables the UART, but enables only the receive interrupt and not the transmit interrupt, which is enabled when the output queue is nonempty by setting *TRDY_EN* which is *UCR1* bit 13. In this example the UART0 is initialized to reduce overhead due to interrupts. Transmit interrupts, when enabled, occur when the hardware output queue has 14 slots for characters, and receive interrupts occur when the hardware input queue has 14 bytes. Then, multiple characters can be input or output upon each interrupt. Alternatively, to ensure that characters are input and output promptly, interrupts can be made to occur when one character or slot is in the queue, by clearing *UOCR1* bits 15, 14, 11, and 10. To output data in *put*, gadfly until the output queue has room, then push the data. In *put*, if the output queue changes from empty to nonempty, enable the transmitter interrupt. To input data from *get*, gadfly until the input queue has data, then pull that data. Upon honoring an interrupt, *handler* checks for a character ready or transmitter ready. When the transmitter is ready (TRDY) and the software output queue has data, up to 14 characters can be pulled from the output queue and pushed onto the hardware output queue. If the software output queue becomes empty, disable the transmitter interrupt. While the software input software queue has room and there is a character ready (CHAR_RDY), pull a character from the hardware input queue and push it into the software input queue. As each data are input, if there is a receive error, set the *error* global variable.

```
unsigned char oQ[8], iQ[8], oTop, oBot, iTop, iBot, error;

#define TRDY_EN 0x2000

void oPush(char data)
    { oQ[oTop = (oTop + 1) & 7] = data; if(oBot == oTop) error = 1;}

char oPull(){if(oBot==oTop)error=1; return oQ[oBot = (oBot+1) & 7];}

void iPush(char i)
    { iQ[iTop = (iTop + 1) & 7] = i; if(iBot == iTop) error = 1; }
```

```
char iPull(){if(iBot==iTop)error=1; return iQ[iBot = (iBot+1) & 7];}

void put(char data) {
    char oldTop = oTop; while(oTop == ((oBot + 7) & 7)) ;
    oPush ( data ); // push data to software queue
    if(oldTop==oBot) U0CR1|=TRDY_EN; // if outQueue empty, enable trns. int.
}
char get() { while(iBot == iTop) ; return iPull(); }

interrupt void handler(){ char oldTop; short i;
    while((iTop!=((iBot+7)&7)) && ((i = U0RX)<0)) // while queue not full,
        {iPush(i);if(i&ERR)error=1;} // UART has data, push it - xfr error
    if((U0SR&TRDY) for(i = 0; i < 14; i++) {// if trns. ready, output up to 14
        U0TX = oPull(); // characters pulled from software queue
        if((oBot==oTop){U0CR1&=~TRDY_EN;break;}// if mpty, disbl. trns. int
    }
}

void main() {
    disableInt(); U0BRGR=208; U0CR1 = 0xdf01;U0PCR = 3;U0CR2 = 0x4020;
    NIER|=(1<<18)|(1<<16); /* enable transmit & receive int. in int. device */
    initInt(0x30000000, 0x30000200, handler, 10); enableInt();
    put(0x55); i = get();
    while(U0CR1 & TRDY_EN ); /* wait for trans. int. to be disabled in handler */
    NIER &= ~((1 << 18)|(1 << 16)); /* disable UART0 trans. and rcv. int. */
}
```

9.3.4   Object-oriented Interfacing to UART0 or UART1

We illustrate object-oriented programming of a UART device with a simple gadfly class UARTg, an interrupt class, UARTi, hardware handshake classes UARTh and UARTah and a software handshake interrupt class UARTsh, each example demonstrating increased complexity. The classes handle UART0 if their constructor parameter id is 0 and UART1 if id is 1. To do this, these classes access ports with offsets from the class's data member port, as defined below, instead of global port declarations. However, we again use the constants enumerated in the earlier section.

```
#define URX *port        #define UTX port[0x20]    #define UCR1 port[0x40]
#define UCR2 port[0x41]   #define UBRGR port[0x42]
#define USR port[0x43]    #define UPCR port[0x45]
```

The simple UARTg class merely implements §8.3.3's gadfly synchronization procedures, and it clears the input queue, should the device be reopened by the constructor after it had been previously closed by the destructor while data was in the input queue. A destructor is needed to wait for the hardware output queue to empty (USR bit 15, TX_MPTY) before the device is turned off, by clearing the control register UCR1.

```
class UARTg : public Port<short> {
    public : UARTg(char id) : Port<short>(0x10009000 + (id << 12) ) {
        UBRGR = 208; /*9600 baud*/ U0PCR = 3; // Rxd, Txd pins used
        UCR1 = UART_EN + TXEN + RXEN; // enable UART, trans., recv.
        U0CR2 = IRTS + WS; // disable hardware RTS, 8-bit frame
        do ; while(URX<0); // test CHAR_RDY, to completely empty input queue
    }
    virtual int get(void){ while((USR&RDRF )==0) ; return URX & 0xff;}

    virtual void put(int d){ while((USR&TDRE)==0 ) ; UTX = d; }

    ~UARTg(){
            while( USR >= 0 ); /* wait for transmitter to become totally empty */
        }   UCR1 = 0; /* turn off UART completely */
} *S;
void main()
 {unsigned char i;S=new UARTg(1);S->put(0x55);i=S->get(); delete S;}
```

UARTi implements §8.3.3's procedures for interrupt synchronization, but it takes advantage of §2-3.3's *Queue* class. Note in *get* and *put*, the disabling and reenabling interrupts to avoid critical section errors. This class correctly keeps track of receiver errors, and turns off when both hardware and software output queues are empty.

```
class UARTi:public Port<short>{protected:Queue<char> *InQ,*OutQ;
    public:
UARTi(char id,unsigned char iSize,unsigned char oSize):Port(id) {
    disableInt(); UCR1 = 0x1301; UCR2 = 0x4020; UBRGR = 208; UPCR = 3;
    OutQ = new Queue<char>(iSize); InQ = new Queue<char>(oSize);
    do ; while ( URX < 0 ); // test CHAR_RDY, to completely empty input queue
    initInt(0x30000000, 0x30000200, handler, 10);enableInt();
}
virtual void put(char data) {
    while(OutQ->size >= OutQ->maxSize) ; // wait for room in the output queue
    disableInt(); OutQ->push(d); enableInt(); // avoid critical section
    if(OutQ->size == 1) UCR1 |= TRDY_EN; // if size becomes 1, en. trns. int.
}
virtual int get(void) { char data;
    while(InQ->size==0);disableInt();data=InQ->pull();enableInt();
    return data;
}
interrupt void handler(){ char oldTop; short i;
    while((InQ->size <= InQ->maxSize) && ((i = URX) < 0))
        { InQ->push(i); if((i & 0x4000) || InQ->errors() ) error = 1; }
    while( OutQ->size) && (USR & 0x2000 ))
        { UTX = OutQ->pull(); if( ! OutQ->size ) UCR1 &= ~TRDY_EN ;}
}
```

~*UARTi ()* { *while (USR >= 0)*; *UCR1 = 0*; /* turn off UART completely */ }
};

Class *UARTh*, derived from *UARTi*, uses *hardware handshaking*. By asserting
clear-to-send (CTS, *UCR2* bit 12), you permit the "other device" (a UART, or
equivalent, sending and receiving serial data to or from this UART) to send characters to this microcontroller, while negating it denies its sending characters. Sensing
request-to-send status (RTSS, *USR* bit 14), tells you if there is room in the "other
device's" input queue. In this class, CTS is generated by, and RTSS is used by, the
class's software; in the next class it will directly control, and be controlled by, the
UART hardware.

Because the "other device's" queue and registers may have data to shift out
when this UART's CTS becomes negated, it is negated when there is still a little
room in the input queue. A constant *THRESHOLD* is defined so as to provide
enough room in the input queue to absorb all of the UART0 device's queue and
registers' data after it is told to stop. CTS is negated in the input portion of the
interrupt handler when the number of characters input queue rises above the queue's
maximum size — *THRESHOLD*. This assures that there will be room in the input
queue to handle an influx of data that might be sent before RTS is negated in the
"other device" and the "other device" stops sending data. CTS is asserted in the get
function member when the number of characters in the input queue drops below
THRESHOLD.

The RTS signal is checked at least once in the *put* function member so that if it is
negated, the transmitter interrupt is disabled, and it is checked repetitively while the
put function is waiting for the output queue to have some room in it, so that if RTS
becomes asserted, the transmitter interrupt is enabled. Note that if the transmitter
interrupt is already enabled, you can try to enable it again, but that won't hurt
anything. The output portion of the interrupt handler also disables the transmitter
interrupt if RTS is found to be negated.

```
enum{ THRESHOLD = 2, RTSS = 0x4000, CTS = 0x1000, TRDY_EN = 0x2000 };

class UARTh : public UARTi { protected : Queue<char> *InQ, *OutQ;

    public:UARTh(char id,unsigned char iS,unsigned char oS) :
        UARTi(id, iS, oS)
        {UPCR = 0xf; *(long *)0x10 = (long)handler + 1;UCR2 |= CTS;}
virtual void put(char d) {
    do { if(USR & RTS) UCR1 |= TRDY_EN; }    // if RTS asserted, en. trn. int.
        while(OutQ->size>=OutQ->maxSize);    // while waiting on full queue
    UARTi::put(d);    // UARTi's put function pushes a byte to the output queue
}
virtual int get(void) {
    if(InQ->size==THRESHOLD) UCR2|=CTS;    // if empty enough, ask more
    return UARTi::get();    // UARTi's get pulls a byte from the input queue
}
```

```
interrupt void handler(){ char oldTop; short i;
    while((InQ->size <= InQ->maxSize) && ((i = URX) < 0)) {
        InQ->push(i); if((i & 0x4000) || InQ->errors())error=1;
            if(InQ->size == (InQ->maxSize - THRESHOLD))UCR2 &= ~CTS;
    }
    while( OutQ->size && ( USR & RTSS ) && (USR & RRDY ) )
    UTX = OutQ->pull();
    if( ! (USR & RTSS) || ! OutQ->size ) UCR1 &= ~TRDY_EN
    }
}*S;
void main()
    {char i;S=new UARTh(1,10,10);S->put(0x55);i=S->get();delete S;}
```

The MMC2001 UART can have RTS and CTS automatically control the hardware without software intervention described above. Class $UARTah$ (UART with automatic handshake) below is essentially class $UARTi$, with $UCR2 = 0 \times 2020$; instead of $UCR2 = 0 \times 4020$; to negate IRTS, bit 13 and assert CCTS. Also UPCR is set to $0xf$; to let the UART use port bits 2 and 3. Finally, the interrupt handler can take care of changes in RTS, which enable and disable transmitter interrupts. To permit such interrupts, set bit 5 of $UCR1$, and to reduce interrupt overhead, interrupt when the queues have 14 characters or slots, by setting bits 15, 14, 11, and 10, as in $UCR1 = 0xfd21$;

```
class UARTah : public Port<int> { protected : Queue<char> *InQ, *OutQ;
    public:UARTah(char id,unsigned char iSize,unsigned char oSize) :
        Port(id) {
        disableInt();UCR1=0xfd21;UCR2=0x2020; UBRGR=208; UPCR=0xf;
        OutQ = new Queue<char>(iSize); InQ = new Queue<char>(oSize);
        do ; while(URX<0);*(long *)0x10=(long)handler+1; enableInt();
    }
    virtual void put(char d) {
        do { if(RTS) UCR1 |= TRDY_EN; } // if RTS asserted, en. trn. int.
            while(OutQ->size>=OutQ->maxSize);// while waiting on full Q
        UARTi::put(d); // UARTi's put function pushes a byte to the output Q
}
virtual int get(void) { char data;
    while( InQ->size == 0 ) ;
    disableInt();flush(); data=InQ->pull();enableInt();return data;
}
interrupt void handler(){ char oldTop; short i;
    if((i=USR)&0x20) {// if there is a change in the RTS input, if RTS is 1 and
        if(( i & 0x4000 ) && (OutQ->size)) UCR1 |= TRDY_EN;
        else UCR1 &= ~TRDY_EN;
    }
    while(OutQ->size)&&(USR&0x2000)) // put output data in hardware queue
        { UTX = OutQ->pull(); if( ! OutQ->size) UCR1 &= ~TRDY_EN;}
    flush(); // get input data into software queue
}
```

```
~UARTah(){ while( USR >= 0 ); UCR1 = 0; /* turn off UART fully */ }
private : void flush(void){ // move data from hardware to software input queue
    while((InQ->size <= InQ->maxSize) && (i = URX) < 0))
        {InQ->push(i);if((i&0x4000)||InQ->errors()) error = 1;}
}
};
```

The class UARTsh, derived from UARTi, uses *software handshaking*. An ASCII character XON (0x11) will be sent by this microcontroller if there is room in its input queue and another ASCII character XOFF (0x13) will be sent by this microcontroller if there is not enough room. These characters are equivalent to changes in the CTS and RTS signals used in hardware handshaking. Relating to hardware handshaking, XON is sent when CTS is asserted and XOFF is sent when CTS is negated. Therefore, when XON is received, this is equivalent to when RTS is asserted, and when XOFF is received, it is equivalent to when RTS is negated. The receiver reconstructs a data member RTS to simulate the RTS signal, responsive to receiving XON or XOFF. XON or XOFF are sent by putting these constants in a data member Msg, which is 0 if no special signal is to be sent. If Msg is nonzero, the transmitter interrupt is enabled, and Msg is sent in place of any data that may be in the output queue. The private function member *send* is used to write its argument in MSG and enable the transmit interrupt. This decreases the delay time from when the input queue state change requires sending an XON or XOFF until the sender reacts by enabling or disabling its transmitter interrupt.

```
class UARTsh : public UARTi { char Msg, RTS;
    public : UARTsh(char id, unsigned char iS, unsigned char oS) :
      UARTi(id, iS, oS)
      { *(long *)0x10 = (long)handler + 1; Msg = 0; RTS = 1;}
}
interrupt void handler(){ char oldTop; short i;
    while((InQ->size <= InQ->maxSize) && ((i = URX) < 0)) {
            if((i&0x)==XON ) RTS=1; // first check if the incoming character
            else if((i & 0x)==XOFF ) RTS=0; // is a special character. If so
            else InQ->push(i);// update simulated RTS; otherwise push it.
            if((i & 0x4000) || InQ->errors() ) error = 1;
            if(InQ->size == (InQ->maxSize - THRESHOLD)) send( XOFF );
        }
        if( OutQ->size && RTS && (USR & RRDY ) ) {
            if(Msg) {UTX=Msg;Msg = 0;} // if special character, send it and clear it
            else UTX=OutQ->pull();// otherwise send a character from output queue
        }
        if( ! RTS || ! OutQ->size ) UCR1 &= ~TRDY_EN
    }
}
virtual int get(void) {
    if(InQ->size == THRESHOLD) send(XON); // if empty enough, ask more
    return UARTi::get();// UARTi's get pulls a byte from the input queue
}
```

```
virtual void put(char data) {
    while(OutQ->size >= OutQ->maxSize) if( RTS ) UCR1 |= TRDY_EN;
    disableInt(); OutQ->push(d); enableInt();
    if( ( OutQ->size >= 1) && RTS ) // if output queue not empty
        UCR1 |= TRDY_EN; // and RTS is 1, enable transmitter interrupt
    private : void send(char c) { MSG = c; UCR1 |= TRDY_EN; }
} *S;
void main()
    {char i; S = new UARTsh(1); S->put(0x55); i = S->get(); delete S;}
```

9.3.5 An Ariel Device Driver for UART0 or UART1

In a multitask system, an MMC2001 UART may well be accessed through a device driver because this device is likely to be used by several of the tasks. This driver should handle special characters, including carriage return, backspace, escape to kill a task, and control-E to start the debugger. It should output (*echo*) characters that have been input to the device, but some characters need to be echoed as two or three characters. This device driver should permit other tasks to run when the device is busy, as in device drivers in §6.2.9.3 and §7.3. This complicates the device driver, because we must return to task (T) state when the device becomes busy, to permit other tasks to become active and use the MPU. On the positive side, the MMC2001 UART's 16-element hardware FIFOs permits the use of gadfly rather than interrupt synchronization, simplifying the code needed to switch between T and S/SE states. On the negative side, the driver should echo input characters, complicating the code needed to switch between T and S/SE states.

Driver declarations and straightforward procedures are described below:

```
#define _EOS_PHYS_FAILURE -10
typedef struct{short rx,d1[31],tx,d2[31],cr1,cr2,brgr,sr,pcr;}
    uart_t;
enum{IRTS=0x4000, CTCS = 0x2000, WS = 0x20, TRDY = 0x2000, RRDY = 0x200,
    ERS=0x4000, RTRF = 0x200, TXEN = 0x1000, RXEN = 0x100, UART_EN = 1 };
enum{RDGO, RDWAIT, WRGO, WRWAIT, BSPACE, ERROR}; /* driver states */
typedef struct param0_s {char *port;  short brgr,cr1,cr2,bcr;}
    param0_t;

param0_t param0 =
    {(short *)0x10009000, 208, UART_EN + TXEN + RXEN, IRTS + WS, 3};
uxid_t i_o_Id; rv_t status;
const drv_t function0={initFun, svcFun, 0, 0, 0, 0, 0};

ucd_t  ucd0={_EXTNAME('P','T','R','O'),&function0,0,&par-
    am0,0,0,{0,0}};
main(){char s[10];
    if(!ISSVCERR(i_o_Id =_pio_create(&ucd0)))i_o(_PIO_READ,s,10,0);
}
```

Initialization executes *initFun()*, which copies constants into control registers. The driver-specific parameters supply the device's base address, which should be *0x10009000* for UART0 and *0x1000A000* for UART1, and the initial values of each of the control registers. While this permits full flexibility in using the device, device drivers often better isolate the device driver user by having separate commands to set the baud rate and frame structure, so the user need not understand the details of MMC2001's UART control registers. The reader can enhance this driver by adding such user-oriented, rather than hardware-oriented, *_pio_cmd()* commands to set up the control registers.

```
int32 initFun( ucd_t *ucp ) {
    param0_t *ip=(param0_t *)ucp->ddp;uart_t *p=(uart_t *)(ip->port);
    p->brgr=ip->brgr; p->cr1=ip->cr1; p->cr2=ip->cr2; return SUCCESS;
}
```

Commands access the driver through *i_o()*; which runs in task (T) state. This procedure waits a tick time using *time_wait()*, essentially whenever the device is found to be busy in function *svcFun()* it calls, which is shown after *i_o()*.

```
void i_o(u_int32 cmd, char *data, short n, char *msg) { cdev_t charDev;
    charDev.aub=data; charDev.lub=n; charDev.axb=msg;
    if(cmd == _PIO_READPR) charDev.spl = WRGO; else charDev.spl = RDGO;
    do {
        _pio_cmd(i_o_Id, cmd, 0L, &charDev, &status, _CONT_NOCOOR);
        if(charDev.lub == 0) break;
        else {time_wait(_TIME_NEXT_TICK); charDev.spl &= ~1;}
    } while(1);
}
```

svcFun() uses subroutines *echo()*, *write()*, and *read()*, all shown below. We discuss these routines' operations after showing them on one page below.

```
void echo(uart_t *port, char c) {while(!(port->sr & TRDY));port->tx=c;}

rv_t read(ucb_t *ucb, cdev_t *dp, short n, int_32 cmd){short c;
    if((ucb->port->sr & RRDY) == 0) return RDWAIT;
    *((*p)++) = c = port->rx; if(c & 0x7c00) {dp->lub = 0; return ERROR;}
    if(cmd == _PIO_READBYTE)return RDGO; /* don't handle sp. chars for raw I/O */
    else if((c &= 0x7f) == 4) {_drv_sig_send(ucb->tcb,0); return BSPACE;}
    else if(c == 0x1b) {_drv_sig_send(ucb->tcb,1); return BSPACE;}
    else if(c == '\r'){ echo(ucb->port, '\r');dp->lub=0; return BSPACE;}
    else if(n == 1){ echo(ucb->port, 7); return BSPACE;}
    else if(c == 0x7f) {
        if(n == 0) echo(ucb->port, 7);
        else {echo(ucb->port, 8); echo(ucb->port, ' ');echo(ucb->port,8);
          dp->aub--; dp->lub++; }
    return BSPACE;
}
```

```
else if((c >= ' ')||(c == '\r')||(c == '\n')||(c == 7))
    {echo(ucb->port,c); return RDGO;}
else { echo(ucb->port, '^'); echo(ucb->port, c + 0x40);return RDGO;}
rv_t write(uart_t *port, char **p)
    {if(port->sr & TRDY){port->tx=*((*p)++);return WRGO;}return
        WRWAIT;}
}


void svcFun(ucb_t *ucb,int fun,cdev_t *dp){ static char p[] = ''?''
    ucd_t *ucd = &ucb->ucd; /* find initFun's argument */
    while(((dp->spl & 1) != 1) && dp->lub){/* loop while state RDGO, WRGO*/
        switch(fun){
        case _PIO_WRITE:
            if(!*dp->aub){dp->lub = 1; dp->spl = RDGO;}
            else dp->spl = write(ucb->port, &dp->aub);
            if(dp->spl==WRGO) dp->lub--; break;
        case _PIO_READ: dp->aub = p;
        case _PIO_READPR:
            if(dp->spl == WRGO)
                {dp->spl = write(ucb->port, &dp->axb);if(*dp->axb)  break}
        case _PIO_READBYTE:
            if((dp->spl=read(ucb,&dp->aub,dp->lub,fun))==BSPACE)
                {dp->lub++; dp->aub--; dp->spl = RDGO;}
            if(dp->spl==RDGO) dp->lub--; break;
        }
} if(dp->lub == 0)
    _drv_pio_done(ucb,
        (dp->spl==ERROR) ? _EOS_PHYS_FAILURE : SUCCESS,0);
}
```

A state stored in _pio_cmd's parameter spl field is used to make i_o() wait whenever the device is busy, and to return to the place in svcFun() exactly where it was stopped. This is similar to restarting an instruction after a memory page fault.

Consider first a command to read a byte vector having a given length. i_o() is called with first parameter _PIO_READBYTE. svcFun() executes its _PIO_READBYTE case, which calls read() to test the device's status register, to see if the receive buffer has data. If no data are available, the return value RDWAIT is put in dp->spl, the state. Otherwise, the received byte is stored using pointer dp->aub which is incremented. read() returns without echoing the character or reacting to special characters. This permits svcFun() to loop, calling read() again, until all data are input, inidicated by dp->lub becoming zero. However, if the state is RDWAIT, svcFun() exits to i_o(), calling time_wait(). After a tick time, svcFun() is called again, calling read() again, to see if data can be transferred to the hardware output queue. Note that if there is data in the hardware queue, this driver transfers data fairly efficiently, but if not, other tasks can run while the device is busy.

Consider next a command to write a null-terminated character string, which is similar to the previous case. *i_o()* is called with first parameter *_PIO_WRITE*. *svcFun()* is called and executes its *_PIO_WRITE* case. If the character that is to be output is the null character, the number of bytes to be sent, *dp->lub*, and the state *dp->spl*, are set to immediately terminate execution of the device driver. Otherwise *write()* is called to test the device's status register, to see if the transmit buffer is available. If the buffer has room in it, the data pointed to by *svcFun()*'s *dp->aub* is output and *dp->aub* is incremented, and the procedure's return value *WRGO* is put into the state *dp->spl*. Otherwise, *svcFun()* exits to *i_o()*, calling *time_- wait()*.

Consider next a command to write a prompt and then read a null-terminated character string. *i_o()* is called with first parameter *_PIO_READPR*. *svcFun()* is called and executes its *_PIO_ READPR* case, which calls *write()* to output the string pointed to by *dp->axb*. Then *svcFun()* calls *read()* to input the string pointed to by *dp->aub*. In this case, the state stored in *dp->spl* distinguishes whether to resume *write()* or *read()* after waiting for a time tick. In *read()*, the received character in *short c* is examined. If the high byte indicates an error, the driver is exited with an error message. If the received data is control-D, a signal is sent to start the debugger, and the character is removed from the input buffer. Likewise, if the received data is the escape code, a signal is sent to start an asynchronous procedure that will respond to escape sequences, and the character is removed from the input buffer. If the received data is the carriage return, *dp->lub* and *dp->spl*, are set to immediately terminate the use of the device driver. If there remains but one character in the input buffer, a bell code is echoed, and the character is deleted from the input buffer. It the character received is delete (0x7f), and the buffer has at least one character, this and the previous character are removed, and a backspace, space, and backspace character are echoed, otherwise a bell is sounded. If the character is printable, it is echoed, but if it is a nonprintable control character, hat "^" is echoed, followed by the control character converted to an equivalent ACSII capital letter.

Echoing uses a simple procedure *echo()* which does not attempt to allow other tasks to execute if the device is busy. Echoing, which occurs when the user types characters, does not usually cause this device to become busy for long. Waiting when the device is busy is not very useful, except when many characters are entered very quickly.

This device driver has a fault that is due to the MMC2001 hardware device. If a task does not use any input, characters are not pulled from the input queue; a control-E character held in the hardware queue will not invoke the debugger. An interrupt-based device driver can pull data from the hardware into the software input queue whenever the former has data in it, whether that data is actually to be input to the task, and at that point, special characters can be detected to generate signals to terminate a task or invoke the debugger. A command similar to BASIC's IN $KEY subroutine can be used to input a character if one is available, but not to wait if there is no character. The reader can easily add this *_pio_cmd()* command to the *svcFun()* procedure. This procedure can be called in tasks that do not otherwise input data, to pull such control characters from the hardware input queue, to generate signals to invoke the debugger.

This section illustrates that serial communication classes and device drivers can be significantly more complex than the simple I/O classes and device drivers. Nevertheless, these classes and device drivers merely combine techniques used for simple I/O classes and device drivers with some specific requirements of communication systems.

9.4   Other Protocols

In addition to the UART protocol, the two most important protocols are the synchronous bit-oriented protocols that include the SDLC, HDLC and ADCCP, X-25, the IEEE-488 bus protocol, and the Small Computer System Interface (SCSI) protocol.

These are important protocols. We fully expect that many if not most of your interfaces will be designed around these protocols. If you are designing an I/O device to be used with a large mainframe computer, you will probably have to interface to it using a synchronous bit-oriented protocol. If you are designing a laboratory instrument, you will probably interface to a minicomputer using the IEEE-488 protocol, so that the minicomputer can remotely control your instrument. We will survey the key ideas of these protocols in this section. The first subsection describes bit-oriented protocols. The second subsection discusses the 488 bus. The final subsection covers the SCSI interface.

9.4.1   Synchronous Bit-Oriented Protocols

Synchronous protocols are able to move a large amount lot of data at a high rate. They are used primarily to communicate between *remote job entry* terminals (which have facilities to handle line printers, card readers, and plotters) and computers, and between computers and computers. The basic idea of a synchronous protocol is that a clock is sent either on a separate wire or along with the data in the Manchester coding scheme. Since a clock is sent with the data, there is little cause to fear that the receiver clock will eventually get out of sync after many bits have been sent, so we are not restricted to short frames as we are in the UART. Once the receiver is synchronized, we will try to keep it in synchronism with the transmitter, and we can send long frames without sending extra control pulses, which are needed to resynchronize the receiver and reduce the efficiency of the channel.

Asynchronous protocols, such as the UART protocol discussed in the last section, are more useful if small amounts of data are generated at random times, such as by a computer terminal. Synchronous protocols would have to get all receivers into synchronism with the transmitter when a new transmitter gain, control of the channel, so their efficiency would be poor for short random messages. Synchronous protocols are more useful when data are sent at once because they do not require the overhead every few bits, such as start and stop bits, that asynchronous protocols need. Bit-oriented synchronous protocols were developed as a result of weaknesses in byte- or character-oriented synchronous protocols when they were used in sending a lot of data at once.

The precursor to the bit-oriented protocol is the binary synchronous *Bisync* protocol, which is primarily character-oriented and is extended to handle arbitrary binary data. This protocol can be used with the ASCII character set. The 32 non-printing ASCII characters include some that are used with the Bisync protocol to send sequences of characters. SYN - ASCII 0x16 is sent whenever nothing else is to be sent. It is a null character used to keep the receiver(s) synchronized to the transmitter. This character can be used to establish which bit in a stream of bits is the beginning of a character. Two Bisync protocols are used, one for sending character text and the other for sending binary data, such as machine code programs, binary numbers, and bit data.

Character text is sent as follows (see Figure 9.8a): A header can be sent, begining with character SOH — ASCII 0x01; its purpose and format are user defined. An arbitrary number of text characters is sent after character STX — ASCII 0x02, and is terminated by character ETX — ASCII 0x03. After the ETX character, a kind of checksum is sent.

To allow any data — such as a machine code program — including characters that happen to be identical to the character ETX, to be sent, a character DLE — ASCII 0x10 is sent before the characters STX and ETX. A byte count is established in some fashion. It may be fixed, so that all frames contain the same number of words, it may be sent in the header, or it may be sent in the first word or two words of the text itself. Whatever scheme is used to establish this byte count, it is used to disable the recognition of DLE-ETX characters that terminate the frame, so such patterns can be sent without confusing the receiver. This is called the *transparent mode* because the bits sent as text are transparent to the receiver controller and can be any pattern.

Bisync uses error correction or error detection and retry. The end of text is followed by a kind of checksum, which differs in differing Bisync protocols. One good error detection technique is to exclusive-OR the bytes that were sent, byte by byte. If characters have a parity bit, that bit can identify which byte is incorrect. The checksum is a parity byte that is computed "at 90 degrees" from the parity bits and can identify the column that has an error. If you know the column and the row, you know which bit is wrong, so you can correct it. Another Bisync protocol uses a *cyclic*

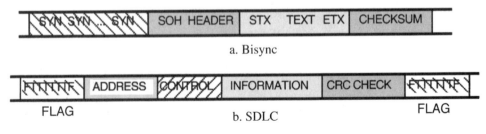

a. Bisync

b. SDLC

Figure 9.8. Synchronous Formats

redundancy check (CRC) that is based on the mathematical theory of error correcting codes. The error detecting "polynomial" X**16 + x**15 + X**2 + 1, called the CRC-16 polynomial, is one of several good polynomials for detecting errors. The CRC check feeds the data sent out of the transmitter through a shift register that shifts bits from the 15*th* stage towards the 0*th* stage. The shift register is cleared, and the data bits to be transmitted are exclusive-ORed with the bit being shifted out of the 0*th* stage, and then exclusive-ORed with some of the bits being shifted in the register at the inputs to the 15*th*, 13*th*, and 0*th* stages. The original data and the contents of the shift register (called the CRC check bits) are transmitted to the receiver. The receiver puts the received data, including the CRC check bits, through the same hardware at its end. When done, the hardware should produce a 0 in the shift register. If it does not, an error (CRC error) has occurred. The Bisync protocol can request that the frame be resent upon a CRC error. If the frame is good, an ACK - ASCII 0x06 is sent, but if an error is detected a NAK - ASCII 0x15 is sent from the receiver back to the sender. If the sender gets an ACK, it can send the next frame, but if it gets a NAK, it should resend the current frame.

Though developed for communication between a computer and a single RJE station, Bisync has been expanded to include *multi-drop*. Several RJE stations are connected to a host computer on a half-duplex line (bus). The host is a master. It controls all transfers between it and the RJE stations. The master *polls* the stations periodically, just as we polled I/O devices after an interrupt, to see if any of them want service. In polling, the master sends a short packet to each station, so that each station can send back a short message as the master waits for the returned messages.

Bisync protocols have some serious shortcomings. It is set up for and is therefore limited to half-duplex transmission. After each frame is sent, you have to wait for the receiver to send back an acknowledge or a negative acknowledge. This causes the computer to stutter, as it waits for a message to be acknowledged. These short-comings are improved in bit-oriented protocols. Features used for polling and multi-drop connections are improved. The information is bit-oriented to efficiently handle characters, machine code programs, or variable width data.

The first significant synchronous bit-oriented protocol was the *Synchronous Data Link Control* (SDLC) protocol developed by IBM. The American National Standards Institute, ANSI, developed a similar protocol, ADCCP, and the CCITT developed another protocol, HDLC. They are all quite similar at the link control and physical levels, which we are studying. We will take a look at the SDLC link, the oldest and simplest of the bit-oriented protocols.

The basic SDLC frame is shown in Figure 9.8b. If no data are sent, either a true bit is continually sent (idle condition) or a *flag* pattern, 0x7E (FTTTTTTF), is sent. The frame itself begins with a flag pattern and ends with a flag pattern, with no flag patterns inside. The flag pattern that ends one frame can be the same flag pattern that starts the next frame.

The frame can be guaranteed free of flag patterns by a five T's detector and F inserter. If the transmitter sees that five T's have been sent, it sends an F regardless of whether the next bit is going to be a T or an F. That way, the data FFTFFTTTTTTF   is   sent   as   FFTFFTTTTTFTTF,   and   the   data

FFTFFTTTTTFTF is sent as FFTFFTTTTTFFTF, free of a flag pattern. The receiver looks for 5 T's. If the next bit is F, it is simply discarded. If the received bit pattern were FFTFFTTTTTFTTF, the F after the five T's is discarded to give FFTFFTTTTTTTF, and if FFTFFTTTTTFFTF is received, we get FFTFFTTTTTFTF. But if the received bit pattern were FTTTTTTF the receiver would recognize the flag pattern and end the frame.

The frame consists of an 8-bit station number address, for which the frame is sent, followed by 8 control bits. Any number of information bits are sent next, from 0 to as many as can be expected to be received comparatively free of errors or as many as can fit in the buffers in the transmitter and receiver. The CRC check bits are sent next. The address, control, information, and CRC check bits are free of flag patterns as a result of five T's detection and F insertion discussed above.

The control bits identify the frame as an information *frame,* or *supervisory* or *nonsequenced* frame. The information frame is the normal frame for sending substantial ammounts of data in the information field. The control field of an information frame has a 3-bit number N. The transmitter can send up to 8 frames, with different values of N, before handshaking is necessary to verify that the frames have arrived in the receiver. As with the ACK and NAK characters in Bisync, supervisory frames are used for retry after error. The receiver can send back the number N of a frame that has an error, requesting that it be resent, or it can send another kind of supervisory frame with N to indicate that all frames up to N have been received correctly. If the receiver happens to be sending other data back to the transmitter, it can send this number N in another field in the information frame it sends back to the transmitter of the original message to confirm receipt of all frames up to the Nth frame, rather than sending an acknowledge supervisory frame. This feature improves efficiency, because most frames will be correctly received.

The SDLC link can be used with multi-drop (bus) networks, as well as with a ring network. The ring network permits a single main, *primary,* station to communicate with up to 255 other *secondary* stations. Communication is full duplex, since the primary can send to the secondary over part of the loop, while the secondary sends other data to the primary on the remainder of the loop. The SDLC has features for the primary to poll the secondary stations and for the transmitting station to abort a frame if something goes wrong.

The SDLC link and the other bit-oriented protocols provide significant improvements over the character-oriented Bisync protocols. Full-duplex communication, allowing up to 8 frames to be sent before they are acknowledged, permits more efficient communication. The communication is inherently transparent, because of the five T's detection feature, and can handle variable length bit data efficiently. It is an excellent protocol for moving large frames of data at a high rate of speed.

The *X*.25 protocol is a three-level protocol established by the CCITT for high-volume data transmission. The physical and link levels are set up for the HDLC protocol, a variation of the SDLC bit-oriented protocol; but synchronous character-oriented protocols can be used so that the industry can grow into the X.25 protocol without scrapping everything. This protocol, moreover, specifies the network level as well. It is oriented to packet switching. Packet switching permits frames of a message to wander through a network on different paths. This dynamic allocation of links to messages permits more efficient use of the links, increases security (because a thief would have to watch the whole network to get the entire message) and enhances reliability. It looks like the communication protocol of the future. While we do not cover it in our discussion of I/O device design, we have been using its terminology throughout this chapter as much as possible.

9.4.2   IEEE-488 Bus Standard

The need to control instruments like voltmeters and signal generators in the laboratory or factory from a computer has led to another kind of protocol, an asynchronous byte-oriented protocol. One of the earliest such protocols was the CAMAC protocol developed by French nuclear scientists for their instruments. Hewlett-Packard, a major instrument manufacturer, developed a similar standard which was adopted by the IEEE and called the IEEE-488 standard. Although Hewlett-Packard owns patents on the handshake methods of this protocol, it has made the rights available on request to most instrument manufacturers, and the IEEE-488 bus standard has been available on most sophisticated instruments, minicomputers, and microcomputers.

Communications to test equipment has some challenging problems. The communications link may be strung out in a different way each time a different experiment is run or a different test is performed. The lengths of the lines can vary. The instruments themselves do not have as much computational power as a large mainframe machine, or even a terminal, so the communications link has to do some work for them such as waiting to be sure that they have picked up the data. A number of instruments may have to be told to do something together, such as simultaneously generating and measuring signals, so they can not be told one at a time when to execute their operation. These characteristics lead to a different protocol for instrumentation busses.

The IEEE-488 bus is fully specified at the physical and link levels. A 16-pin connector, somewhat like the RS232 connector, is prescribed by the standard, as are the functions of the 16 signals and 8 ground pins. The 16 signal lines include a 9-bit parallel data bus, three handshaking lines, and five control lines. The control lines include one that behaves like the system reset line in the MMC2001 microcomputer. Others are used to get attention and perform other bus management functions. However, the heart of the bus standard is the asynchronous protocol used to transmit data on the bus.

An asynchronous bus protocol uses a kind of expandable clock signal, which can be automatically stretched when the bus is longer or shortened if the bus is shorter. This happens because the "clock" is sent from the station transmitting the data to the station that receives the data on one line, then back to the transmitter on another line. The transmitter waits for the return signal before it begins another transmission. If the bus is lengthened, so are the delays of this "clock" signal. The IEEE-488 bus uses this principle a few times to reliably move a word on an 9-bit bus from a transmitter to a receiver. (See Figure 9.9.)

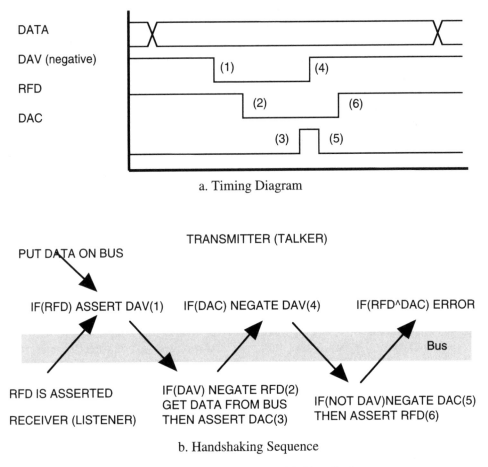

a. Timing Diagram

b. Handshaking Sequence

Figure 9.9. IEEE-488 Bus Handshaking Cycle

The handshake cycle is like a clock cycle. Each time a word is to be moved, the bus goes through a handshake cycle to move the word, as shown in Figure 9.9. The cycle involves (negative logic) *data available* (DAV), sent by the transmitter of the data, and (positive logic) *ready for data* (RFD) and (positive logic) *data accepted* (DAC), sent by the receiver of the data.

If the receiver is able to take data, it has already asserted RFD (high). When the transmitter wants to send a data word, it first puts the word on the bus, and then begins the handshake cycle. It checks for the RFD signal. If it is asserted at the transmitter, the transmitter asserts DAV (low) to indicate that the data are available. This is step 1 in Figures 9.9a and 9.9b. When the receiver sees DAV asserted, it negates RFD (low) in step 2 because it is no longer ready for data. When the processor picks up the data from the interface, the receiver asserts DAC (high) to indicate that data are accepted. This is step 3. When the transmitter sees DAC asserted, it negates DAV (high) in step 4 because it will soon stop sending data on the data bus. When the receiver sees DAV negated, it negates DAC in step 5. The data are removed sometime after the DAV has become negated. When it is ready to accept new data, it asserts RFD (high) in step 6 to begin a new handshake cycle.

The IEEE-488 bus is designed for some special problems in busing data to and from instruments. First, the bus is asynchronous. If the receiver is far away and the data will take a long time to get to it, the DAV signal will also take a long time, and the other handshake signals will be similarly delayed. Long cables are automatically accounted for by the handshake mechanism. Second, the instrument at the receiver may be slow or just busy when the data arrive. DAC is asserted as soon as the data get into the interface, to inform the transmitter that they got there; but RFD is asserted as soon as the instrument gets the data from the interface, so the interface will not get an overrun error that a UART can get. Third, although only one station transmits a word in any handshake cycle, a number of stations can be transmitters at one time or another. Fourth, the same word can be sent to more than one receiver, and the handshaking should be able to make sure all receivers get the word. These last two problems are solved using open collector bus lines for DAV, RFD, and DAC. DAC, sent by the transmitter, is negative logic so the line is wire-OR. That way, if any transmitter wants to send data, it can short the line low to assert DAV. RFD and DAC, on the other hand, are positive logic signals so the line is a wire-AND bus. RFD is high only if all receivers are ready for data, and DAC is high only when all receivers have accepted data.

The IEEE-488 bus is well suited to remote control of instrumentation and is becoming available on many of the instruments being designed at this time. You will probably see a lot of the IEEE-488 bus in your design experiences.

9.4.3   The Small Computer System Interface (SCSI)

The microcontroller has made I/O devices quite intelligent, such as hard disks or CD-ROMs. Commonly, a personal computer communicates with such an intelligent I/O device using the *Small Computer System Interface* (*SCSI*) protocol. Up to eight devices can be on a SCSI bus, and they may be *initiators* (e.g., microcontrollers) or *targets* (e.g., disk drives). The SCSI protocol permits multiple initiators to *arbitrate* for the bus, granting one of them its use, and for that initiator to *select* one of the devices as its target and give the target a command. The two devices then transfer data. This process is called a *nexus*. This section considers the two lowest layers of the SCSI protocol, which are the transfer of a byte of data, and the execution of a nexus.

A SCSI byte transfer is quite similar to the IEEE-488 byte transfer. This movement involves an 9-bit (8-data plus odd parity) parallel bus, a handshake protocol involving a direction signal (IO) and a request (REQ) sent by the target, and an acknowledge (ACK) sent by the initiator. (See Figure 9.10.) Signal IO is low when the initiator gets data from the bus, and is high when the initiator puts data on the bus.

If the initiator is to receive a byte, the target puts the byte and parity on the bus and drops REQ low. When the initiator sees REQ drop, it picks up the byte, verifies

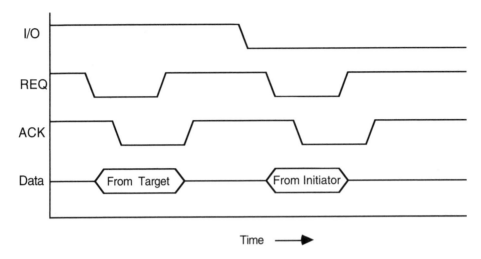

Figure 9.10. SCSI Timing

parity, and drops ACK low. When the target sees ACK low, it raises REQ high, releasing data, and when the initiator sees REQ rise, it raises ACK, so the next transfer can take place.

If the initiator is sending a byte, the target drops REQ. When the initiator sees REQ drop, it outputs the byte and odd parity and drops ACK. When the target sees ACK drop, it picks up the data from the data bus, and it raises REQ. When the target sees REQ rise, it releases data and raises ACK high in order for the next transfer to take place.

```
#define myId 7
#define hisId 5
#define t(n) (n / 62) /* for 32 MHz clock */
enum{selIn=1, cdIn=2, reqIn=4, ioIn=8, rstOut=0x10, selOut=0x20,
    busyOut = 0x40, dataIn = 0x80, attnIn = 0x100, attnOut = 0x200,
    ackDirIn=0x400, ackOut=0x800, ackIn = 0x1000, busyIn = 0x2000,
    rstIn=0x4000, msgIn=0x8000, hiIds=~((1<<(myId+1))-1) & 0xffff};
enum{ Arbitration_d=t(2400), Assertion_d=t(90),
    Bus_clear_d=t(800),
    Bus_free_d = t(800), Bus_set_d = t(1800), Bus_settle_d = t(400),
    Cable_skew_d = t(10), Data_release_d = t(400), Deskew_d = t(45),
    Disconnect_d = t(200000), Hold_t = t(45), Negation_p = t(90),
    Selection_abort_t=t(200000), Reset_to_selection_t=t(250000000),
    Reset_hold_t = t(250000) }; // timing constants, in nanoseconds
volatile unsigned short
KDDR@0x10003004,KPDR@0x10003006, EPDDR@0x10007002, EPDR@0x10007004;
char dataBuffer[512], statusBuffer[4];
const char Inquiry[6] = { 0x12, 0, 0, 0, 36, 0 };
```

```
void wait (long t) { asm{
    setc
L: loopt r2,L ; delay 62 ns per value of t
}}

int isLow(short v) { return (v & (~KPDR)) == v; }

int isHi (short v) { return (v & KPDR) == v; }

void raise (short v) { KPDR |= v; }

void drop (short v) { KPDR &= ~v; }

void initialize() {
    KDDR=rstOut|selOut|busyOut|attnOut|ackOut|dataIn|ackDirIn;
    KPDR = 0xffff; EPDDR = 0; drop(rstOut); wait(Reset_hold_t);
    raise(rstOut); wait(Reset_to_selection_t);
}
void nexus( const char *commandPtr, char *dataPtr, char *statusPtr ) {
    char outmsg = 8;
    do{ // arbitration phase
      EPDDR = 0; raise( ackDirIn ); raise( busyOut ); // wait for bus free
      do if(isLow(rstIn)) return; while(isLow(selIn)||isLow(busyIn));
      do { wait( Bus_settle_d ); if(isLow(rstIn)) return; }
        while(isLow(selIn) || isLow(busyIn));
      wait( Bus_free_d ); EPDR = ~ (1 << myId); drop( dataIn );
      EPDDR = 0x; drop( busyOut ); wait( Arbitration_d );
    } while((EPDR ^ 0xff) & hiIds);
    drop( selOut ); wait( Bus_clear_d + Bus_settle_d );
    EPDR=~((1<<myId)|(1 << hisId));drop(attnOut|ackDirIn|dataIn );
    wait( Deskew_d * 2 ); raise( busyOut ); wait( Bus_settle_d );
    do if(isLow(rstIn)) return; while(isHi(busyIn));
    wait(Deskew_d*2); raise(selOut); EPDDR=0; raise(dataIn);

    do{ // information transfer phase
        do if(isLow(rstIn)) { outmsg = 5; break; } while(isHi(reqIn));
        if(isLow(ioIn)){ // input operations
          EPDDR = 0; raise( dataIn );
          if(isLow(msgIn)) { outmsg = ~EPDR; break; }
          else if(isLow(cdIn)) *statusPtr++ = ~EPDR;
          else *dataPtr++ = ~EPDR;
        }
    else { // output operations
        if(isLow(msgIn)) EPDR = ~0x80 raise( attnOut ); }
        else if(isLow(cdIn)) EPDR = ~ *commandPtr++;
        else EPDR = ~*dataPtr++;
        drop( dataIn ); EPDDR = 0xff;
      }
```

```
    wait( Deskew_d + Cable_skew_d); drop( ackOut ); // confirm acceptance
    do if(isLow(rstIn)) { outmsg = 5; break; } while(isLow(reqIn));
    EPDDR = 0; raise( dataIn | ackOut );
  } while(isLow(busyIn));
  *statusPtr++ = outmsg; /* can examine outmsg to see why we quit */
}
main(){ initialize(); nexus(Inquiry, dataBuffer, statusBuffer); }
```

The reader can get the flavor of a protocol, specifically the SCSI-2 protocol, from the program above. The specifications on sequencing and timing control signals are given in the ANSI document, *Information Technology - Small Computer System Interface - 2*, reference number ISO/IEC 9316-1:199x, §6. Parallel ports are used as in indirect I/O (§5.3.1) but delays are inserted using real-time programming (§6.1.1).

Six other control signals use the 50-pin connector shown to the left of Figure 9.11 to execute a nexus. The program above shows a nexus for the identify command, used to obtain the identity of a SCSI device. The arbitration protocol, coded in the first `do while` loop of `nexus`, assures that two initiators will not use the SCSI bus at the same time. The winner of arbitration, which becomes the initiator, selects its target and requests that the target ask for a message. The arbitration phase begins right after the above mentioned `do while` loop. After the target is selected, this target controls the transfer of bytes. The remaining do while loop transfers bytes, which may be messages, commands, status, or data, using the protocol discussed earlier in this section. The first sends the "identify" message 0x80, and then it sends a command. In the program above, the command in the 6-byte char vector `Inquiry` requests that the target send an identification packet, which includes the ASCII character string names of the device's manufacturer and product. The first byte, 0x12, is the command and the fifth byte, 36, is the size of the buffer that is able to receive the identification packet, so the target will not overflow the buffer.

We will extend this protocol in §10.2.4, to write a file on a ZIP drive. It can be used as well to access a hard disk or a CD-ROM drive. However, as it is wired and written, this example is really only suited to an SCSI bus with just one processor and one I/O device, and it does not respond to errors other than a reset condition. To extend this example, the reader needs to consult *Information Technology - Small Computer System Interface - 2*, which is a document as thick as this textbook, and requires very careful reading. Nevertheless, if the reader implements the hardware and program in this section, he or she has much of the knowledge needed to extend the design to get a practical SCSI interface, which allows a microcontroller such as the MMC2001 to control a large number of SCSI devices. The MMC2001 can then be used to store data on, or read data from, removable "floppy" disks including ZIP drives, hard disks, and CR-ROMS, and the MMC2001 can use scanners and printers that have a SCSI interface.

The SCSI bus is specially designed for communication between a small but powerful computer like a personal computer, and an intelligent I/O device. Many systems that you build may fit this specification and thus may use an SCSI interface.

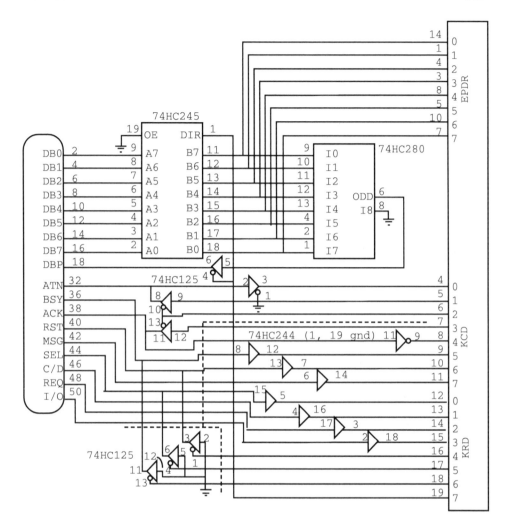

Figure 9.11. An SCSI interface.

9.5 Conclusions

Communication systems are among the most important I/O systems in a micro-computer. The microcomputer communicates with keyboards, displays, and type-writers, as well as with remote control stations and other microcomputers, using the UART protocol. The microcomputer can be in a large computer system and have to communicate with other parts of the system using the SDLC protocol. It may be in a laboratory and have to communicate with instrumentation on an IEEE-488 bus. It may have to talk with other systems in a different protocol.

This chapter covered the main concepts of communication systems at the phy-sical and link control levels. You should be aware of these concepts so you can understand the problems and capabilities of specialists in this field. You should be able to handle the UART protocol — the simplest and most widely used protocol — and its variations. You should be able to use the UART0 system in the MMC2001. You should be able to use the UART, in hardware designs. You should be able to write initialization rituals, interrupt handlers, and gadfly routines to input or output data using such hardware. Hardware and software tools like these should serve most of your design needs and prepare you for designing with the SDLC, IEEE-488, or SCSI interface protocol systems.

Do You Know These Terms?

See the End of Chapter 1 for Instructions.

coordinated	hardware	answer phone	poll stations
movement	handshake	switch hook	synchronous data
level of abstraction	bit level	ring indicator	link control
peers	frame level	data terminal ready	(SDLC)
end-to-end	message level	clear to send	flag pattern
network control	protocol	break	information frame
link control	handshake protocol	universal	supervisory frame
physical control	stack	asynchronous	nonsequenced
medium	structure	receiver transmitter	frame
frequency shift	store and forward	UART protocol	primary station
keying	circuit	start bit	secondary station
frequency	governed	stop bit	x.25
multiplexing	centralized	parity error	data available
time multiplexing	distributed	framing error	ready for data
channel	master slave	double buffering	data accepted
simplex	differential line	overrun error	small computer
half-duplex	rs442 standard	echo	system
full duplex	active	remote job entry	interface (SCSI)
bit time period	passive	bisync	initiator
baud rate	modem	transparent mode	target
bit rat	originate modem	cyclic redundancy	arbitrate
synchronous	answer modem	check	select
asynchronous	data coupler	multi-drop	nexus
manchester code	request-to-send	software handshake	information frame

Problems

Problem 1 is a paragraph correction problem; for guidelines, see the problems at the end of Chapter 1. The guidelines for software problems are given at the end of Chapter 2, and those for hardware problems are at the end of Chapter 3. Special guidelines for problems using the MMC2001 counter/timer modules and M6840 counter/timer chip are presented at the end of Chapter 7.

1. * To avoid any ambiguity, the peer-to-peer interfaces in communications systems are specified in terms of all lower level interfaces. The physical level, which is at the lowest level, is concerned with the transmission of signals such as voltage levels, with multiplexing schemes, and with the clocking of data if the clock must be sent with the data or on a separate line. The baud rate is the number of bytes of user data that can be sent per second. A channel is a data path between entities at the link control level. It is half duplex if every other bit is a data bit and the remainder are clock bits. Protocols are conventions used to manage the transmission and reception of data at the link control level and to negotiate for and direct communications at the network level. A handshake protocol is one in which congratulations are extended whenever a frame is correctly sent, but the receiver is silent if the data do not appear to be correct. A store-and-forward network is one that sends frames, called packets, from node to node and stores a frame in a node before negotiating to send it to the next node closer to its destination. The bus is a particularly common half-duplex store-and-forward network.

2. Design a 1-bit input/output port using the SN75119 differential transceiver that is connected to a full-duplex differential line. Reading *KCD* bit 7 will read the input data, and data written in bit 6 will pass through the transmitter for 8 cycles (1 µs) after the MMC2001 writes in the output register. Use *KCD* bit 5 to enable the transmitter.

a. Show the logic diagram of this device. Do not show pin numbers.
b. Show a self-initializing procedure `send(char d)` that outputs the least significant bit of *d*.

3. A "null modem" is a simple module permitting one computer's RS232-C plug to connect to another computer's RS232-C plug that has the same pin connections. Suppose the computer uses only transmitted and received data, request to send, data set ready, and signal ground. Show socket connections in the "null modem" that can correctly interconnect the two computers so that each looks like a terminal to the other.

4. Use the MMC2001 ISPI to implement a modified Pierce loop, which is an inter-microcontroller circular shift register, with, say, 8 bits of the shift register in each microcontroller. One microcontroller, the "master," supplies the Pierce loop clock, the others are "slaves." Frames circulate in the shift register synchronously and continuously. A simplified frame consists of a 3-bit source address (most significant bits), a 3-bit destination address (middle bits), and a 2-bit data field (least significant bits). When a frame is available to be used, the source and destination address are made TTT (111); no microcontroller sends to itself. Any microcontroller sends data to another by putting its addresses and data into such an available frame. When the frame shifts by its destination microcontroller, its data are saved there in global *char dataIn*. When a frame shifts by its source, it is rewritten as "available" again.

a. Show the logical design of a system using the ISPI module to shift the data among three MMC2001s, denoted microcontrollers 0 to 2. Show signal names but not pin numbers. Assume master microcontroller 0 supplies the ISPI clock. The clock SCLK and slave selects EN are tied in common, but MISO and MOSI form a ring.

b. Write the initialization in *main()*, procedures *send(char address, char data)*, *receive()*, and interrupt handler *handler()* so that *main()* outputs the message 3 to destination 1, receives incoming data into local variable *i*, and sends the message 0, to destination 1; to be used in the master micro-controller 0.

c. Write the initialization in *main()*, procedures *send(address, data) char address, data; receive()*, and interrupt handler *ccspi()* so that *main()* outputs the message 1 to destination 0, receive incoming data into local variable i, and send the message 2, to destination 0, to be used in either slave microcontroller 1 or 2.

5. Write a program to output 9600 baud UART signals, having 8 data, no parity, and 1 stop bits, using MMC2001 PWM device 0. *char* global variables *UartOut* hold output data, and *outBits* hold the output bit count. Write *main()* to in-itialize the PWM, output 0x55, and turn off the device. Output should be general, so that it can be repetitively done. Write *handler()* to interrupt each bit time, to send *UartOut*.

6. Write a program to input 9600 baud UART signals, having 8 data, no parity, and 1 stop bits, using MMC2001 edge port bit 0 and PWM device 0. *char* global variables *UartIn* hold completely shifted input data, *inBits* hold the input bit count, *shiftRegister* to hold bits shifted in, and *flag* is set when data have arrived. Write *main()* to initialize the edge prot and PWM, input a byte, output the byte that was input, and terminate the use of the device. Input should be general, so that it can be repetitively done. Write *handler0()* to detect rising and falling edges that indicate the widths, and therefore the bit values, of input data, which should be shifted into *UartIn*. *handler0()* should also detect each start bit, and enable device 1 when it has been found. Write *handler1()* to interrupt at the time of the middle of the stop bit to complete received data in *UartIn*, set *flag*, and disable its interrupt Hint: call a *service()* procedure from either handler to shift input bits.

7. Show a logic diagram of an I/O device using an IM6403 connected to an MMC2001. Use MMC2001 chip select CS0 to write data to be sent at 0x2d020000, read data that was received at 0x2d020000, write control at 0x2d020002, and the OE, FE, PE, DR, and TBRE status bits can be read at location 0x2d020002 as bits 0 to 4 respectively. Connect control and status so that the lower-numbered pins on the

IM6403 are connected to lower-numbered data bits for each I/O word, and use the lower-number data bits if fewer than 8 bits are to be connected. Show signal names of all pin connections to the MMC2001, and the signal names and the numbers of pins on the IM6403.

8. Show the logic diagram of a remote station that uses an IM6403 UART and a 74HC259 addressable latch so that when the number $I + 2N$ is sent, latch N is loaded with the bit I. Be careful about the timing of DR and DRR signals, and the G clock for the 74HC259. Show only signal names, and not pin numbers.

9. Show initialization rituals in *main()* to initialize the MMC2001 UART0 module (using a 32-MHz CLK clock) for:

a. 8 data, 1 stop bit, 9600 baud, all interrupts disabled, enable receiver and transmitter

b. 9 data, 1 stop bit, 300 baud, enable only receiver interrupt, enable receiver and transmitter

c. 8 data, 1 stop bit, 9600 baud, all interrupts disabled, enable transmitter only, send break

d. As in part a, but interrupt (wake up) when the line is idle

e. As in part a, where the CLK clock is set to 16 MHz

10. Write an UART0 device 1 background teletype handler to feed characters to a slow teletype using interrupt synchronization, so you can continue working with the computer as the printing is being done. A 0x100 byte *queue* contains characters yet to be printed. Part b will fill *queue* when your program is ready to print something, but part c's interrupt handler pulls words from the queue as they are sent through the UART0.

a. Write *main()* to initialize UART0 device 1 for 8 data, no parity, and 1 stop bit, and 1200 baud (the MMC2001 has a 32-MHz CLK clock). Only the transmitter is enabled.

b. Write a procedure *put(char *s, char n)* that will, starting at address *s*, output *n* words by first pushing them on the queue (if the queue is not full) so they can be output by the handler in part c. If the queue is full, wait in *put* until it has room for all words. Write the code to push data into the queue in *put* without calling a subroutine to push the data.

c. Write handler *handler()* that will pull a word from the queue and output it, but if the queue is empty it will output a NULL character (0). Write the code to pull data from the queue in the handler without calling a subroutine to pull the data.

11. Implement a Newhall loop, using interrupt synchronization on UART0 device 1. Such a Newhall loop is a ring of modules (microcontrollers) where each module's

TxD1 is connected to the next's *RxD1*, to circulate messages. The message's first byte's most significant nibble is a module address to which the message is to be sent, and its least significant nibble is a count of the number of data bytes left in the message, which is less than 0x10. If the message address is this module's address, input data are stored into an input buffer, otherwise input data are pushed onto a 16-byte queue. Transmitter interrupts are enabled only when the queue is nonempty or output buffer data is to be sent. Upon a transmitter interrupt, if the queue is nonempty, a byte is pulled and output. Otherwise, the output buffer data is sent highest index first to simplify counting. If neither data are to be sent, transmitter interrupts are disabled. Write *main()* to initialize the UART0, send a 4-byte message 1,2,3,4 in the 16-byte output buffer to module 12, wait until all data are moved, and disable UART0 device 1. Write *ccUART01()* to store an input message with address 5 into the module's input buffer, and move other messages through it around the loop. If no data are sent through the module, the handler will send the module's outgoing message. *push* and *pull* statements should be written into the handler code rather than implemented in separate subroutines.

12. Write a C *main()* to initialize, and interrupt handlers *handler0()* to locate bit boundary times, and *handler1()* to collect Bisync data bits, which arrive most significant bit first at 1K baud without a clock, and which are read from *EPDR* bit 0. PWM device 0 emulates a phase-locked loop. Whenever any edge appears on EPDR bit 0, PWM device 1 is set to interrupt half a bit-time later, and after that, device 1 repeatedly interrupts each bit time later, to collect a data bit from *EPDR* bit 0. When each bit arrives, if it is a SYNC character, bytes are picked up after each 8 bit times. When a byte matches STX, until ETX is met, bytes are stored in a buffer, unless it is full. The arithmetic sum of bytes from STX to ETX, inclusive, should be zero.

13. Write gadfly synchronized C procedures *main()*, *put(char *v)* and *get(char *v)* to send and receive 9-frame, 9-bit frames, in which the 9th bit is even parity for the row, and an even parity is developed for each column, at 4800 baud, and perform single-bit error correction using the 2-dimensional parity protocol. *main()* initializes UART1, sends, and then receives these 9-frame, 9-bit frames. *get* corrects any single error by observing which row and column position has a parity error.

14. Rewrite §8.3.4's class *UARTi* as a derived class of *Port* that works with §7.2's real-time interrupt scheduler, so that member functions *get* and *put* go to sleep indefinitely when the input queue is empty or output queue is filled, and saves the thread number, so that when a character arrives in the UART's input queue, or there is room to put data in the output queue, the thread is awakened by setting its sleep time to zero.

15. Write a class *UARTdd* as a derived class of *Port* that uses §8.3.5's device driver to actually perform the I/O. User calls to constructor, destructors, or member functions of *UARTdd* will be used for all I/O operations.

16. Rewrite §8.3.5's device driver to use interrupt synchronization. The input device should cause an interrupt whenever one character is present in its buffer. In the interrupt handler, each character should be checked for control-D (0x4) sending a signal to activate the debugger immediately when it arrives, rather than pushing it on a queue.

17. Rewrite §8.3.5's device driver to use software handshake. Each time the output interrupt occurs, when the input buffer is more than 3/4 fill and XOFF (0x13) has not been sent, XOFF is sent, when the input buffer is less than 1/4 fill and XON (0x11) has not been sent, XON is sent. 18. Write a C procedure main(), using EPDR bit 0 for clock and bit 1 for data input (determinate when the clock rises), and a gadfly loop to receive 100 bits and compute the CRC check value for the polynomial $X**16 + X**15 + X**2 + 1$. When is data valid?

18. Write a C procedure `main()`, using `EPDR` bit 0 for clock and bit 1 for data input (determine when the clock rises), and a gadfly loop to receive 100 bits and compute the CRC check value for the polynomial $X**16 + X**15 + X**2 + 1$. When is data valid?.

19. Write a real-time C procedure `main()` to initialize the devices and output the stream of data bits in `char buffer[0x100]`, most significant bit of lowest-addressed word first, checking for five T's and inserting F, as in the SDLC protocol. Send the data at 100 baud, gadflying on timer device 0 to time the bits. The clock is sent on *KCD* bit 0, and data are sent on *KCD* bit 1 (to be determinate when the clock rises).

20. Write a gadfly routine to handshake on the IEEE-488 bus. Data can be read or written in *KCD*, and *DAV*, *RFD*, and *DAC* are *KRD* bits 2 to 0, respectively.

   a. Show a C procedure `send(char i)` to initialize the ports, send `i`, and perform the handshake for a transmitter (talker).

   b. Show a C procedure `receive()` to initialize the ports, perform the handshake, and return the received word (listener).

21. Write gadfly C procedures to initialize the ports and handshake on the SCSI bus. Eight-bit data can be read or written in *KCD* (ignore parity), and I/O, REQ, and ACK are *KRD* bits 2 to 0, respectively. When not in use, all control lines should be high.

   a. Show `initateS(char c)` to send `c` from an initiator through its SCSI device.

   b. Show `targetR();` to return the data obtained from the SCSI device in a target.

   c. Show `targetS(char c)` to send `c` from a target through its SCSI device.

   d. Show `initateR()` to return the data obtained from the SCSI device in an initiator.

10

Display and Storage Systems

The previous chapter discussed the techniques by which microcomputers can communicate with other computers. They may also have to communicate with humans using LCD displays covered in Chapter 5 or using more complex CRT displays. We now cover CRT display technology. Also, a microcomputer may have to store data on a magnetic tape or disk. This stored data can be used by the microcomputer later or it may be moved to another computer. Thus, on an abstract level, a magnetic storage medium can be an alternative to an electronic communication link.

This chapter covers both the CRT display and the magnetic storage device. We discuss display systems first and then storage systems. In display systems, we will use a single-chip MMC2001 with only an additional transistor and its resistors etc., to implement a primitive device. We follow this with a small bit-mapped display. In storage systems, we use a special chip, the Western Digital WD37C65C, to implement a very useful floppy disk controller. We then illustrate a ZIP-100 disk SCSI interface. First, using the surprisingly powerful MMC2001 alone lets us show the principles of these devices and provides the opportunity to experiment with them without much expense. However, the special-purpose floppy disk controller and SCSI interface are quite easy to use and to be designed into real systems.

In this chapter, we spend quite a bit of time on the details of video and disk formats. We also present some rather larger system designs and refer to earlier discussions for many concepts. We have somewhat less space for the important notion of top-down design than in previous chapters because the design alternatives for CRT and disk systems are a bit too unwieldly to include in this short chapter. They are important, nevertheless, and are covered in some of the problems at the end of the chapter.

Upon completing this final chapter, you should have gained enough information to understand the format of the black-and-white National Televeision System Committee (NTSC) television signal and implement a single-chip or a bit-mapped CRT display. You should have gained enough information to understand the floppy disk and be able to use a floppy disk controller chip or SCSI interface to a ZIP drive, to record and play back data for a microcomputer, or use it to move data to or form it from or to another computer. Moreover, you will see a number of fairly complete designs similar to those you will build.

10.1   Display Systems

A microcomputer may be used in an intelligent terminal or in a personal computer. Such systems require a display. Any microcomputer, requiring the display of more than about hundred digits an LED or LCD display can handle, can use a CRT display.

This section describes CRT display systems concepts. We present the format of the NTSC black-and-white signal and then show a program that enables the MMC2001 to display a checkerboard block on a white screen. We then illustrate a useful bit-mapped display, and show object-oriented functions for elementary graphics applications.

10.1.1   NTSC Television Signals

A *National Television System Committee* (*NTSC*) signal is used in the United States and Canada for all commercial television. A computer display system consists of the CRT and its drive electronics — essentially a specialized TV set — and hardware and software able to send pulses to time the electron beam, which is a stream of bits to make the TV screen black or white at different points. Figure 10.1 diagrams the front of a TV screen. An electron beam, generated by a hot cathode and controlled by a grid, is deflected by electromagnets in the back of the CRT and made to move from left side to right side and from top to bottom across the face of the CRT. More electrons produce a whiter spot. The traversal of the beam across the face is called a *raster line*. The set of raster lines that "paint" the screen from top to bottom is a field. NTSC signals use two fields, one slightly offset from the other, as shown in Figure 10.1a, to paint a picture *frame*.

In NTSC signals, a frame takes 1/30th s and a field takes 1/60th s. The raster line takes 1/15,750th s, a field has 262 1/2 raster lines and a frame has 525 raster lines. As the beam moves from side to side and from top to bottom, the electron beam is

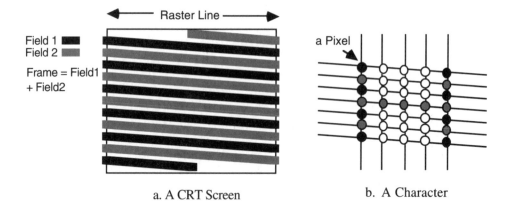

a. A CRT Screen   b. A Character

Figure 10.1. The Raster Scan Display Used in Television

controlled to light up the screen in a pattern. A *pixel* is the (smallest controllable) dot on the screen; a clear circle represents a pixel having no light, and a dark circle (black for field 1 and gray for field 2) shows a lighted pixel.

Figure 10.1b shows how H is written in both fields of a frame. ASCII characters will be painted in a 7- by 12-pixel rectangle, 80 characters per line (Figure 10.2).

The *NTSC composite video signal* is an analog signal, diagramed in Figure 10.3a. The displayed signal is an analog signal where a maximum voltage (about 1/2 V) produces a white dot, a lower voltage (3/8 V) produces gray, and a lower voltage (1/4

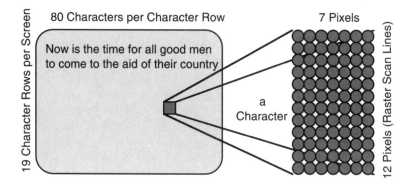

Figure 10.2. Character Display

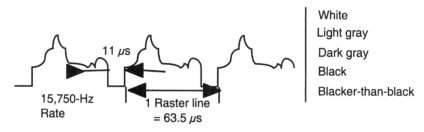

a. Video Signal and Sync Levels

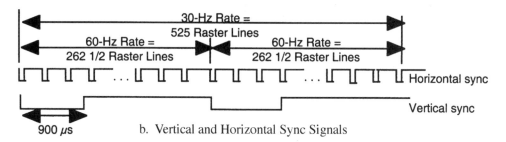

b. Vertical and Horizontal Sync Signals

Figure 10.3. The Composite Video Signal

V) produces a black dot. The part of the signal corresponding to the time when the electron beam is moved to the left side (*horizontal retrace*) or to the top (*vertical retrace*) occurs between the displayed parts. At these times, *horizontal sync* and *vertical sync* pulses appear as lower voltage (0 V) or "blacker-than-black" pulses. See Figures 10.3b. A CRT uses a *sync separator* circuit to extract these pulses so it can derive the horizontal and vertical sync pulses, which are used to time the beam deflections on the screen. This signal is called the composite video signal because it has the video signal and the sync signals composed onto one signal. If this signal is to be sent over the air, it is modulated onto a radio frequency (r.f.) carrier (such as channel 2). Alternatively, the separate video, horizontal, and vertical sync signals can be sent over different wires to the CRT system, so they do not need to be separated; this gives the best resolution, such as is needed in 1024 by 1024 pixel CRT displays in engineering workstations. The composite video is used in character displays that have 80 characters per line. The r.f. modulated signals are used in games and home computers intended to be connected to unmodified home TV sets, but are capable of only ~ 51 characters per line.

The frequency of the vertical sync pulses, which corresponds to the time of a field, is generally fixed to 60 Hz, to prevent AC hum from making the screen image have bars run across it, as in inexpensive TVs. It is also about the lowest frequency at which the human eye does not detect flicker. American computer CRTs often use this vertical sync frequency. The horizontal sync frequency in computer CRTs is usually about 15,750 Hz, as specified by the NTSC standard, but may be a bit faster to permit more lines on the screen yet keep the vertical sync frequency at 60 Hz. The magnetic beam deflection on the CRT is tuned to a specific frequency, and the electronics must provide horizontal sync pulses at this frequency, or the picture will be nonlinear. The pulse widths of these horizontal and vertical pulses are specified by the electronics that drive the CRT. Thus, the CRT controller must be set up to give a specific horizontal and vertical frequency and pulse width, as specified by the CRT electronics.

10.1.2   An MMC2001 ISPI Display

We are fortunate that the MMC2001 has a built-in counter and shift register able to generate the synchronization pulses and the bit stream to implement a primitive CRT display. The MMC2001 pulsewidth modulator, described in Chapter 7, is capable of generating the vertical and horizontal sync pulses and Chapter 5's Interval (Mode) Serial Peripheral Interface (ISPI), has the capability of generating a CRT display having poor, but useful, resolution. The upcoming C procedure *main( )* should produce a checkerboard pattern as shown in Figure 10.4, using the simple hardware diagrammed in Figure 10.5 with a single-chip MMC2001. It is quite useful for explaining the principles of CRT display systems, because it uses familiar MMC2001 peripherals. It might be useful for multicomputer systems as a diagnostic display available on each microcomputer. We have found it useful in testing some "bargain basement" CRTs when we did not have specifications on the permissible range of horizontal and vertical sync pulse widths and frequencies. This little pro-

Figure 10.4. Screen Display

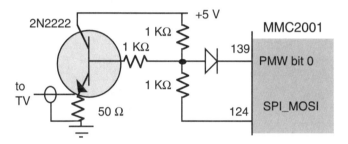

Figure 10.5. Circuit Used for TV Generation

gram lets us easily test these systems to generate the specifications. We now describe how MMC2001's "built-in" CRT generator can produce a CRT display.

A combined sync signal is generated which is the exclusive-OR of the vertical and horizontal sync signals. The CRT's sync separator outputs its high-frequency component to the horizontal oscillator and a low-frequency component to the vertical oscillator. By inverting the horizontal sync signal during vertical retrace, the signal's low-frequency component has a pulse during this period. The high-frequency output of the sync separator continues to synchronize the horizontal oscillator during vertical retrace, while the low-frequency component synchronizes the vertical oscillator during vertical retrace.

Figure 10.5 shows a simple circuit for the generation of composite video. If your CRT requires separated video and sync signals, the combined sync can be taken directly from pin 139 and the data from pin 124, respectively, and the circuit diagrammed in Figure 10.5 is not needed. The program is just a pair of PWM square-wave generator programs, as described in §7.1.1, with an ISPI routine. A stand-alone PWM0 generates the horizontal sync pulse. A PWM1 interrupt *handler* generates the output signal to display a checkerboard, shown in Figure 10.4, by sending a pattern to the ISPI. Using a primitive "character generator" between row *VPOS* and row *VPOS+12*, an element of *char pattern[12]* is put into the SPI data port, which is the video signal shifted out. *VPOS* determines which row the pattern appears int. PWM1 is started just a little after PWM0, and has the same period. This delay, *HPOS*, positions the output pattern, relative to the horizontal sync pulse, which determines in which column the checkerboard appears. *handler* sets and clears

PWM0's *POL* bit to invert its outputs during the vertical sync time, which combines
the horizontal and vertical sync on pin 139.

```
#define HPOS 127 * 2 /* horizontal location of square - half way across screen */
#define VPOS 260 /* vertical location of square */
enum { PWM_IRQ = 0x400, IRQ_EN = 0x200, LOAD = 0x100, DATA = 0x80, DIR
    = 0x40, POL = 0x20, MODE = 0x10, COUNT_EN = 8};

short lineNo;
char pattern[12] =
    {0x55,0xaa,0x55,0xaa,0x55,0xaa,0x55,0xaa,0x55,0xaa,0x55,0xaa};

interrupt void handler(){short i; /* read contol-status to clear int. */
    if((lineNo > VPOS) && (lineNo <= (VPOS + 12)))
        { SPDR = pattern[lineNo - VPOS]; }
    i= PWMCR1; // clear interrupt
    lineNo += 2; // interlaced display, output every other line
    if (lineNo == 524) // generate Hsync by altering polarity of Vsync output
        { lineNo = 1; while(PWMCTR0 < (127*2)) ; PWMCR0 |= POL; }
    else if (lineNo == 525) { lineNo = 0; PWMCR0 |= POL; }
    ;else if (lineNo == 13) { while(PWMCTR0 < (127*2)) ;PWMCR0 &= ~POL;}
    else if (lineNo == 12) PWMCR0 &= ~POL; // terminate horizontal sync
}
main(){ disableInt();
    PWMPR1 = PWMPR0 = 127 * 4; /* hor. period = 63.5 µs * 8 */
    PMWWR0 = (127 * 4) - 88; /* Hsync pulse width is 11 µs * 8 */
    PMWWR1 = 127*2; /* pwl duty cycle: can be anything < 127 * 4 */
    PWMCR0 = MODE + COUNT_EN; /* pwm0 to produce horizontal sync */
    while(PWMCTR0 < HPOS) ; // delay horizontal position
    PWMCR1 = IRQ_EN + MODE + COUNT_EN; /* start pwm 1 */
    NIER |= (1 << 11); /* enable normal interrupt for PWM device 1 */
    SPCR = 0x4807; /* spi for 8 bits, no interrupt */
    initInt(0x30000000, 0x30000200, handler, 10); enableInt();
    do ; while(1);
}
```

The use of two PWMs removes jitter from the display. One generates the horizontal sync pulse while the other, by means of the handler, feeds the ISPI to generate
the video signal. The two PWMs keep in sync because the hardware increments their
counters simultaneously and they have the same period. Without sync, the screen
image would appear torn and would "dance" around because the interrupt would
start on completion of the current instruction in any of several cycles after an output
compare occurred, and this would result in a mismatch between the timing of the
sync signal and the data signal.

For a 7 by 12 pixel character form with 1 pixel between characters, data is
shifted at 4 Mbits/s, so about 20 characters can be put on a line. The image is
interlaced, so 40 lines of characters appear on the screen. While the pattern we used

is a checkerboard, a different pattern can be written for each letter, and the pattern can be chosen by the ASCII representation of the letter. Thus, characters in words can be written across the screen to implement a useful display. Lines can also be drawn, to draw rectangles and other geometric figures. However, the SPI's shift rate is too slow to generate finer pixels needed to represent letters more satisfactorily. What is mildly surprising, though, is that the MMC2001, with very little external hardware, has the ability to generate marginally useful CRT signals. Motorola can therefore claim that the MMC2001 has a built-in CRT controller.

10.1.3 Bit-Mapped Display

Whereas the previous section showed how to output a single character on the screen at a location that can be selected at random, this section essentially expands the same technique to handle a window for a window display class introduced in the next section. MMC2001's PWM-ISPI system can have an ISPI clock as fast as 4 MHz, which outputs a bit every 250 ns. To make the pixel fairly square, we output the same data for four consecutive lines. Therefore the window display will be 160 horizontal by 120 pixels vertical. The SRAM's size is sufficient to hold a bit image for this window.

A more realistic bitmappled display can be achieved using an external 16-bit shift register in place of the SPI (Figure 10.6). However, the MMC2001's PWM-ISPI system provides a bit-mapped display that adequately teaches how software can write into this bit-mapped display to draw various figures and characters. This poor-

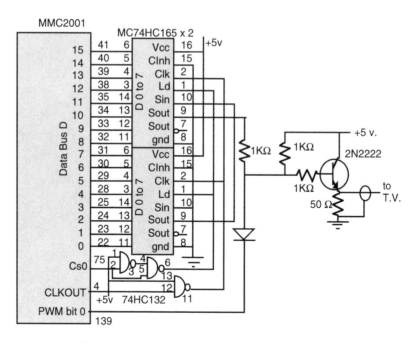

Figure 10.6. Hardware for a More Realistic Display

resolution display will be used in the next section to make concrete the ideas of graphics processing.

```
#define HPOS 10      /* horizontal left location of rectangle */
#define VPOS 80      /* vertical top location of rectangle */
#define WIDTH 10       /* rectangle horizontal width, in 16-bit words - 160 pixels */
#define HEIGHT 120    /* rect vert height, in multiples of 4 scan lines - 480 lines */
enum { PWM_IRQ = 0x400, IRQ_EN = 0x200, LOAD = 0x100, DATA = 0x80,
DIR = 0x40, POL = 0x20, MODE = 0x10, COUNT_EN = 8};
short lineNo, dummy, buffer[HEIGHT][WIDTH], *bPtr = buffer;
#pragma NO_EXIT
void scanout(short **ptr){asm{
run: lrw     r7,[portAddr] /* get current buffer pointer *
     movi    r6,9            /* 10 16-bit outputs */
     ld.w    r5,(r2,0)      /* get pointer */
     setc
run1: ld.h r4,(r5,0)      /* 2 get bytes from memory */
      st.h r4,(r7,0)       /* 2 put into ispi */
      addi r5,2            /* 1 next word */
      mfcr r3,cr0          /* 1 save condition for loopt */
      movi r4,58           /* 1 wait for (128 - (10 + 2))/2 clocks */
w1: loopt r4,w1           /* 2 less number of clocks in rest of program */
    mtcr r3,cr0           /* 1 restore condition for outer loopt */
    loopt r6,run1         /* 2 repeat for line */
    jmp r15
portAddr: dc.w 0x10008000;    /* location of SPI shift register */
}}
interrupt void handler(){short i;    /* read contol-status to clear int. */
    i = PWMCR1;    /* clear interrupt */
    bPtr = buffer + ((lineNo - VPOS) >> 2);
    if((lineNo > VPOS) && (lineNo <= (VPOS + (HEIGHT * 4))))
        scanout(&bPtr); else SPDR = 0;
    lineNo += 2; // interlaced display, output every other line
    if (lineNo == 524) { // generate Hsync by altering polarity of Vsync output
        lineNo = 1; while(PWMCTR0 < (127 * 2)) ;
        bPtr = buffer + WIDTH; PWMCR0 |= POL;
    }
    else if (lineNo == 525)
        { lineNo = 0; bPtr = buffer; PWMCR0 |= POL; }
    else if (lineNo == 13)
        { while(PWMCTR0 < (127 * 2)) ; PWMCR0 &= ~POL; }
        else if (lineNo == 12) PWMCR0 &= ~POL; // terminate horizontal sync
}
```

```
main(){ short i, j;
     disableInt(); initInt(0x30000000, 0x30000200, handler, 10);
     PWMPR1 = PWMPR0 = 127 * 4;   /* hor. period = 63.5 μs * 8 */
     PMWWR0 = (127 * 4) - 88;   /* hor sync pulse width is 11 μs * 8 */
     PMWWR1 = 127 * 2;   /* pwl duty cycle is half of width */
     PWMCR0 = MODE + COUNT_EN;   /* pwm0 to produce horizontal sync */
     while(PWMCTR0 < HPOS) ;   // delay horizontal position
     PWMCR1 = IRQ_EN + MODE + COUNT_EN; /* start pwm 1 */
     SPCR = 0x480f; // spi for 16 bits
     enableInt(); NIER |= (1 << 11);   /* enable normal int for PWM 1 */
     display *S = new display((short *)buffer); S->point(30, 20);
     S->line(5, 10, 1, 1, 5); S->triangle(30, 40);
     for(j = 0; j < 10; j++) for(i = 0; i < 13; i++)
       S->put(patternA, i * 9, j * 10);
     do ; while (1); delete S; // asm trap #3
}
```

The above C program is essentially the same as that used in §10.1.2 except that an assembler language subroutine *scanout* outputs ten 16-bit words from memory through the ISPI to send bits from the screen buffer to write a horizontal line of the display. Assembly language real-time programming is used in the procedure *scanout* to output a display line. A 16-bit word should be output in 128 memory cycles, but two extra memory cycles are required between outputting consecuitve 16-bit words, so that at every 16th pixel along a row is stretched by 25%. Using the gadfly or interrupt techniques to feed data to the ISPI yields longer times between outputting 16-bit words, which results in wider gaps in the display.

10.1.4 An Object-oriented Display

The previous section's program displays the bit-mapped image stored in the frame buffer, *short buffer[HEIGHT][WIDTH]*. In this section, we illustrate an object-oriented class *display* that can draw geometric figures and characters in this frame buffer.

```
const unsigned char
    patternA[8]={0x10,0x28,0x44,0x82,0xFE,0x82,0x82,0x82};
const short points[16] = {0x8000, 0x4000, 0x2000, 0x1000, 0x800,
    0x400, 0x200, 0x100, 0x80, 0x40, 0x20, 0x10, 8, 4, 2, 1};
class display{ short *dPtr;
    short inScreen( unsigned short h, unsigned short v )
      { return (v < (WIDTH * 16)) && (h < HEIGHT ); }

    unsigned short *address( unsigned short h, unsigned short v )
      { return (unsigned short *)dPtr + (v * WIDTH) + (h >> 4); }

    public: display(short *ptr) { dPtr = ptr; /* constructor */ }
    virtual void point(unsigned short h, unsigned short v)
      {if(inScreen( h, v)) *address(h, v) |= points[h & 0xf];}
```

```
virtual void line(unsigned short h, unsigned short v,
   unsigned short dh, unsigned short dv, unsigned short n)
      { /*point( h, v);*/ while(--n) point( h += dh, v += dv); }

virtual void triangle(unsigned short h, unsigned short v) {
   line(10 + h, 10 + v, 2, 1, 20); line(10 + h, 10 + v, 1, 2, 20);
   line(50 + h, 28 + v, -1, 1, 20);
}

virtual void put(const unsigned char *pattern, unsigned short h,
   unsigned short v) {
   unsigned char row; unsigned short *p, hi, lo;
   unsigned long bits;
      if(!inScreen(h, v) || !inScreen(h + 8, v + 8)) return;
      for(p=address(h, v), row=0; row<8; row++, p += WIDTH) {
      bits = pattern[row] << (24 - (h & 0xf));
   }
}
};

main(){ short i, j;
   display *S = new display((short *)buffer); S->point(30, 20);
   S->line(5, 10, 1, 1, 5); S->triangle(30, 40);
   S->put(patternA, i * 9, j * 10);
   do ; while (1);
}
```

The class' constructor merely sets *dPtr* to the location of the frame buffer. The function member *point(int h, int v)* sets a bit to display a point at horizontal row *h* and vertical pixel column *v* provided the point is inside the display area as determined by *inScreen()*. The vertical coordinate and the horizontal co-ordinate's high-order bits determine which 16-bit word in the buffer is to be changed, and the horizontal coordinate's low order bits determine which bit in that word is to be changed. This function member sets the bit chosen by the function's parameters; other variations of it can clear or complement the indicated bit.

The function member *void line(int h, int v, int dh, int dv, int n)* draws an *n*-point line from point *h, v*. Each time a pixel is drawn, it adds *dh* to *h* and *dv* to *v*. This simple algorithm is only suitable for lines where increment *dh* or increment *dv* is one, and the other increment is between -2 and +2. It can draw a rectangle (see Problem 8) and some triangles. The *Bresenham algorithm* is commonly used for general line drawing (see Problem 9). The basic algorithm works only for the octant in which both *dh* and *dv* are positive, and *dh* is greater than *dv*. As points are drawn in consecutive columns, the algorithm keeps track of an error *e* whose sign indicates whether a point should be drawn in the same row as the last point or the next higher numbered row. The calculation of *e* is based on the line differential equation and is explained by most textbooks on computer graphics. The general algorithm determines which octant a line is in, and calls the function member that implements the basic algorithm with operands interchanged or negated as needed, so that the basic algorithm operates in its preferred octant. A variation of this algorithm can draw ellipses and circles. The function member $triangle(int \; h, \; int \; v)$ draws a triangle with upper left vertex at h, v. A more general triangle can be drawn using the Bresenham line drawing algorithm (see problem 10).

The function member $put(char \; *pattern, \; int \; h, \; int \; v)$ writes a character whose pattern is defined by the vector $pattern$ (such as $patternA$ above) so its upper left pixel is at h, v. If the character is entirely within the display area, the pattern is ORed into the buffer a row at a time, at a 16-bit word offset determined by v and the high-order bits of h, using a shift offset determined by the low order bits of h. This function member is suited only for characters whose maximum width is 8 pixels. A slightly more general function member put can draw characters whose maximum width is 16 pixels (see problem 11). Calling put with different character patterns and offsets can write words on the screen (see problem 12). A more general function member could write null-terminated strings of characters on the screen, keeping track of the position of the last drawn character as a data member so that the calling routine need not pass this parameter to the function member.

The routines for drawing lines and characters are described here as object-oriented function members for a class $display$. While this example does not seem to warrant the use of object-oriented programming, a simple extension of this class will utilize object-oriented capabilities. Consider the display of multiple separate *windows*, each of which occupy a separate portion of the buffer and therefore of the screen. The class constructor can have an origin, which initializes the data member $dPtr$, and a horizontal and vertical range of pixels, as arguments, and points and characters within that range can be drawn using the offset indicated by $dPtr$. Overlapping windows are drawn from the farthest to the nearest windows, and later drawn windows overwrite the earlier drawn windows. Each window has its own horizontal and vertical axis, and when a window is moved, by modifying $dPtr$, all the line and text items are drawn relative to the new origin.

This class of graphics objects can be significantly improved and extended. Rather than drawing each window from farthest to nearest, portions of windows that will be overwritten can be *clipped*. The windows can be linked in a hierarchy so that if a parent window is moved its offspring will move. This class of graphics objects is essentially what is used in the Macintosh and Microsoft Windows operating systems. Graphics is one of the most common applications of object-oriented programming.

10.2 Storage Systems

Most microcomputers require either a communications system to move data into and out of them or a storage system to get data from or to save data in. The latter, called *secondary storage*, generally uses some form of magnetic medium. Floppy disk systems have become so inexpensive that they are likely to be used in many microcontrollers. This section describes techniques for data storage on floppy disks. We discuss a floppy disk format, then use a Western Digital WD37C65C chip, which is particularly easy to interface to the MMC2001, to show a floppy disk interface and an object-oriented class to read and write files in a 3 1/2" PC disk.

10.2.1   Floppy Disk Format

We now describe the 3 1/2" double-density floppy disk *format*. Data can be stored on
the disk using either of two popular formats. Figure 10.7 shows how a bit and a byte
of data can be stored on a disk, using FM (single density) and MFM (double-
density) formats. The FM format is just Manchester coding, as introduced in §10.1.
Figure 10.7a shows a bit cell and Figure 10.7c shows a byte of data, in the FM
format. Every 8 µss there is a clock pulse. If a 1 is sent, a pulse is put in between the
clock pulses, and if a 0 is sent, no pulse is put between the clock pulses. MFM format
provides half the bit cell size as FM format; it does this by using minimal spacing
between pulses in the disk medium: MFM format has at most one pulse per bit cell.
It is thus called "double density" storage. The idea is that a 1 cell, which has a pulse
in it, doesn't need a clock pulse, and a 0 cell only needs a clock pulse if the previous
cell is also a 0 cell. Figure 10.7b shows a byte of data in the MFM format. High
density merely doubles the density for the MFM format. For the remainder of this
section, we discuss the "high-density" MFM format. Every 2 µss there is a data bit.
If a 1 is sent, a pulse is put near the end of the bit time; if a 0 is sent after a 1, no pulse
is put between the clock pulses; and if a 0 is sent after a 0, a pulse appears early in the
bit time. Note that data must be read or written at the rate of 1 byte per 16 µs, which
is 128 memory cycles for the MMC2001 using a 16-MHz CLK clock.

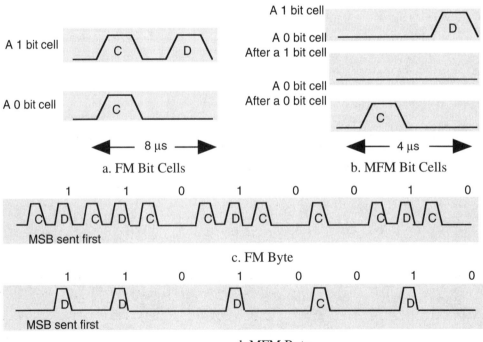

Figure 10.7. Bit and Byte Storage for FM and MFM Encoding

Data read from the disk are separated by a phase-locked loop (PLL), which synchronizes to the bit cell rather like a flywheel. Once the bit cell is locked on to, the data bits can be extracted from the input analog signal. The PLL must be designed to lock into the bit cells within 48 bit cell times.

A disk drive may have one or more disks, stacked pancake-style, and each disk may have one or two *surfaces*. Figure 10.8a shows a surface of a disk; a *track* is shown, and tracks are numbered — track 0 on the extreme outside, and track $i + 1$ next towards the center to track i. The track spacing density is the number of tracks per inch and is generally 48 or 96 tracks per inch. Floppy disks have diameters of 8", 5 1/4", or 3 1/2", and these typically have 77, 35, and 80 tracks, respectively. Although disks exist which have a head on each track, generally disks have a single head per surface — used to both read and write the data on that surface — that is moved by a stepper motor to a track that is to be read or written. In a multiple

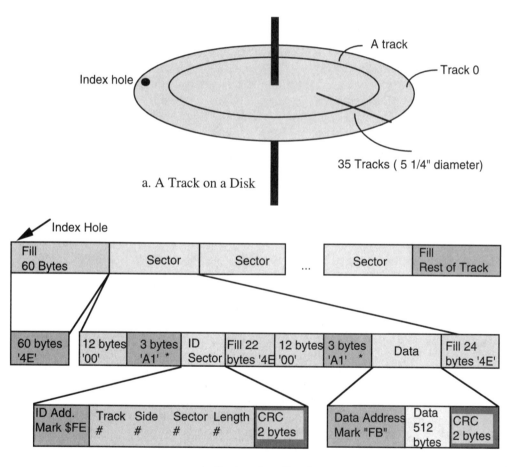

a. A Track on a Disk

b. Sectors in a Track

Figure 10.8. Organization of Sectors and Tracks on a Disk Surface

surface disk, the same tracks are accessed on each surface by a comb-like mechanism holding the read-write head for each surface; the collection of tracks accessed at the same time is called a *cylinder*. We soon describe an example of a single-sided 3 1/2" disk's track format. The formats for other types of disks are similar.

Relative to later discussions of the operation of the floppy disk controller, timing of head movements significantly affects the disk system's performance. The *step rate* is the rate at which the stepping motor can be pulsed to move the head from track i to track $i+1$ (step in) or to track $i-1$ (step out). There is also a *settling time*, which is the time needed to allow the mechanical parts to stop bouncing (see contact bounce in §6.2.7.1). Floppy disk drives have stepping rates from 2 to 30 ms and settling times of about 30 ms. If a drive has a 3-ms stepping rate and a 30-ms settling time, the time to move from track i to track j is $3*|i-j| + 30$ ms. The average time to position from track 0 to a random track is the time to move over half of the (80) tracks of the disk. There is some additional time needed to get to the data on the track, as discussed soon. Thus, on the average, about 80 ms would be used to move the head, and no data is transferred during that time.

The problem with a disk is that to record data, a head must be energized, and the process of energizing or deenergizing a head erases the data below the head. The track is thus organized with *fill* areas where data are not stored and where the head may be energized to begin or deenergized to end writing, and the data between these fill areas, called (disk) *sectors*, are written in their entirety if they are written at all. A disk's indivisible storage objects thus are sectors. Figure 10.8b shows the breakdown of a typical track in terms of sectors and the breakdown of a sector in terms of its ID pattern and data. (Later, we discuss a "logical sector"; when we need to distinguish a "logical sector" from what we describe here, we call this a "disk sector.") There is an *index hole* on the disk (Figure 10.8a) that defines a track's beginning; it is sensed by an optical switch that provides an *index pulse* when the hole passes by the switch. The track first contains a 60 byte fill pattern. (Each fill pattern is 0x4E.) There are then 18 disk sectors on each track, as described soon. The remainder of the track is filled with the fill pattern.

Regarding the timing of disk accesses, after the head moves to the right track, it may have to wait 1/2 revolution of the disk, on the average, before it finds a track it would like to read or write. Since a floppy disk rotates at 10 revolutions per second, the average wait would be 50 ms. If several sectors are to be read together, the time needed to move from one track to another can be eliminated if the data are on the same track, and the time needed to get to the right sector can be eliminated if the sectors are located one after another. We will think of sectors as if they were consecutively numbered from 0 (the *logical sector number*), and we will position consecutively numbered sectors on the same track, so consecutively numbered sectors can be read as fast as possible. Actually, two consecutively read disk sectors should have some other sectors between them because the computer has to process the data read and determine what to do next before it is ready to read another sector. The number of disk sectors actually physically between two "consecutively numbered" logical sectors is called the *interleave factor*, and is generally about four.

We need to know which disk sector is passing under the head as the disk rotates, since sectors may be put in some different order, as just described, and we would also like to be able to verify that we are on the right track after the head has been moved. When the read head begins to read data (it may begin reading anywhere on a track), it will examine this address in an ID pattern to find out where it is.

There is a small problem in identifying the beginning of an ID pattern or a data field when the head makes contact with the surface and begins to read the data on a track. To solve this, there is a special pattern whose presence is indicated by the deletion of some of the clock pulses that would have been there when data are recorded in MFM format, and there are identifying patterns called the ID address mark and data address mark. The special pattern, shown in Figure 10.9, is said to have a data pattern of 0xA1 and a missing clock pulse between bits 4 and 5. The ID address mark 0xFE is used to locate the beginning of an ID pattern on a track. The data address mark similarly identifies the beginning of data in the sector but is 0xFB rather than 0xFE.

The ID pattern consists of a 1-byte ID address mark (0xFE), a track number, side number, sector number and sector length (each is 1 byte and is coded in binary), and a 2-byte CRC check. The track number, beginning with track 0 (outermost), and the sector number, beginning with either sector 0 (zero-origin indexing) or 1 (one-origin indexing), is stored in two of the bytes. A simple method of mapping the logical sector number into a track and zero-origin indexing disk sector number is to divide the logical sector number by the number of sectors per track: the quotient is the track number, and the remainder is the sector number. The side number for a single-surface drive is 0, the sector length for a 256-byte sector is 1, and the sector length for a 512-byte sector is 2.

A sector is composed of a pattern of 12 zeros and 3 bytes of 0xA1 (in which a clock pulse is missing), followed by an ID pattern as just described, a 22-byte fill, and another pattern of 12 zeros, followed by 3 bytes of 0xA1 (in which a clock pulse is missing). The 512 bytes of data are then stored. The ID pattern and the data in a sector have some error detection information called a CRC, discussed in §9.4.1, to ensure reliable reading of the data. The track may have a total capacity of about 13 K bytes, called the *unformatted capacity* of the track, but because so much of the disk is needed for fill, ID, and CRC, the *formatted capacity* of a track (the available data) may be reduced to 9216 bytes.

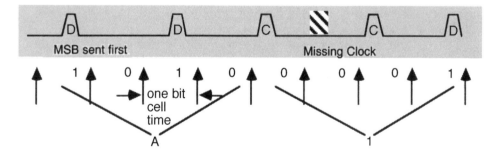

Figure 10.9. A Special Byte (data = 0xA1, clock pulse missing between bits 4,5)

The *format* of a disk is the structure just described, disregarding the content of the data field. To *format* a disk is to write this structure. Later, when data are written, only the data part of the sector, together with its data address mark and CRC check, are written. The ID pattern is not rewritten. If it is altered, data cannot be read or written because the controller will be unable to find the id part of the sector to locate the sector.

10.2.2   The Western Digital 37C65C Floppy Disk Controller

We now examine a hardware-software system for reading and writing a double-sided high-density (HD) 3 1/2" floppy disk using the Western Digital WD37C65C chip (the '65C). We first catalog pin connections, and ports in the organization of the chip. Then we list the status ports and functions that the controller executes. In the next section, we will present and describe the software used to control the chip (See Figure 10.10a.)

The 34-conductor cable connects the controller chip to the drive. Grounded odd numbered pins reduce noise pick-up, and all signals are in negative logic. Motor controls MOT1 and MOT2 are asserted true (grounded) to make the motor run continuously in this example, although the '65C has means to control the motor. Output Step causes the drive's stepper motor to move to another cylinder, and Dirc specifies which direction to move the head. DS1 selects drive 1, DS2 selects drive 2, and Hs selects the head used. Wd is the write data, and We is the write enable. Input Rdd is the read data, write protect Wp is asserted if a disk tab is positioned to prevent writing in the disk, Tr00 is the track 0 signal, and Ix is the index pulse signal; inputs use a pull-up resistor.

Positive logic *KRD* signals communicate to '65C control and status pins. Bit 4 resets the '65C, bit 3 senses an "interrupt" that signals completion of a '65C operation, and bit 2 (TC) terminates counting, to stop an operation.

The '65C's organization has five ports, of which two read ports and two write ports are used herein (Figure 10.10b). The control port's two least significant bits specify disk density. The master status port has a request bit (Req) a data direction bit (Out) an execution phase bit (Exec) and five busy bits indicating the status of the control chip and up to four drives. The data port transfers data into and out of the '65C, and also sends commands into the device (as listed in Figure 10.10d) and gets more detailed status from the chip (by reading the status bytes St0, St1, St2 and St3 as shown in Figure 10.10c).

The '65C commands read and write sectors and perform auxilliary operations. (See Figure 10.10d.) The user may wish to read one or more bytes from the disk. Because it is only possible to read whole sectors, the sector or sectors that the data are at are read, and the data are extracted once the sectors are read. To read a sector, the user must *seek* the track first, then *read* the desired *sector*. The two commands, seek cylinder and read sector, are given to the floppy disk controller. Seek cylinder puts the read/write head in the disk drive over the desired cylinder, and the read sector command causes the data from a sector on the track to be read into primary memory. If the read/write head is definitely over the right cylinder, the seek com-

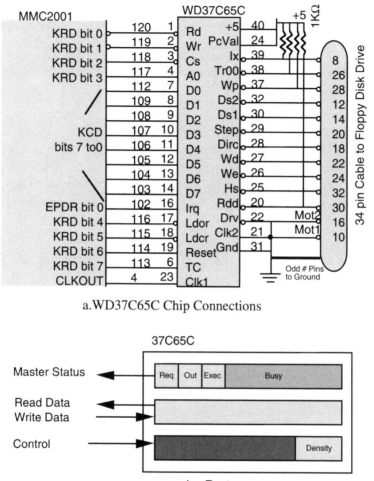

a.WD37C65C Chip Connections

b. Ports

Figure 10.10. The Western Digital WD37C65C

mand may be omitted. Also, in some floppy disk controllers having intelligence in them, the user only gives a read command regardless of where the read/write head is and an *implied seek* may be automatically generated as a result, without the user giving it if the head is in the wrong place.

The user may wish to write one or more bytes into the disk. To write a sector, the commands to seek cylinder and *write sector* are given as for the read operations above. Good programs often read a sector each time right after it is written to be sure there are no errors in writing the sector. A disk can be formatted by executing the *format* command on each track. Finally, when a disk is being initialized, the position of the read/write head must be moved to track zero (zeroed) to later establish how to move it to a desired track. This operation is called *restoring* or *recalibrating* the drive.

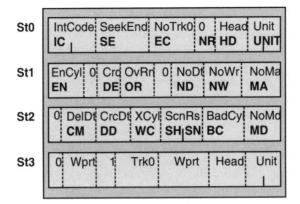

c. Status Port Fields

	Specify	Seek Cylinder		Read Sector	Write Sector	Format Track
Command	0000 0011 StepRate Unload LoadTime NoDMA	0 0 0 0 1111 xxxxx Hs Unit[2] Cylinder	Command	Mt Mf Sk 0 0110 xxxxx Hs Unit[2] C Cylinder H Head R Sector N Size EOT End GLP Gap DTL DLen	Mt Mf Sk 0 0101 xxxxx Hs Unit[2] C Cylinder H Head R Sector N Size EOT End GLP Gap DTL DLen	0 Mf 0 0 1101 xxxxx Hs Unit[2] Bytes/Sector Sectors/Track Gap Data
	Read Id	**Recalibrate**				
Command	0 0 0 0 1010 xxxxx Hs Unit[2] Results: See Read Sector	0 0 0 0 0111 xxxxx 0 Unit[2]	Execution	Read 2**(Size+7) Bytes	Write 2**(Size+7) Bytes	Supply (Sectors/Track)* Cylinder Head Sector Size
	Sense Drive	**Sense Status**				
Command	00000100 xxxxx Hs Unit[2]	0 0 0 0 1000	Results	St0 St1 St2 C Cylinder H Head R Sector N Size	St0 St1 St2 C Cylinder H Head R Sector N Size	St0 St1 St2 C Cylinder H Head R Sector N Size
Results	St3	St0 Cylinder				

d. Command Summary

Figure 10.10. (continued)

The '65C has additional commands. *Specify* writes the step rate, head load, and head unload times to initialize the '65C, *read id* will read the first valid id sector found on the track, *sense drive* will indicate the drive's status, and *sense status*, often used after an interrupt or a gadfly loop, will indicate the cause of the interrupt or termination of the loop. We will develop procedures to execute these functions and write an object-oriented class to handle these operations, as we did for other I/O devices.

10.2.3 Floppy Disk Interfacing Procedures

This section illustrates some simple software that implements the '65C commands discussed in the previous section. We begin with the innermost procedures used to write commands and to read status from the '65C. We then illustrate the initialization procedure for the controller chip, drive, and disk in the drive. Then we discuss procedures to read and write a sector on the disk.

A handshake protocol is used to issue commands and garner status information. This protocol checks the master status register and inputs and outputs through the data port. We give these declarations for the ports, for connections shown in Figure 10.10a and the map in Figure 10.10b. Comments indicate the use of the variables.

The innermost procedure *w* writes command bytes into the '65C, and the procedure *r* reads status bytes from the '65C.

```
char cs(void){ char s;   /* read master status */
    KRD &= -0xD; s = KCD; KRD = 0xBF; return s & 0xc0;
}
void w(char c){   /* write data */
    while( cs() !=0x80) ; KRD&=-6;KCDD=0xff;KCD=c;KRD=0xBF;KCDD=0;
}
char r(){ char d; /* read data */
    while (cs() != 0xc0) ; KDR |= 1; d = KDC ; KDR &= ~1; return d;
}
```

Note that before reading or writing, we gadfly on the master status register. Within the '65C, a microcontroller changes this status register when it is ready to move data through its data port.

The software will use error message numbers to report errors in a manner that assists the user in locating and fixing errors. These error numbers are declared as follows:

```
enum{illOp=1,chpErr,drvError,dskErr,addrErr,seekErr,rdErr,wrErr};
```

The initialization of the '65C and drive hardware is shown in the procedure *init65* below. This procedure initializes the MMC2001, resets the '65C, recalibrates the drive, and reads the ID of the disk in the drive. The MMC2001 is configured to access the '65C. We need an 8-bit data bus to connect the '65C, and we need CS0 and CS1 to enable the '65C's Rd and Wr pins. We set *KRDD* to output *KRD* control signals RESET and TC, then we pulse RESET high, then initialize RESET to L and TC to H. Compare each of the following program segments with the sequences shown in Figure 10.10d. The specify command sets up the '65C, and the recalibrate command sets up the drive hardware. The sense interrupt status command is used after recalibrate command, and will later be used after each seek operation, and the sense drive status command is used to verify the success of each operation. The last program segment which reads a sector I.D., uses global variables introduced after this procedure is described, so it will be discussed later.

```
char init65(char drive, char step, unsigned char unload, char load){int i;
    EPPAR=1; EPFR=1; KDDR=0xFF; KRD=0xBF; KRD=0xFF; KRD=0xBF;
    w(3); w(((-step) << 4) | (unload >> 4)); w(load | 1); /* specify */
    w(4); w(drive & 7); i = r(); /* sense drive status */
    if((i & 0xa0) != 0x20) return error = chpErr;

    w(7); w(drive & 3); /* recalibrate (restore) */
    while( ! (KRD & 8)) ; /* gadfly on drive being in execution mode */

    w(8); i = r(); C = r(); /* sense interrupt status */
    if(i != 0x20) return error = drvError;

    control=0; w(0x4a); w(0); while( ! (EPFR & 1)) ;EPFR=1;check();
    if((S[0] & 0xc0) != 0) return error = dskErr;
    return 0;
}
```

Procedures that operate on sectors use global data declared below. The variables C, H, and R are the first three bytes in an id sector field of a sector of the disk (Figure 10.8b). C is the cylinder number, which is the same as the track number. H, 0 or 1, indicates which head is used. R is the disk sector number to be accessed. *seek*, called by *read* and *write*, is passed a logical sector number (discussed in §10.2.1) and computes C, H and R. In disks formatted for the **IBM PC**, and used in the next section, sector numbers on a track begin with one rather than zero (one-origin indexing). Four status bytes $ST[4]$ are read from the controller (Figure 10.10c), *error* is nonzero when an error occurs, *verify* is 1 if we will read after writing a sector to verify it was saved. is Input data is put or output data is taken from Buffer B (*spacer* places B to be read in a memory dump at a 16-byte boundary, at location 0x810).

```
char  C,                    // cylinder
      H,                    // head
      R,                    // sector number on track
      ST[4],                // status
      error,                // nonzero indicates error
      verify = 1,           // TRUE means will verify
      spacer[7], B[512];    // buffer for sectors
```

The *seek* procedure, called up at the beginning of the *get* and *put* procedures, converts the logical sector number to head, track and disk sector numbers for the disk controller. It checks that the sector number is within range (a double-sided 80 track disk, with 18 sectors per track, has 2880 sectors), and then, dividing by the number of sectors per track, essentially obtains the disk sector number (remainder) and a combined head-cylinder number (quotient). Recall that in IBM disks, disk sector numbers begin with 1 so the remainder is incremented to get the disk sector number. The head-cylinder number is divided by two to get the disk cylinder number

C (quotient) and the head H (remainder). The '65C is given a seek command and the status register $S[0]$ is examined to determine if the seek was successful. If the drive is already on the desired track, the seek command completes quickly, but this command can take ms to execute.

```
char seek(int sectorNumber) {
    if(sectorNumber>2880)return error=addrErr;/* check for sec out of range */
    R=(sectorNumber%18)+1; /* get sector # (PC disks use one-origin indexing) */
    C=sectorNumber/18; /* get combined cylinder-head number, temporarily in C */
    H = C & 1; C >>= 1; /* separate into cylinder C (high bits) and head (lsb) */
    w(0xf); w(H << 2); w(C); /* give command to seek cylinder */
    while( ! (EPFR & 1)) ; /* gadfly on drive seeking cylinder */
    w(8); ST[0] = r(); /* get interrupt status */
    if((r()!=C)||((ST[0]&0xf8)!=0x20))return error=seekErr;
    return 0;
}
```

Procedures `get` and `put` input (read) a sector and output (write) a sector. Procedures `check` and `setup` are used in these methods to "factor out" some common code from them. `setup` writes the first eight parameters of a read or write command. `check` reads out the status information of the command result phase.

```
void setup(char cmd)
    {w(cmd); w(H<<2); w(C); w(H); w(R); w(2); w(18); w(0x1b); w(0xff);}

char check() {
    if((ST[0] = r()) & 0x80) { error = illOp; return 0; }
    ST[1] = r(); ST[2] = r(); C = r(); H = r(); R = r(); N = r();
    return (ST[1] & 0x7f) || (ST[2] & 0x33);
}

char get(int sector, register char *buffer) { int i; char *end;
    if(error||seek(sector))return 1;end=buffer+(i=512);KRD=0;
        setup(0x46);
    do {
        while(!(cs()&0x80)) ; if(!(cs()&0x20)) break; *buffer++=r();
        break;*buffer++=rData;
        } while( --i );
        if( check() || (buffer != end) ) return error = rdErr; return 0;
}
```

The procedure `get` reads a sector. It calls `seek`, which computes C, H, and R, and executes a seek command. First, during a *command phase*, nine bytes are written to the data port using the *setup* procedure. The command $0x46$ requests reading an MFM sector. Then the surface, which is the head and drive number, is sent. The cylinder, head, sector, and size bytes are sent exactly as they should appear in the sector's id (figure 10.8b), and the last sector, format gap size and data length are sent to complete the command phase. The *execution phase* reads each byte of the sector's data.

Asserting TC (*KRD* bit 4) terminates this phase. Finally, the *result phase* uses the *check* procedure to read each byte of the result. The status ports S[0] to S[2] are read, then the sector's cylinder, head, sector, and size bytes are read back exactly as they appeared in the sector's id (Figure 10.8b), so software can *verify* that the right sector was read.

The procedure *main ( )* below reads the first sector of a file that stores the floppy disk program's source code, which is logical sector 57.

void main ( ) { init65 (0, 3, 240, 16); get (57, B); }

The seek procedure calculates the cylinder number as 1, the head as 1, and the sector number as 4. Then a dump (figure 10.11) shows the data in logical sector 57.

Similarly, the *put* procedure writes a sector, and if *verify* is nonzero, the sector that was just written is read again using the *get* method to verify that it is properly stored and can likely be read later without errors. Rather than actually using *get* to verify the sector as in this example, which destroys *B,* a procedure exactly like *get* can be used, except that it writes read data into a dummy local variable. The CRC check verifies success. But reading the sector in the *verify* step slows down the writing of sectors two because a sector is written and a whole disk revolution later the same sector is read. Verification is indispensable when the data being written might be lost forever if it is not written correctly. But there are times when this verify step should be omitted. When copying a whole disk, by clearing the *verify* instance variable, this verify step is omitted to speed up writing of sectors. The destination disk is first fully written without verifying each sector, and then each sector on the disk is read just to verify it.

```
char put (int sector, register char *buffer) { int i;
    if (error || seek (sector)) return 1; i = 512;
    setup (0x45);
    do {
        while (!(cs ()&0x80)) ; if (!(cs ()&0x20)) break;w (*buffer++);
    } while ( --i );
if (i||check ()) {if (ST[1]&2) return error=wrProtErr;return
    error=wrErr; }
    if (verify) if (get (sector,B)) return error=wrErr;
    return 0;
}
```

```
      00 01 02 03 04 05 06 07 08 09 0A 0B 0C 0D 0E 0F
0810 23 64 65 66 69 6E 65 20 55 54 42 55 47 0D 23 69  #define UTBUG.#i
0820 6E 63 6C 75 64 65 20 3C 36 38 31 32 2E 68 3E 0D  nclude <MMC2001.h>.
0830 23 69 6E 63 6C 75 64 65 20 22 44 69 73 6B 2E 68  #include "Disk.h
```

Figure 10.11. File Dump

We deferred a discussion of *init65's* read id command, which reads any valid id sector. It uses the same status reporting mechanism that *get* and *put* to check the disk in the drive for readability.

From these examples, the reader should see that movement of data to or from the '65C always uses handshaking, such as gadflying on some master status bits. The read sector (also read id, write sector, and format) commands go through three phases: command where control values are sent to the chip, execution where data are read or written, and result where status is read from the chip. Other operations (sense drive, sense status) have only a command and a result phase, and the remainder (specify, recalibrate, and seek) have only a command phase.

The example above can be significantly improved. For instance, it used gadfly synchronization; by connecting '65C's interrupt request to a *EPD* bit rather than *KRD* bit 3, a key wakeup interrupt at the end of a seek could permit other programs to run while a long seek is in progress. See Problem 27 at the end of the chapter.

10.2.4   An SCSI Interface to a ZIP Drive

Rather than build and program a disk interface, an SCSI interface (§9.4.3) can interface to an external SCSI disk such as a ZIP-100 drive. Figure 10.12 lists the SCSI commands used in place of *const char Inquiry[6]* in §9.4.3's procedure *nexus*. *Start* is used before any other commands are give, and *Eject* is used at the end. *RequestSense* is given after an error is reported. The others obviously read or write sectors on the disk.

10.2.5   Personal Computer Disk Data Organization

Each operating system organizes its disks differently; herein we first briefly describe the organization of the popular IBM PC 3 1/2" HD disk, which we formatted on a Macintosh for a PC. Later we look at the orgainzation of a ZIP drive that is accessed as described in §10.2.4, using an SCSI interface. Important data on logical sector 0, called the *boot sector*, are listed in Figure 10.13a. Key logical sectors are arranged as in Figure 10.13b.

Start	0x1B	0	0	0	1	0				
Eject	0x1B	0	0	0	2	0				
RequestSense	3	0	0	0	36	0				
Read10	0x28	0	START SECTOR			0	# sectors	0		
Write10	0x2a	0	START SECTOR			0	# sectors	0		
WriteVerify	0x2e	0	START SECTOR			0	# sectors	0		

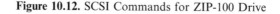

Figure 10.12. SCSI Commands for ZIP-100 Drive

Hex Address	Length Bytes	Meaning/Contents
0	3	0xe9xxxx or 0xebxx90
3	8	OEM Name
b	2	# byte/Sector
d	1	# Sectors/Cluster
e	2	# Sectors in Boot Record
10	1	# FATs
11	2	# Root Directory Entries
13	2	# Logical Sectors
15	1	Medium Descriptor = 0xf0
16	2	# Sectors/FAT
18	2	# Sectors/Track
1a	2	# Heads
1c	2	# Hidden Sectors
1e	...	Boot Program

Logical Sectors	Contents
0	Boot Sector
1 - 9	FAT
10 - 18	Duplicate FAT
19 - 32	Root Directory
33 +	Data Files

a. The Boot Sector (logical sector 0)   b. Logical Sectors

Figure 10.13. PC Disk Organization

The boot sector, which is logical sector 0, or disk sector 1 in track 0 using head 0, is shown in Figure 10.13a. Besides storing a boot program that loads and starts an operating system, this sector stores parameters that can be used to unambiguously read the disk. Figure 10.14 shows a dump of the boot sector of the disk we are using. You should examine it to locate Figure 10.13a's parameters.

A *cluster* is a fixed number of sectors (one in this case) that are allocated or deallocated as a unit. The two bytes at location 0xb indicate the number of bytes per sector and the byte at location 0xd indicates the number of sectors per cluster. The two bytes at location 0xe indicate the number of sectors, starting at sector 0, that store these parameters and the boot program used to install an operating system, and is one. The number of FATs is two, so that a duplicate is available if one is corrupted when there is a disk crash. Any 32-byte entry in a *directory*, the *root directory* being in logical sectors 19 to 32, describes one file listed in the directory, as illustrated in Figure 10.15a. (Other directories are subdirectories.) The number of root directory entries, each of which is 32-bytes (Figure 10.15a), is always a multiple of 16 to use up an integral number of sectors, and is 224. The number of logical sectors on the disk is

```
     00 01 02 03 04 05 06 07 08 09 0A 0B 0C 0D 0E 0F
0810 EB 34 90 50 43 58 20 32 2E 30 20 00 02 01 01 00  .4.PCX 2.0 .....
0820 02 E0 00 40 0B F0 09 00 12 00 02 00 00 00 00 00  ...@...........
```

Figure 10.14. Dump of a Boot Sector

stored in two bytes at 0x13. Figure 10.15b shows how the *file allocation table* (FAT), stored in logical sectors 1 to 9, and duplicated in logical sectors 10 to 18 to aid in file recovery after a disk crash, is associated with sectors storing data in the file. The number of sectors in each FAT is in the two bytes at location 0x16, and the number of sectors in each track is in the two bytes at location 0x18. The number of heads, or surfaces, is in the two bytes at location 0xa. The number of hidden sectors is in the two bytes at location 0x1c. The boot program is stored after location 0x1e.

Incidentally, each two-byte value, and all multibyte values stored on the disk, are in the Intel format: least significant byte first. For instance, the number of bytes per sector may be 0x12 at location 0xb and 0 at 0xc. The value is not 0x1200, but rather 0x0012.

Boot sector parameters unambiguously determine the root directory's beginning logical sector. If the boot record length at location 0xb is 1, the number of FATs stored in 0x10 is 2, and the number of sectors per FAT stored in 0x16 is 9, then the root directory begins at 1 + (2 * 9) = 19. EACH 32-byte directory entry, shown in Figure 10.15a, begins with an eight character name and a three character extension, each in ASCII, and each entry location 0x1a stores a two-byte "data area logical sector number" (*DALSN*).

After the boot sector is loaded into buffer *B*, key locations needed in locating the file are computed by the following program segment.

```
void getFat(){ int fat, dir, entries;
    fat = B[0xe] + (B[0xf] << 8); /* get boot record size, this is lsn of FAT */
    dir=fat+(B[0x10] * (B[0x16]+(B[0x17] << 8))); /* is base of directory */
    entries = B[0x11] + (B[0x12] << 8); /* is # directory entries */
    base=dir+(entries >> 4)-2; /* is base of data sectors ("2" is discussed later)*/
}
```

Hex Address	Length Bytes	Meaning/Contents
0	8	Name
8	3	Extension
b	1	Attribute
c	10	Reserved
16	2	Time of Modification
18	2	Data of Modification
1a	2	Start Cluster
1c	4	File Length

a. Directory Entry   b. FAT Mapping

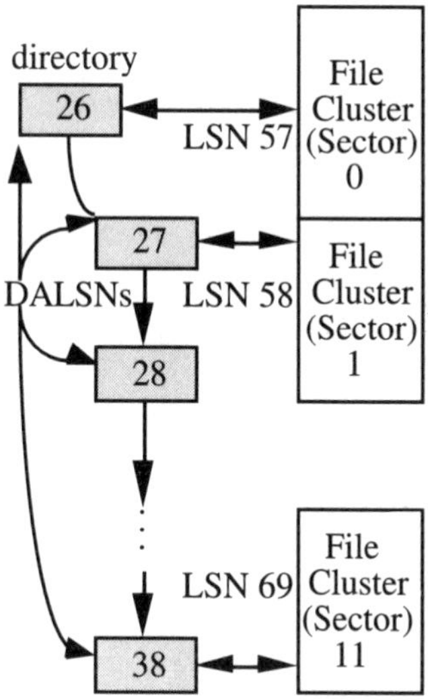

Figure 10.15. PC File Organization

To read the first byte of a file having a given name, compare the desired name with each directory entry name until a match. Our disk's root directory dump is shown in Figure 10.16. The file name DISK.C is in the seventh entry. This file's beginning DALSN is in the two bytes in this entry's location 0x1a, and its value is 0x1a or 26. Its length is 0x00001770 bytes.

After the first directory sector is loaded into buffer B, it can be searched to get the starting *firstDALSN* and `length` by the following program segment. This program segment will only search a root directory with fewer than 32 files. Problem 24 expands this program segment to a 14-sector root directory.

```
void Search(char fileName[]){ int i, j, k, length;
    for(i=0; i<512; i+=32){ /* should search 14 sectors; here we search 1 sector */
    for(j = k = 0; j < 11; j++)
        if((B[i + j] & 0xff) != (fileName[j] & 0xff)) { k = 1; break; }
    }
    if(k == 0) { firstDALSN = B[i + 0x1a] | (B[i + 0x1b] << 8); }
    length=B[i +
        0x1c]|(B[i+0x1d]<<8)|(B[i+0x1e]<<16)|(B[i+0x1f]<<24);
}
```

Since the data area follows the root directory (Figure 10.13b), its logical sector number is essentially the logical sector number of the beginning of the directory, computed previously, added to the length of the directory in sectors. The 2-byte value in boot sector location 11 is the number of directory entries; this divided by 16 gives the number of sectors in the root directory. In our example the data area base is sector

```
      00 01 02 03 04 05 06 07 08 09 0A 0B 0C 0D 0E 0F
0810  46 49 4E 44 45 52 20 20 44 41 54 22 00 00 00 00  FINDER  DAT"....
0820  00 00 00 00 00 00 3C 77 15 23 02 00 B8 02 00 00  ......<w.#......
0830  44 45 53 4B 54 4F 50 20 20 20 20 22 00 00 00 00  DESKTOP    "....
0840  00 00 00 00 00 00 41 77 15 23 00 00 00 00 00 00  ......Aw.#......
0850  53 61 6D 70 6C 65 20 20 20 20 20 28 00 00 00 00  Sample     (....
0860  00 00 00 00 00 00 D4 76 15 23 00 00 00 00 00 00  .......v.#......
0870  52 45 53 4F 55 52 43 45 46 52 4B 12 00 00 00 00  RESOURCEFRK.....
0880  00 00 00 00 00 00 D5 76 15 23 03 00 00 00 00 00  .......v.#......
0890  46 49 4C 45 49 44 20 20 44 41 54 22 00 00 00 00  FILEID  DAT"....
08A0  00 00 00 00 00 00 9B A4 26 23 05 00 80 00 00 00  ........&#......
08B0  E5 44 45 53 4B 54 4F 50 46 4F 4C 10 00 00 00 00  .DESKTOPFOL.....
08C0  00 00 00 00 00 00 97 A4 26 23 06 00 00 00 00 00  ........&#......
08D0  44 49 53 4B 20 20 20 20 43 20 20 20 00 00 00 00  DISK    C   ....
08E0  00 00 00 00 00 00 54 2F 26 23 1A 00 70 17 00 00  ......T/&#..p...
08F0  E5 52 41 53 48 20 20 20 20 20 20 10 00 00 00 00  .RASH       .....
0900  00 00 00 00 00 00 99 A4 26 23 08 00 00 00 00 00  ........&#......
```

Figure 10.16. Dump of a Directory

31. If our directory search located a file whose DALSN is 26, then the beginning of the file is in logical sector 31 + 26 = 57. The first 512 bytes of the file are stored therein.

Each consecutive logical sector is associated with a consecutive 12-bit DALSN in the FAT and its duplicate (Figure 10.17). Each sector's corresponding FAT entry gives the DALSN of the file's next consecutive sector, or gives a value greater than 0xff7 when the file doesn't have a next sector. This is effectively a linked list used to find the remainder of the file. The total number of valid bytes in the file is the four-byte number at the end of the directory entry for the file; it must be less than the file's number of DALSN entries times 512. DALSNs herein are 12-bit binary numbers in Intel format; for every three consecutive bytes in the FAT such as 0x12 0x34 0x56, there are two DALSNs, 0x412 and 0x356. Peculiarly, each FAT's first three bytes are 0xf0 0xff 0xff, and the first DALSN is numbered 2. This initial 3-byte pattern has an advantage of assuring that the sector being examined is actually the beginning of the FAT. Correspondingly, because the first two FAT entries do not correspong to logical sectors in the data area, the DALSNs are converted to LSNs by adding the base address which is the address of the beginning of the data area less two sectors. This is accommodated in the calculation of *base* in the program segment eariler in this section. We suspect that small PC disks had a boot and a directory sector which were included in the "data area", and this FAT "signature" pattern was kept, as larger disks had larger boot files and root directories, to retain this convenient distinctive pattern for verification purposes.

The file DISK.C is stored beginning at DALSN 0x01a = 26. The next segment of the file is determined by the "left half" of the 13th triple-byte entry in dump bytes 0x837 to 0x839. The value of this left half is 0x01b. The next segment of the file is determined by the "right half" of the 13th triple-byte entry in dump bytes 0x837 to 0x839. The value of this right half is 0x01c. Our file's DALSNs are: 0x01a, 0x01b, 0x01c, 0x01d, 0x01e, 0x01f, 0x020, 0x021, 0x022, 0x023, 0x024, and 0x025. These can be dumped to examine the contents of the file.

After a file's *firstDALSN* is determined by the directory search and the first FAT sector is loaded into buffer B, LSNs for the file can be stored in *int list[64]*. This program segment is able to build the list of LSNs if all the disk's DALSNs are held in only the first sector. Problem 26 expands this program segment to a 9-sector FAT. These program segments are built into the object-oriented class in the next section.

```
void FindSector(){ int i, j, curDALSN, list[10];
    for(i=0,curDALSN=firstDALSN; (curDALSN<0xff0) && (i<64); i++)
        list[i] = curDALSN + base; j= (curDALSN >> 1) * 3;
        if(curDALSN & 1)
            curDALSN=(((B[j+1]>>4)&0xf) | (B[j+2]<<4)) & 0xfff;
        else curDALSN = (B[j]|((B[j + 1] << 8) & 0xf00)) & 0xfff;
    }
}
```

```
     00 01 02 03 04 05 06 07 08 09 0A 0B 0C 0D 0E 0F
0810 F0 FF FF 07 F0 FF 0B F0 FF 00 F0 FF 00 00 00 00   ................
0820 C0 00 0D E0 00 0F 00 01 11 20 01 13 40 01 15 60   ......... ..@..`
0830 01 17 80 01 FF 0F 00 1B C0 01 1D E0 01 1F 00 02   ................
0840 21 20 02 23 40 02 25 F0 FF FF 0F 00 00 00 00 00   ! .#@.%.........
```

Figure 10.17. Dump of an Initial FAT Sector

10.2.6   Object-oriented Disk I/O

Object-oriented programming encapsulates data members for a file so that multiple files can be easily read from or rewritten in the same procedure. We show two examples, in which files are only read from, or only rewritten, to simplify the functions.

In this section's first example, using a Western Digital chip disk controller, the class *File* below is made a derived class of *Port* < *char* > so that bytes can be read or written by the *put* and *get* function members and overloaded assignment, cast, < <, and > > operators. The constructor executes the program segments of the previous section, leaving the list of LSNs for the file in the data member *list[64]*. Procedures *init65*, *get*, and *put* from §10.2.3 are also called in this class. Subsequent member functions use the *list* prepared by the constructor to read a sector into a data member buffer *B* in preparation for reading each 512 bytes, or write a sector into *B* into sectors after each 512 bytes are written into it. The destructor writes out *B* into the last sector if it is partially written into.

```
enum{ rd = 1, wr};

class Drive { char C, H, R, N, ST[4]; public : int errors;

    Drive(char drive, char step, unsigned char unload, char load){int i;
        KRD = 0x14; KRD = 4; /* positive logic reset of '65 */

        w(3); w(((-step) << 4)|(unload>>4)); w(load|1); /* specify */
        w(4); w(drive & 7); i = r(); /* sense drive status */
        if((i & 0xa0) != 0x20) { errors = chpErr; return; } errors = 0;

        w(7); w(drive & 3); /* recalibrate (restore) */
        while( ! (KRD & 8)) ; /* gadfly on drive being in execution mode */

        w(8); i = r(); C = r(); /* sense interrupt status */
        if(i != 0x20) { errors = drvErr; return; }

        control=0;w(0x4a);w(0);while(!(KRD&8)); check(); /* read i.d. */
        if((ST[0] & 0xc0) != 0) errors = dskErr;
        return;
    }
char cs(void){ char s; /*read master status */
    KRD &= -0xD; s = KCD; KRD = 0xBF; return s & 0xc0;
}
```

```
void w(char c) {/*write data */
    while(cs()!=0x80) ;KRD&=-6;KCDD=0xff;KCD=c;KRD=0xBF;KCDD=0;
}
char r(){ char d; /*read data */
    while(cs()!=0xC0;KDR&=-5;d=KCD;KDR=0xBF; return d;
}
    void setup(char cmd)
        {w(cmd);w(H<<2);w(C);w(H);w(R);w(2);w(18);w(0x1b);w(0xff);}
    char check() {
        if((ST[0] = r()) & 0x80) { errors = illOp; return 0; }
        ST[1] = r(); ST[2] = r(); C = r(); H = r(); R = r(); N = r();
        return (ST[1] & 0x7f) || (ST[2] & 0x33);
    }
    char seek(int sectorNumber) {
        if(sectorNumber>2880) return errors=addrErr;/* sec outOfRange? */
        R=(sectorNumber%18)+1; /* get sector # (PC disks use one-origin indexing) */
        C=sectorNumber/18; /* get combined cylinder-head number, temporarily in C */
        H = C & 1; C >>= 1; /* separate into cyl C (high bits) and head (lsb) */
        w(0xf); w(H << 2); w(C); /* give command to seek cylinder */
        while( ! (KRD & 8)) ; /* gadfly on drive seeking cylinder */
        w(8); ST[0] = r(); /* get interrupt status */
        if((r()!= C)||((ST[0]&0xf8)!=0x20))return errors=seekErr;
        return 0;
    }
    char get(int sector, register char *buffer) { int i; char *end;
        if(errors||seek(sector))return 1;
        end=buffer+(i=512); KRD=0; setup(0x46);
        do {
        while(!(cs()&0x80));if(!(cs()&0x20)) break;*buffer++=r();
        } while( --i );
        KRD = 4;
        if(check()||(buffer!=end)) return errors = rdErr; return 0;
    }
    char put(int sector, register char *buffer, char verify){ int i;
        if(errors || seek(sector)) return 1; i = 512;
        setup(0x45);
        do {
          while(!(cs()&0x80));if(!(cs()&0x20)) break;
          (*buffer++);
        } while( --i );
        if(i||check())
        {if(ST[1]&2) return errors= wrProtErr;return errors=wrErr;}
        if(verify && get(sector, buffer)) return errors = wrErr;
        return 0;
    }
};
```

```
class File : public Port<char>{ char B[512], mode; Drive
    unsigned int curDALSN, firstDALSN, base, fat, dir, entries, list[64];
    long length, position; public : int errors; char verify;
    File(char * fileName, Drive &D, char mode):Port(0x200){int i,j,k;
        errors = position = 0; verify = 1; this->D = D;

        // locate fat and read directory
        if(D.get(0, B)) return; /* read boot sector */
        fat=B[0xe]+(B[0xf]<<8); /* get boot record size; it is LSN of FAT */
        dir=fat+(B[0x10]*(B[0x16]+(B[0x17]<<8)));/*this is directory base*/
        entries = B[0x11] + (B[0x12] << 8); /* is # directory entries */
        base = dir + (entries >> 4) - 2; /* is data sector base */

        // search directory for file name
        if(D.get(dir, B)) return; /* read directory */
        for(i = 0; i < 512; i += 32){ /* search one sector of directory */
            for(j = k = 0; j < 11; j++)
                    if((B[i+j]&0xff)!=(fileName[j]&0xff)){k=1; break;}
            if(k==0){ firstDALSN=B[i+0x1a]|(B[i+0x1b]<<8); break;}
        }
        length = B[i + 0x1c] | (B[i + 0x1d] << 8);

        // get list of sectors
        if(D.get(fat, B)) return; /* read FAT */
        for(i=0, curDALSN=firstDALSN; (curDALSN<0xff0)&&(i<64);i++){
            list[i] = curDALSN + base; j= (curDALSN >> 1) * 3;
            if(curDALSN & 1)
                    curDALSN=(((B[j+1]>>4)&0xf)|(B[j+2]<<4)) & 0xfff;
            else curDALSN=(B[j]|((B[j+1]<<8)&0xf00))
        }
    }
    virtual char get(void){ // input
        if(!(mode&rd)||(position>(length-1))) {error =1; return; }
        if((position & 0x1ff) == 0) D.get(list[position >> 9], B);
        return B[((position++) & 0x1ff)];
    };
    virtual void put(char data) {// output
        if(!(mode && wr)||(position>=length)) {errors=1; return;}
        B[((position++) & 0x1ff)] = data;
          if((position & 0x1ff) == 0)
                D.put(list[(position >> 9) - 1], B, verify);
    };
    void seek(long position) { // seek a location in the file
        if(position >= length) {errors = 1; return; }
        if((mode == wr) && ((position & 0x1ff) == 0))
            D.put(list[(position >> 9) - 1], B, verify);
        this->position = position; D.get(list[position >> 9], B);
    }
```

```
File &operator = (File &f); // copy file
      {do put(f.get()); while(!(errors|=f.errors)); return *this;}
   ~File(){if(position&0x1ff) D.put(list[position>>9] B,1);} // de-
      structor
};
Drive d1(0,3,240,16); // delare the object; call the constructor to initialize the disk
void main() {
   File f1(''F1'',d1,wr), f2(''F2'',d1,rd); // delare objs; call constr to open files
   f1 = f2; // copy files
   f1.~File; f2.~File; // close the files
}
```

Our second example, shown below, illustrates an object-oriented class *File* for use with an SCSI interface, to a ZIP-100 drive. It uses the SCSI commands in Figure 10.12, with the SCSI procedure *nexus* in §9.4.3. ZIP disks have 100 MByte capacity, so their FAT structure uses 16-bit, rather than 12-bit, DALSNs, and each DALSN corresponds to four 512-byte sectors. Also, the boot sector is the first sector on track 1 (logical sector 0x20), rather than the first sector on track 0.

Like the first *class* above, the second *class* below gives a class *File* for reading or rewriting a file that has been written on the ZIP disk by a PC that has the capability for writing the directory and FATs. Member functions *get* and *put* which transfer a byte at a time use a buffer to hold a sector, and those that transfer a vector at a time, will try to directly transfer all the data in one nexus if the location and the vector size are multiples of 512, otherwise they internally transfer one byte at a time. Member function permits reporting errors, seeking a location in the file, reporting that location, enabling or disabling write verify capability, and determining the sector-in-the-track location.

```
enum{ fromDisk=0,fromFile,NO_START=1, NO_BOOT, BAD_BOOT, NO_DIREC-
      TORY,
   NO_FILE_NAME,TOO_MANY_FATS,NO_READ,NO_WRITE,EOF,START_STOP=0x1b,
   START=0x100, EJECT=0x200, REQUEST_SENSE=3, READ=0x28, WRITE=0x2A,
   WRITE_VERIFY=0x2e };

struct { unsigned char valid:1, code:7, segNum:8, FileMark:1, EOM:1,
      ILI:1, :1, snsKey:4, information[4], length, cmdInfo[4], auxSnsCode,
      auxQual, Ucode, Sksv:1, specific0:7, specific1,specific2;
} SenseData;

#define PAGESIZE 512
char page[PAGESIZE];
void main(){ short j, k;
   initializeSCSI();F=newFile(''10   '');
   if(F->ERROR) return; /* faulty constructor execution */
   for(k = 0; k < 16; k++) { for(j = 0; j < PAGESIZE; j++)
      {page[j] = F->get(); if(F->ERROR) return;} }
   delete F;
}
```

```
class File {
    long fileLength, location, curFileSectorNum, beginSegment, segmentSize,
        index, sectorsPerCluster, residual, dataBase, segmentCount;
    unsigned char error, validData, dirty, verify,
        buffer[512], statusBuffer[32], msgBuffer[32], packet[10];
    unsigned short lowFatList[32], highFatList[32];
    long swap(unsigned char *ptr){ return (*(ptr+1) << 8) | *ptr; }
    unsigned short findSegment( long fileSectorNumber){
        for(beginSegment = index = segmentSize = 0;
            (beginSegment + segmentSize) <= fileSectorNumber;
                segmentSize = (highFatList[index] + 1 -
                    lowFatList[index]) * sectorsPerCluster;
        }
        residual=fileSectorNumber + segmentSize - beginSegment; index-;
        return dataBase + residual + ((lowFatList[index] - 2) *
            sectorsPerCluster);
    }
    short moveData(char command, char *b, long length){
        long oldLength = length, size = 1, available, sector;
        if((length & 0x1ff) || (location & 0x1ff)){ // if not mult of 512
            while(!error && length-) // move single bytes
                if(command == READ) *b++ = get(); else put(*b++);
            return oldLength - length; // return amount moved
        }
        while(!error && length){ // if 512X, move whole sectors
            if(location > fileLength) { error = EOF; return 0; }
            size = length >> 9; sector = findSegment(location >> 9);
            available = segmentSize - residual;
            if(size>available)size=available;//don't move > than is contiguous
            if(sendPacket(command, b, sector, size, fromDisk))
                error = (command == READ) ? NO_READ : NO_WRITE;
            location += size * 512; length -= size * 512;
        }
        return oldLength - length;
    }
    void flush(){
        if(dirty) if(sendPacket( verify ? WRITE_VERIFY : WRITE,
            buffer, location >> 9, 1, fromFile))
    }
    int check(){
        if((statusSize == 1) && (*statusBuffer == 0)) return 0;
        sendPacket(REQUEST_SENSE, (char *)&SenseData,
            sizeof(SenseData) << 8, 0, 0);
        if(SenseData.snsKey == 6) return 0; /* warning: media changed */
            return 1;
    }
```

```
public:File(char *name){
    unsigned short i, j, noMatch, sectorsPerBoot, numFats,
      numSectorsPerFat, numDirEntries, newFatSector, curFatSector,
      firstDalsn, curDalsn, index, directorySector, fatSectorBase;
    segmentCount = curFatSector = location = error = validData =
      curFileSectorNum = dirty = verify = 0;
    if(sendPacket(START_STOP, 0, START, 0, 0))
        { error = NO_START; return; }
    if(sendPacket(READ, buffer, 0x20, 1, fromDisk))
        { error = NO_BOOT; return; } // read boot
    if((!((buffer[0]==0xeb)&&(buffer[2]==0x90)))&&(buffer[0]!=0xe9)
    ||(buffer[0xb]!=0)||(buffer[0xc]!=2)){error=BAD_BOOT;return;}
    sectorsPerCluster=buffer[0xd]; sectorsPerBoot=swap(buffer+0xe);
    numFats = buffer[0x10]; numDirEntries = swap(buffer + 0x11);
    numSectorsPerFat = swap(buffer + 0x16);
    fatSectorBase = sectorsPerBoot + 0x20;
    directorySector=fatSectorBase + ( numFats * numSectorsPerFat );
    dataBase = directorySector + ((numDirEntries) >> 4);
    if(sendPacket(READ, buffer, directorySector, 1, fromDisk))
        { error = NO_DIRECTORY; return; }
    for(firstDalsn=fileLength=i=0; i < 512; i += 32){ // search directory
      for(noMatch = j = 0; j < 11; j++)
            if(name[j] != buffer[j + i]){ noMatch = 1; break; }
        if(noMatch == 0) {
            firstDalsn = swap(buffer + 26 + i);
            fileLength=(swap(buffer+30+i)<<16)|swap(buffer+28 + i);
            break;
      }
    }
    if(noMatch) { error = NO_FILE_NAME; return; }
    j = ((fileLength-1) / (sectorsPerCluster*0x200))+1;
    if(j>256)j=256;
    lowFatList[0] = highFatList[0] = curDalsn = firstDalsn;
    for(i = 1; i < j; i++){
      newFatSector = (curDalsn >> 8) + fatSectorBase;
      if(newFatSector != curFatSector)
 sendPacket(READ,buffer,curFatSector=newFatSector,1,fromDisk);
        curDalsn = swap(buffer + ((curDalsn & 0xff) << 1));
        if(curDalsn == 0xffff) break;
        if((curDalsn - 1) == highFatList[segmentCount])
            highFatList[segmentCount]++;
 elselowFatList[++segmentCount]=highFatList[segmentCount]=curDalsn;
        if(segmentCount > 32) {error = TOO_MANY_FATS; return;}
    }
    highFatList[index]++;
}
```

```
char get () { short newFileSectorNum;
    newFileSectorNum = location >> 9; if (location > fileLength)
       { error = EOF; return 0; }
    if (!validData || (newFileSectorNum != curFileSectorNum))
       if (sendPacket (READ, buffer, curFileSectorNum =
             newFileSectorNum, 1, fromFile)) error = NO_READ;
    validData = 1; return buffer[(location++) & 0x1ff];
}
void put (unsigned char c) { short newFileSectorNum;
    if (location > fileLength) {
       flush (), error = NO_WRITE; error = EOF; return;
    }
    newFileSectorNum = location >> 9;
    if (!validData || (newFileSectorNum != curFileSectorNum))
       if (sendPacket (READ, buffer, curFileSectorNum =
             newFileSectorNum, 1, fromFile)) error = NO_READ;
    dirty = validData = 1; buffer[(location++) & 0x1ff] = c;
    if ((location & 0x1ff) == 0)
       {sendPacket ( verify ? WRITE_VERIFY : WRITE, buffer,
       (location - 1) >> 9, 1, fromFile); validData = dirty = 0; }
}
int sendPacket (char command, char *b, long sector, short size,
    char fromFile) {
       if (fromFile) sector = findSegment (sector);
       packet[0] = command; packet[1] = packet[6] = packet[9] = 0;
       packet[2] = sector >> 24; packet[3] = sector >> 16;
       packet[4] = sector >> 8; packet[5] = sector;
       packet[7] = size >> 8; packet[8] = size;
       nexus (packet, b, statusBuffer, msgBuffer);
       return check ();
}
long get (char *b, long length) { return moveData (READ, b, length); }
long put (unsigned char *b, long length)
    { return moveData (verify ? WRITE_VERIFY : WRITE, b, length); }
int option (int code = 0, int value = 0) { unsigned int v;
    if (code == 0) { v = error; error = 0; return v; }
    if (code == 0x10) { // seek
       if (dirty) sendPacket ( verify ? WRITE_VERIFY : WRITE, buffer,
             location >> 9, 1, fromFile);
       location = value; return dirty = 0;
    }
    if (code == 0x11) return location; // get location
    if (code == 0x12) {verify = v; return 0; } // enable/disable write verify
    if (code == 0x13) {return findSegment ( location << 9) & 0x1f; }
}
~File () {flush (), sendPacket (START_STOP, 0, EJECT, 0, 0); }
} *F;
```

Either class for the floppy or ZIP disk can be used to read or write data from several files, as the following program illustrates. This program copies file "F1" to the file "F2".

```
File f1(''F1''), f2(''F2''); // delare the objects; call the constructor to open the files
void main() {do f1=f2;while(!f1.error);~f1();~f2();} // copy until eof error
```

Either object-oriented disk access class *File* makes the disk appear like any other I/O device described in this book. Using device independence, a disk can be substituted for another I/O device at run time. For instance, if a monitoring device has a serial port through a modem and a ZIP drive, data it collects can be sent through the modem if it is operational, but can be saved on the ZIP disk if the modem is not available. The ZIP drive can be substituted for the serial port at run time, by blessing the object to be an object of class *File* rather than *Uart*. Using I/O independence, a disk can be substituted for another I/O device at compile time. For instance, the program can be compiled with the first or second *File* class in this section, depending on whether the system uses floppy or ZIP disks. Objects encapsulate all the data and functions needed to access a disk, so that two or more files can be accessed in the same program. These objects have many of the advantages of operating system device drivers, but with much lower overhead. These capabilities illustrate the advantages of object-oriented I/O.

10.3   Conclusions

This chapter introduced two common interfaces: CRT display and secondary storage. These rather complete case studies give a reasonably full example of common interface designs. They also embody the techniques you have studied in earlier chapters. Besides presenting these important interfaces, this chapter serves to complete the book by showing how the techniques in the other chapters will find extensive application in almost any interface design.

For further reading on floppy disks, we strongly recommend the data sheets for the '65C from Western Digital. Harold Stone's *Microcomputer* Interfacing has additional general information on the analog aspects of storage devices. These can be consulted for further examples and inspiration.

This text has been fun for us. Microcomputers like the MMC2001 are such powerful tools that it challenges the mind to dream up ways to use them well. We sincerely hope you have enjoyed reading about and experimenting with the MMC2001 microcomputer.

Do You Know These Terms?

See the End of Chapter 1 for Instructions.

National Television Broadcast Bresenham   logical sector   specify
  System Committee   algorithm   number (LSN)   read id
  (NTSC)   window   interleave factor   sense drive
raster line   clip   unformatted   sense status
frame   secondary storage   capacity   command phase
pixel   surface   formatted capacity execution phase
NTSC composite   track   format   result phase
video signal   cylinder   seek   verify
horizontal retrace   step rate   read sector   boot sector
vertical retrace   settling time   implied seek   cluster
horizontal sync   fill   write sector   directory
vertical sync   sector   format   root directory
sync separator   index hole   restore   file allocation
                           index pulse   recalibrate   table (FAT)

Problems

Problems 1 and 13 are paragraph correction problems; see guidelines at the end of Chapter 1. Programming guidelines are given at the end of Chapter 2, and hardware design guidelines are at the end of Chapter 3.

1.* A TV screen is a series of fields; and in the NTCS format, a field takes 1/30 s. There are about 500 raster lines in a field, each line scanning from top to bottom of the screen, and each raster line takes about 60 µs. Sync pulses are incorporated into the composite video signal as gray level signals, and these are used to synchronize the horizontal and vertical oscillators that cause the electron beam to scan the screen. CRT controllers use either character or graphics display modes at any time. The former can use an independent mode, where the CRT gets characters from the primary memory of the processor using DMA; or the shared mode, where the processor writes into a separate display memory only du'ring the horizontal retrace periods.

2. Rewrite `char pattern[12]` in §10.1.2 to display:

    a. a solid black 8 by 12 square.
    b. an 8 by 12 black outlined white square. The outline is 1 pixel wide.
    c. a horizontal line two pixels high across the top of the 8 by 12 black square.
    d. a vertical line two pixels wide on the left of the 8 by 12 black square.
    e. a letter A on the top 8 lines of the 8 by 12 black square, with the right column blank, with four blank lines on the bottom of the 8 by 12 black square.

3. Rewrite *#defines* in §10.1.2 to display the square: (give approximate values ±10%)

    a. at the top left corner of the screen.
    b. at the top right corner of the screen.
    c. at the bottom left corner of the screen.   d. at the bottom right corner of the screen.

4. Rewrite the program in §10.1.2 that outputs the same TV picture as in Figure 10.4, using gadfly synchronization rather than interrupt synchronization, to implement horizontal and vertical sync pulses, using the same counter and SPI modules.

5. Write a C procedure *border ()* which will draw a border that is four pixels wide around §10.1.3's display, rather than two pixels wide, and fill the display with gray, rather than black. Use §10.1.3's constants such as HEIGHT, HWIDTH, and WIDTH.

6. When displaying on a low-bandwidth CRT, so that horizontal and vertical lines have equal brightness, the video signal should be "chopped" by ANDing it with the shift clock because horizontal lines have more low-frequency signal than vertical lines. The 74HC132's remaining NAND gate can "chop" the video signal, requiring two hardware changes and changes in the accompanying software to display the same white border on black background as in §10.1.3. Write a paragraph accurately describing these changes.

7. The program in §10.1.3 displays 256 lines of 512 pixels per line. By displaying the same line in both fields, for instance so that location 0x8000 appears on the top left of the first scan line of the first field and again on the top left of the first scan line of the second field, our (8K, 16) SRAM can display a 496 by 512 screen image. Show the program needed to display this 496 by 512 screen, wherein each memory location is displayed twice, in the same relative location of each field.

8. Write a function member *rectangle(int h, int v, int w, int ht)* for §10.1.4's class *screen* that will draw a rectangle whose top left corner is at row *v*, column *h*, and whose width is *w* and height is *h*, using the function member *line* given in §10.1.4.

9. The function member *line(int h, int v, int dh, int dv, int n)* for §10.1.4's class *screen* can only draw lines where either *dh* or *dv* is 1, and the other is between −2 and +2. Use the Bresenham algorithm to make a practical line drawing function.

    a. Write a function member *line1(int h1, int v1, int h2, int v2, int dh, int dv)* where *dh = h2 − h1*, *dv = v2 − v1*, *dh > dv* and *dh > 0*, to draw a continuous line from row *v1* column *h1* to row *v2* column *h2*.

b. Write a function member *lineto(int h1, int v1, int h2, int v2)* which draws a continuous line from row *v1* column *h1* to row *v2* column *h2*. Use a modification of part a's function member *line1*, with a fifth argument *reverse*, so that if *reverse* is 0, *line1* calls *point* with unsubstituted *h* and *v*, if 1, *line1* calls *point* interchanging *h* and *v*, if 2, *line1* calls *point* negating *h*, and if 3, *line1* calls *point* interchanging *h* and the negative of *h*.

10. Using problem 9's function member `lineto,` write a function member `triangle(int h1, int v1, int h2, int v2, int h3, int v3)` that will draw a triangle with a vertex in row *v1* column *h1,* a vertex in row *v2* column *h3,* and a vertex in row *v3* column *h3* .

11. Write a function member *wchar(int *a, int h, int v)* for §10.1.4's class *screen* that draws a character whose pattern is pointed to by *a,* whose top left corner is at row *v,* column *h,* and whose width is up to 16 pixels wide and 16 pixels tall. Use only `long` variables and pointers. Also write a vector to draw a letter 'A.'

12. Write a program *main ( )* that writes MISSISSIPPI in the middle of the screen. Show the vectors *patternM, patternI, patternS*, and *patternP* analogous to the §10.1.4 vector *patternA*, to draw the letters M, I, S, and P, in a 7 pixel wide, 8 pixel high, font. Write *main( )* to bless a pointer *SCN* to an object of the class *screen*, and then write the word MISSISSIPPI in the middle line and around the middle column of the display area.

13. A surface of a typical floppy disk is divided into concentric rings called sectors, and each sector is divided into segments. A sector may be read or written as a whole, but individual bytes in it may not be read or written. The format of a sector has only some zeros, a 0xA1 flag pattern, data address mark, the data, and a CRC check; counters are used to keep track of the track and sector. To read (write) a sector, it is necessary to first give a command to seek the track, then give a command to read (write) the sector.

14. Trace the pattern for the following bytes: 0x80, 0x55, 0xcc, 0xca, 0x1 (assume the bit previous to this byte is a 0).

    a. FM encoding   b. MFM encoding.

15. Show timing diagrams of the middle special byte 0xA1 in Figure 10.9 and relevant parts of the beginning and end of the previous and following special bytes as they shift through a byte-sized window. Show that the shifting byte appears to match the required pattern exactly once, which defines the byte boundary.

16. Determine how many total bits are in a sector, and how long it takes to read it.

17. The operations register OR is written when LDOR, attached to address line A3, is asserted and WR is asserted.

a. Into which locations is this operations register and only this register written?

b. OR bit 4, output in negative logic on pin 33, runs the drive 0 motor. OR bit 5 similarly on pin 34 runs the drive 1 motor. Show drive cable connections so that asserting OR bit 4 runs the drive 0 motor, and asserting OR bit 5 runs the drive 1 motor.

18. For the logic diagram of Figure 10.10, determine which addresses can be used to uniquely select the ports. Identify *all* addresses in which the following can be done.

a. Read the master status register

b. Read the data ragister

c. Write the data register

d. Write the control register

19. Deleted data has a different Delete Data Mark $FD in place of the Data Mark $FB in Figure 10.8 and a sector with this mark is skipped when reading data; one writes this mark by the command, write deleted data (command 0xc), when one finds defective media in the sector. Such a deleted sector can still be read using the command, read deleted data (command 0x9). Otherwise commands to write and read deleted data are the same as the commands, write data (command 0x7) and read data (command code 0x6). Show pictorial descriptions that can be added to Figure 10.10d to describe these two commands.

20. Write a multithread scheduled procedure *seek* and a handler *handler* for the 65C IRQ pin attached to *KRD* bit 0. When a seek cylinder operation is begun (see the centronics printer, §7.2.2), if the '65C asserts IRQ within 12 µs, *seek* exits, but otherwise *seek* puts the thread to sleep. When this operation is complete a key wakeup interrupt executes handler *handler* to awaken the sleeping thread. Assume thread 1 is used.

21. Write procedures that try five times to read or write a sector, until the sector is read or written correctly. When writing, if verify is 1, verify the written sector without destroying *B*. After the second attempt, and if the head is not on cylinder 0, move the head to the next lower cylinder, then read or write the sector. After the fourth attempt, and if the head is not on cylinder 80, similarly move it out a cylinder and back in. "Jiggling" the head this way facilitates reading or writing a misaligned cylinder.

a. Write a *get* procedure to read a sector.   b. Write a *put* procedure to write a sector.

22. Write a procedure *format (int c, char h)* to format cylinder *c* on side *h* of drive 0. Before this procedure is executed, the drive should have been initialized using *init65*, but at the beginning of this procedure, a seek cylinder command is given. Note that errors will occur during initialization and seeking that should be ignored. To format a track for an HD disk, the 65C is given the command 0x4d, a byte containing the head (bit 2) and drive number (bits 1 and 0), the number N of bytes per sector, the number of sectors per track, a gap width, and the byte used to fill the data portion of each sector. For an HD disk, N is 2, there are 18 sectors per track, the gap width is 0x54, and the data portion of each sector is filled with 0x46. The execution phase waits for the index pulse, then formats an entire track, and then asserts IRQ. While the 65C formats the track, the MMC2001 writes each sector bytes C, H, R, and N into the 65C data port as in the write sector command. That is, for the 18 sectors, write 72 bytes. Your procedure should write the track with an interleave factor of four. The status phase returns the status bytes ST[0], ST[1], and ST[2], which should be checked for errors, and four bytes (C, H, R, and N), which have no significance in this case. Use the §10.2.3 variables.

23. In Figure 10.14, determine the DALSN and length of the files:

  a. FINDER.DAT, b. DESKTOP, c. RESOURCE.FRK

24. Write a program segment to search a 14-sector root directory, analogous to the program segment below Figure 10.14.

25. From Figure 10.17, determine the next (hex) DALSN when the current DALSN is:

  a. 2   b. 3   c. 4   d. 0xb   e. 0xc   f. 0xd

26. Write a program segment to construct the file DALSN *list* from a 9-sector FAT, analogous to the subroutine `FindSector`. Note that DALSN of consecutive sectors in a file are not necessarily consecutive, and may even be nonmonotonic (they may skip around). Take care to handle the special case where a three-byte sequence containing two DALSN overlaps a sector boundary.

27. Write a function member `seek(long a)` for class `File,` which will position the read or write in *position a* so the next byte read by `char get()` or written by `put(char)` is the ath byte of the file. If a sector needs to be output to save bytes written before *seek* is executed, do so and then read in the sector in which byte a appears.

28. The class `File` can be modified to permit either reading of data in the file, writing of data in the file, or reading and writing of data in the same file (called updating the file), but we have to be careful about putting back sectors that may have been partially overwritten when we read the data, and about putting a sector into the buffer before writing a byte into it, in case we will read bytes from this sector later.

  a. Write a function member `char get()` for class `File,` which will output the next byte of the file (at location `position`). However if this requires reading in another sector, the sector previously stored in the buffer is written back.

  b. Write a function member *put(char c)* for class *File*, which will write *c* into the next byte of the file (at location *position*). However if this requires writing into another sector, that sector is read into the buffer.

c. Explain why an object of class *File* should be declared or blessed with "permissions" read-only, write-only, or update to make the file both readable and writeable at the same time. In particular, comment on how long reading or writing can take in the worst case for each case. How can our class *File* be modified to permit this capablility to be declared in the constructor and used in the function members.

Appendix
Using the HIWARE CD-ROM

This appendix helps you use the accompanying CD-ROM to simulate your programs and to download and debug them on EVB Boards and other target microcontrollers.

A-1   Loading HIWARE Software

Open the CD ROM, check "installation", and choose the M·CORE target. If you have 60 Megabytes of disk space, load all parts of the tool chain.

A-2   Running the Simulator

You can use the software on the CD-ROM to simulate your programs on a PC running Windows·95 or later, or Windows NT, without using any extra hardware. Using Acrobat Reader 3.0 or later, run the \hiware\docu\mcore\demomc.pdf file. This file provides a tutorial guide on how to load and run the compiler, linker, and simulator. Following this guide, compile, link, and simulate the program Fibo.c.

A very simple way to experiment with other programs is to modify the file Fibo.c. Using any text editor, such as NOTEPAD, rewrite the Fib.c file with a program that you wish to study. Compile, link, and simulate the modified program Fib.c. You can rewrite Fibo.c each time you wish to study a new program. You can use more sophisticated techniques, but this simple technique can get you started with minimal effort.

A-3   Running Examples from This Book

Note that the folder EXAMPLES on the CD-ROM has files in it such as Em2.txt. These files contain examples from this textbook, which you can copy-and-paste into Fibo.c, so that you can run these examples on the Hiware simulator. The file Em2.txt contains all the examples in Chapter 2 of this textbook, and the file Em5.txt contains all the examples in Chapter 5 of this textbook, and so on. Copy this folder into your hard disk; most conveniently, put it into your HIWARE folder.

A-4   Downloading to an Axiom M·CORE Board using the EBDM

You can use the HIWARE software and Motorola's extended background debug module (EBDM), to connect to the Axiom M·CORE board (called the target) to run experiments. Begin by simulating Fibo.c. on the HIWAVE simulator, as described in the foregoing. After you are comfortable with the simulator operation, substitute the following text for the contents of fib.prm and run the linker. In HIWAVE, select the target component Motoesl instead of sim. You should be running your program on the target. You should always apply the 5-V power after all connections are made, and you should never change a connector while power is applied to the board.

```
LINK Fibo.abs
NAMES Fibo.o Strtmco.o END
SECTIONS
     MY_RAM = READ_WRITE 0x30000200 TO 0x30003FFF;
     MY_ROM = READ_ONLY 0x30004000 TO 0x30007FFF;
PLACEMENT
DEFAULT_ROM    INTO MY_ROM;
DEFAULT_RAM    INTO MY_RAM;
END
STACKSIZE 0x600
```

At the time of the writing of this book, we used the Motorola EBDI interface, but we expect that other interfaces will become available and might be more suited to many of our readers. However, the other interfaces are probably going to closely resemble this one.

A-5   Techniques for HIWARE Tools

We have had some experiences with HIWARE tools, which might help you use them more efficiently. We add a note here on our suggestions, to help you with this powerful software.

   A problem with the current version is that when you change project files, the compiler/linker/hiwave debugger may read or write the wrong files, or fail to find the files it needs. We found that by shutting down all HIWARE programs, and starting them up again, the problem goes away. But you do not have to restart the computer. If you have verified that the paths to the files are correct, but you are unable to access them through the compiler/linker/hiwave debugger, then try restarting all HIWARE programs "from scratch". The same remedy is suggested when the HIWAVE simulator or debugger fails to execute single-step commands, or breakpoints, correctly.

   When dealing with different environments such as your own PC running Windows 95, workstations running Windows NT, and a PC running Windows 98 in the laboratory, keep separate complete project folders for each environment, and copy the source code from one to another folder. In that way, you will spend less time readjusting the paths to your programs and HIWARE applications when you switch platforms.

We hope that the CD-ROM supplied through HIWARE makes your reading of this book much more profitable and enjoyable. We have found it to be most helpful in debugging our examples and problem solutions.

INDEX